The Structure and Function of Skin

THIRD EDITION

DEDICATION

to

ALL OF OUR TEACHERS

THE STRUCTURE AND FUNCTION OF SKIN

THIRD EDITION

William Montagna and Paul F. Parakkal

OREGON REGIONAL PRIMATE RESEARCH CENTER

BEAVERTON, OREGON

1974

ACADEMIC PRESS   New York   San Francisco   London

A Subsidiary of Harcourt Brace Jovanovich, Publishers

LMLML K

ACADEMIC PRESS, INC.
111 Fifth Avenue, New York, New York 10003

United Kingdom Edition published by
ACADEMIC PRESS, INC. (LONDON) LTD.
24/28 Oval Road, London NW1

Library of Congress Cataloging in Publication Data

Montagna, William.
     The structure and function of skin.

     Previous editions by W. Montagna.
     Includes bibliographies.
     1.   Skin.   I.   Parakkal, Paul F., joint author.
II.   Title.   [DNLM: 1.   Skin—Anatomy and physiology.
2.   Skin—Physiology.   WR101 M758s 1974]
QM481.M6 1974   611'.77   73-0439
ISBN 0–12–505263–4

PRINTED IN THE UNITED STATES OF AMERICA

Contents

Chapter 8.   Effects of Malnutrition on the Morphology of Hair Roots

Chapter 9.   Nails

Chapter 10.   Sebaceous Glands

Chapter 11.   Apocrine Glands

Chapter 12.   Eccrine Sweat Glands

Preface

This book was written by two biologists, a teacher and his former pupil, each of whom is unaccountably but deeply devoted to all matters pertaining to skin. Written primarily for dermatologists, whether seasoned veterans or neophytes, it should be useful to all biologists who are interested in biomedical disciplines. If it is incomplete, it is because our knowledge of human skin is still limited and fragmented; if it contains some errors, it is because most of the available facts, morphological or biochemical, are not yet fully or exactly known. Finally, if it reflects our bias, it is because the subject of cutaneous biology has more than its share of controversial areas on which, like most other authors, we have taken a stand. We have not limited our treatment exclusively to human skin, but it is our point of reference. We regard human skin as a complex and complicated template that shares in unique fashion the biological principles of other organ systems. Many well-known data on human skin and much information from comparative cutaneous biology were deliberately omitted lest this book grow to encyclopedic proportions.

We must declare at once that there are many more authors than the two listed on the case. In a very real sense, this book is the product of a family effort. For example, Drs. Kenji Adachi, Michael Im, Dennis Knutson, Robert Kellum, Alan Shalita, and, in particular, Mary Bell have contributed enormously to the chapter on sebaceous glands. The chapter on the dermis was written almost entirely by our former student Dr. Jeffrey Pinto and that on eccrine sweat glands by our friend and former colleague Dr. Richard Dobson. Dr. Robert Bradfield has written the remarkably lucid and concise account of the effects of protein deficiency on hair growth. Dr. Kenji Adachi and our former student Dr. Michael Im have written most of what we have included on metabolism, and Dr. Giuseppe Moretti has done yeoman work on the chapter on hair. Our friend Dr. Vicente Pecoraro has contributed the results obtained by his group using the trichogram technique.

But our indebtedness does not end here. All of the delineations were executed by our master artist Joel Ito and much of the photography by our own "Steichen" Harry Wohlsein. Nearly all of the superb electron photomicrographs were prepared by Dr. Mary Bell and some of the scanning electron micrographs by our colleague Dr. Wolf H. Fahrenbach. Many of the histological materials used reflect the incomparable technical skill of Nickolas Roman. Elizabeth Macpherson helped us with the bibliography and some illustrations. Cathy Taylor was helpful with many of the illustrations. Margaret Shininger not only typed, but organized, supervised, checked for accuracy, and devoted herself unstintingly to every detail of the book. And finally, running throughout is the unmistakable evidence of Margaret Barss's perceptive knowledge of and love for the English language and her keen sense of logic and clarity. To all of these selfless friends and colleagues and to other contributors too numerous to name, we are forever grateful.

WILLIAM MONTAGNA
PAUL F. PARAKKAL*

* *Present address:* Division of Research Grants, National Institutes of Health, Bethesda, Maryland.

Introduction

The first edition of this book, published in 1956, was a modest effort that reflected mainly the limited results of the author's own morphological studies of skin. Six years later a more ambitious second edition was published. Now, after more than ten years, the third edition represents both individual and team work, claiming two authors—a teacher and his former student—and notable contributions from many former students, colleagues, and friends. For example, the chapter on sebaceous glands is a digest of the knowledge of nearly all modern authorities on that subject. This collaboration is as it should be, for in the short span of seventeen years studies in cutaneous biology have advanced at such speed and in such volume that it is impossible for any one person to keep pace with them.

The extension and deepening of our knowledge of skin is evident at all levels, from the purely morphological to the biochemical. Aided by the vast advances in technology, skin research has surged forward and publications have mushroomed. However, such rapid progress has sometimes created new problems. As new instrumentation and techniques became available, investigative dermatologists were able to look at old details in many different ways; however, less able investigators began to dabble into research that was quasi foreign to them and to inundate the literature with floods of studies, many of them sadly uncritical. For example, in the early days of the evolution of the electron microscope, that redoubtable instrument was sometimes used with more exuberance than skill. Fortunately, however, standards have risen, and with minor exceptions the results obtained today in the electron microscopy of skin compare well with those obtained in any other organ.

Our intention is far from deliberately discouraging the young explorer from venturing into new territory, whether for the sheer love of learning or for more practical reasons. But the scientific adventurer should proceed only if he is equipped, first with an idea and then with basic knowledge, an absence of bias, keen senses, and a readiness to accept facts. Those who have ignored these cardinal principles in studying skin have spewed out an enormous amount of worthless conjecture, too often dangerously disguised as fact.

In the hierarchy of the biological sciences, none are more exalted than those which use numbers because somehow numerical results convey the impression of being "exact." Theoretically, such exactitude is the goal of every scientist, but sometimes these mathematical disciplines are manipulated by scientists who are hopelessly inept in them. Nevertheless, such triflings in physiology, pharmacology, and biochemistry have sometimes been published as fact. Numbers, to be magic, must carry authority.

Another pitfall in skin research has been the danger of losing sight of the whole in a sincere, if somewhat misguided, search for the parts. With the advent of electron microscopy, the cutting of skin into as thin sections as possible has become a fetish. This procedure certainly is indispensable for achieving cytological details, but for studying the general architecture and pathways of such structures as nerves, vessels, and fibers, it is the worst possible one to follow. Indeed, for such purposes the thinner the sections, the more worthless they are. Years ago the most important lesson taught by the esteemed Professor Howard B. Adelmann at Cornell University was his insistence that beginning students in histology collect areolar connective tissue from a freshly killed animal, tease it out on slides, just as Bichat had done in the eighteenth century, study it and then, and only then, study the details by fixation and staining. We would do well to follow his lead.

Yet another common disaster is the comparative approach and use of animal models as substitutes for the human integument. Many dermatologists fail to appreciate the fact that the resemblance of human skin to that of various breeds of hairless mice and dogs and sparsely haired pigs is only superficial and does not constitute a real similarity. Still implicit in the reports of many workers is the mistaken conviction that human skin is largely glabrous, that is, without hair. No wonder it comes as a shock to some to learn that man has more hair than the gorilla! If we knew all there is to know about the skin of most extant animals, we would be in the enviable position of knowing exactly which species to select for specific experimental purposes pertinent to human medicine. Unfortunately, most studies in comparative cutaneous biology are poorly documented and we know next to nothing about the skin of some animals. This is a serious problem since no skin, not even that of the great apes, is quite like that of man.

In retrospect, when the first edition of this book was published, scarcely any other book on skin could be found in the English language except Rothman's pioneer work "The Physiology and Biochemistry of Skin." In the interval between that first edition and this one, many symposia and books dealing with one or another aspect of skin have been published. But the danger remains that in pursuing minutiae and grubbing away at the particular, the investigator can easily lose sight of its relation to the whole. To remedy this myopia in regard to skin, we have drawn upon the accumulated data derived from embryology, histology, anatomy, physiology, biochemistry, and pharmacology. This volume then is an attempt to integrate and synthesize and provide as total a view of this organ as can be found in a small book. This view is so important to the survival of the species that to neglect it is to court disaster. Our emphasis has been on human skin because despite the similarities between it and other mammalian skin which are clearly delineated in the comparative studies reported here, man's skin is in the last analysis *unique*.

WILLIAM MONTAGNA
PAUL F. PARAKKAL

1

An Introduction to Skin

Comel's (1953) description of human skin as the "monumental facade of the human body" though somewhat poetic captures most of its essential aspects. Certainly, skin envelops the entire body, is the chief means of identifying individual human beings, and is the major organ of sexual attraction. Its relatively hairless condition and its flawless conformity to the contours of the body, especially in youth, add an esthetic quality that is uniquely human. It is customary to speak of it as the largest organ of the body, but this must be accepted with some reservation. Even when we acknowledge that skin is the heaviest single organ, we cannot assert that it is the "largest" organ of the body, at least not in surface area. Despite its wide extent, it is only slightly plicated and, depending on the size of the individual, is only some two yards square. Compare this with the entire surface of the gastrointestinal tract or the 404 million alveoli in the lungs with a total surface of about five square yards when collapsed during expiration and 19 or more square yards when distended during deep inspiration (Testut, 1945).

Functionally, skin is a most extraordinary organ. Among its numerous functions, none are more important than preventing the organism from desiccating and protecting it from its environment even while maintaining it in uninterrupted communication with the environment. The

1

remarkable ability of skin to adapt to its surroundings accounts not only for the seemingly endless structural and functional differences between the various species but for certain basic patterns common to all.

In mammals, the most striking structural differences exist between the glabrous and the hairy skin. For example, with some singular exceptions (burrowing animals are a case in point), heavily haired skin has considerably less tactile sensibility than skin with few or sparse hairs because hair buffers the epidermis from contact with stimuli. Animals with extremely sensitive skin receive continuous external stimuli to which they must respond. Thus cutaneous sensitivity is intimately linked with reaction, which in turn depends on a brain that is adequate to receive and respond to messages. Thus, superlative cutaneous sensibility places the responsibility for action on the brain, which in most large animals is probably too small to assume it. The exceptions to this general pattern are equally important.

If, for example, the vast surface area of such bulky animals as whales, elephants, hippopotami, and rhinoceroses were equipped with nerve receptors numerically comparable to those in man, these animals would probably require a larger brain than they have to respond to the volleys of stimuli transmitted to it. On the other hand, exquisite cutaneous sensibility would be both useless and detrimental to such mammoth creatures; their bodies are so large and rigid that they would be helpless to rid themselves of annoyances signaled by the skin. The very tough skin of feral rhinoceroses, for example, is generally "pock marked" with open raw wounds mostly suffered in combat; still, neither the wounds nor the tick birds that keep them from healing by their constant picking seem to bother the animals. When the skin of these animals is compared with the heavily furred skin of horses, which are phylogenetically close relatives, the differences in adaptation to and protection against the environment become apparent. Despite his size, the horse is nimble and mobile, with a long free-moving neck and head that enable the recumbent animal to reach any part of his body. The switching of his long tail serves to shoo off annoying vermin from his posterior. On the other hand, the stocky rhino, relatively short-legged and short-necked, lacks such defenses against his rugged environment and relies instead on his nearly impenetrable skin, no doubt endowed with high thresholds of sensibility. To be sure, lacking experimental data, no one knows exactly how sensitive the skin of these animals is.

Two distinct patterns emerge from the study of hairy and glabrous skin. First, the epidermis is always thicker in glabrous than in hairy skin. Second, unlike the relatively simple dermal–epidermal junction of hairy skin, the epidermis of naked or sparsely haired skin always has a variably complex understructure depending on the number of hairs. Specifically, the glabrous epidermis in rodents, most carnivores, ungulates, and others has an elaborate, often labyrinthine, understructure unlike that of their hairy skin which is relatively thin and has a somewhat flat dermo–epidermal interphase anchored down, for the most part, by hair follicles and glands (Fig. 1). The thick epidermis of the elephant, with its sparse small hairs, has a highly complex, honeycomb-patterned underside (Fig. 2) which presumably enables it to adhere better to the dermis and helps to facilitate the flow of nutrients from the dermis.

Basic similarities as well as profound differences characterize the structure and function of skin in the major classes of vertebrates. In every order, suborder, or even closely related species of mammals, skin

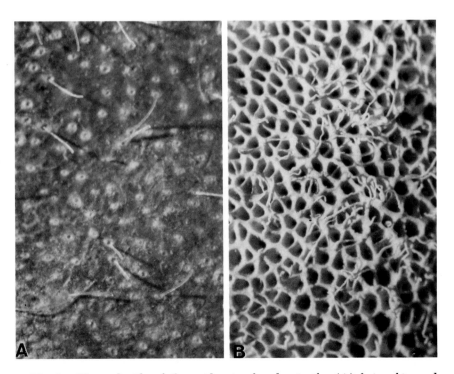

Fig. 1. The underside of the epidermis of a dog in the (A) hairy skin, and (B) toe pad. Split skin preparations.

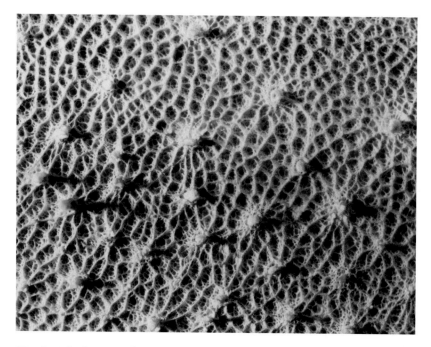

Fig. 2. The honeycomb pattern on the underside of the epidermis of an elephant (split skin preparation).

often is adapted so uniquely that only broad generalizations can be made. Even this, however, is a dangerous practice because of the many exceptions that come to mind. First, skin protects the body from injury and is largely responsible for preventing body fluid from escaping and external fluids from penetrating. But at the same time skin must in various degrees, depending on the animal, enable it to communicate with and to respond and adapt to its environment.

Skin also has numerous and varied specific functions. It becomes part of the locomotory devices in bats and aquatic mammals and its glands secrete substances that attract or repel. The pelage and other structures that grow from skin protect, warn, camouflage, or attract attention. Think, for example, of the spots on a fawn and the white warning hair on the buttocks of many deer. Some cutaneous structures are of value to the organism because of their role in the social or sexual aspects of its existence.

According to types and numbers, cutaneous structures in the different vertebrates are almost infinitely varied. Scales and spines abound in fish, and amphibian skin has copious and various kinds of glands about which we know very little. When snakes and lizards shed their outer horny layer, intact or in patches, another already keratinized layer in reserve beneath assumes the protective role. Reptilian skin is covered by various keratinous structures—scales, spines, and scutes (many of which consist of tough β-keratin). Feathers, cutaneous structures with an architecture of incomparable complexity and beauty, maintain the homeostasis of birds in hot or cold environments, on the earth, in the burrow, the water, or the air. As ornamental sensory mechanisms, they are also devices of communication and the principal means of locomotion; without them, birds could not fly, float in water, or recognize each other. Just as precisely structured are their keratinous beaks, feet and tarsi covered with horny plates, and toes that terminate in claws of various sizes and shapes. The toes of swimming birds are webbed or lobed and like the other cutaneous structures are uniquely tailored to enable these birds to adapt to their particular environments and to follow their specific patterns of life.

Among vertebrates, only mammals have hair, which varies in density, shape, thickness and length, color, patterns of growth, and texture. Some mammals also have spines, spurs, scales, horns, hooves, claws, and nails. Today the number and variety of mammalian skin glands have become fascinating objects of study because of the social and sexual implications of their secretions. However, the composition, mechanisms of control, and significance to the organism of these secretions—watery or viscid, fatty or proteinous, colorless or pigmented, odorless or fetid—remain for the most part unknown. Doubtless, though, the rising interest in the role of pheromones in communication will result in a better understanding of their value to the well-being and survival of the species. To acquire a better insight into the biological properties of human skin, scientists have assiduously studied the skin of other animals, but the knowledge derived from these studies has not always shed light on the paradoxes of man's skin, the most important single feature of which is its relative nakedness.

The significance of man's nakedness is much more profound than at first appears. Rather than an impediment, as many believe, nakedness is probably man's distinctive asset. Had he been covered with a heavy fur, a substantial and effective barrier against the environment, he probably would never have achieved his present eminence on earth. The unique quality of human life was attainable only by a naked or scantily furred mobile animal, whose large brain has many areas designed to interpret the numberless stimuli relayed by the skin and whose free skilled hands enable him to reach all parts of his body. Ironically, this very uniqueness has posed and still poses problems for the investigator, who is increasingly aware that not a single mammal except himself can be used as an effective model for *all* research needs. Parenthetically, however, it cannot be denied that with an increasing knowledge of the skin of other animals scientists are progressively better able to select more appropriate experimental animals than in the past.

Man's skin has many and varied topographic differences. Smooth over some parts of the body, it is rough and furrowed over others. Glabrous in some areas, it is downy or hairy in still others. Sometimes it is thick, horny, or taut; at others, thin, translucent, and lax. Attached to bony structures, it is usually firm, but it glides and is flaccid over joints and the ventral parts of the body. In any given area, the thickness and characteristic properties of the superficial horny layer and the amount and nature of the secretion of the skin glands make its surface rough or smooth, dry or moist. Tensile strength and resiliency vary not only in the different regions of the body and in different individuals but even in the same individual with aging. Differences in the amount of pigment, in vascularity, and in the thickness of the horny layer affect the color of skin. Knowledge of normal skin involves a familiarity with these and other variables, an understanding of the factors that brought them about, and an awareness of the basic similarities in all skin regardless of apparent differences.

A veneered or stratified tissue, skin is composed of the epidermis at the outer surface and of the connective tissue layer, dermis or corium, under it (Fig. 3). Below the dermis is the variably thick fatty layer, or panniculus adiposus, under which is a discontinuous flat sheet of skeletal muscle, the panniculus carnosus, which separates the rest of the body tissues from the integument. Well-developed in most mammals, the panniculus carnosus is vestigial in man, the platysma of the neck being about all of it that remains. Bundles of smooth muscle fibers, musculi arrectores pilorum, have a scattered origin in the upper dermis and a more discrete insertion on the bulge of hair follicles. The smooth muscle fibers of the tunica dartos in the dermis of the scrotum are oriented perpendicular to the creases on the surface and when contracted cause the ridges to be accentuated. Numerous smooth muscle

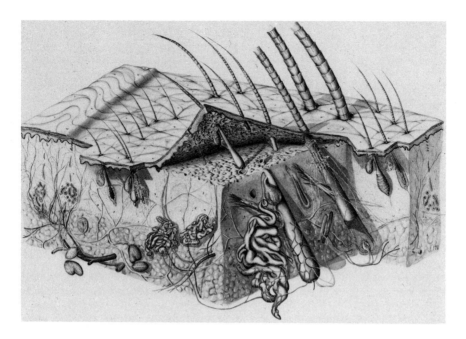

Fig. 3. Diagram of human skin, in which all of the various structures discussed in this book have been assembled. Nowhere in normal skin are these structures found all together.

fibers radiate like the vanes of a fan from the nipple and the areola of the breast; contraction of these muscles causes the areola to pucker and the nipple automatically to rise. Smooth muscle fibers are also numerous in the skin of the perineum. A bed of loose areolar tissue, or tela subcutanea, binds the skin to the fasciae of superficial skeletal muscles and bones and other connective tissues. Both the epidermis and the dermis are stratified in their own right. The successive connective tissue layers of the dermis are oriented at variable angles to one another; in some primitive mammals this arrangement is so precise that the layers form an orthogonal pattern. Five functional and mechanical frameworks compose the dermis: layers of collagen fibers and elastic fibers, nerves, blood vessels, and lymphatics.

To adapt to the numerous and nearly continuous movements of the body, skin must be a mobile tissue, capable of extension and relaxation. When increasing tension is applied to an area of relaxed skin, very small increments in load increase extension until a "stage" is reached when even great increases produce little further extension. Such elasticity depends partly on the mechanical properties of the dermal fibers themselves and partly on the pattern in which they are woven (Gibson and Kenedi, 1970). Other factors are also involved: the thickness, intrinsic elasticity, and number of folds of the skin, the firmness of fixation by the tela subcutanea, and the age, sex, and genetic makeup of the individual. The extensibility of skin, measured in different directions and at different points of the body, usually shows considerable variability (Gibson and Kenedi, 1968). These differences can be plotted on ellipses, like those removed by a surgeon, the long and short diameters of which indicate the direction of maximum and minimum extensibility. The directional variations of skin extensibility fall into patterns that seem to follow Langer's lines, which were plotted by puncturing the skin of cadavers with an awl. Because of uneven tension, the holes become elliptical along the direction of greater tension. Collagen fibers generally seem to run parallel to the long axis of the ellipse and serve to show the coincidence of direction of minimum extensibility. The skin of the abdomen has an unusual capacity for distension, but if extended beyond certain limits, unique in each individual, becomes damaged. During pregnancy, for example, striae gravidarum (Fig. 4) which are really tears appear as red streaks; during

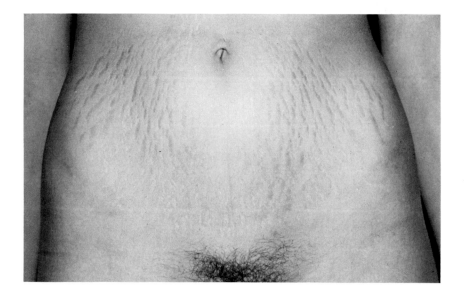

Fig. 4.   Striae gravidarum in a young woman (primipara).

postpartum these wounds are repaired but in some women remain as permanent white scars. Similar tears generally appear in the skin of some women who during adolescence experience a too rapid acquisition of excessive muscles or fat or on the sides of large breasts immediately after adolescence, when fat replaces connective tissues and causes these appendages to lose their turgidity. Permanent scars do not appear in all individuals probably because of genetic factors, since the quality of skin is unique to individuals. Although the skin of infants and children is free of creases except over joints, it nonetheless has all the congenital markings and all the true, congenital flexure lines. These lines are fixed creases or "skin-joints," which indicate a surface registering of mobile parts and the folding point of the skin and subcutaneous tissues. With maturity and aging, the direct pull of muscles on the integument, particularly on the face, and certain habitual facial expressions begin to emphasize the sharpness of the congenital lines and also cause non-congenital but permanent wrinkles to appear. What causes aged skin to lose its tone and elasticity is not altogether known. In some parts of the body, the apparent decrease in the fat content of the tela subcutanea, loss of dermal matrix, and changes in connective tissue fibers may cause the skin to sag and become deeply furrowed by wrinkles (Fig. 5). The more loosely anchored skin on either side of the congenital flexure lines is folded passively toward them during movement. The architecture of the connective tissue in wrinkles and flexure lines is still largely unknown.

The entire outer surface of the skin is imprinted with patterned inter-secting lines, which, like the dermatoglyphics, are unique to the area where they are found and are never exactly alike in any two individuals. The shape and orientation of these patterns reflect the particular stress to which the skin is subjected in a certain area of the body. If one compares the creases in the skin over the elbow, for example, which when the forearm is flexed must stretch over the conical olecranon process of the ulna, with those over the knee, which are mostly stretched on one direction, the point becomes very clear. It is also evident that the creases of the antecubital fossa and those of the popliteal fossa are similar (Fig. 6).

Surface inscriptions on the epidermis are found in pigs, elephants, and a few others, but in most other animals the surface of the skin is smooth, even in nonhuman primates unless the skin is redundant.

Flexure lines, ridges, furrows, and folds are formed in definite pat-terns during fetal life. The first ridges are formed in 13-week-old fetuses on the palmar and plantar surfaces of the tips of the digits (Hale, 1952)

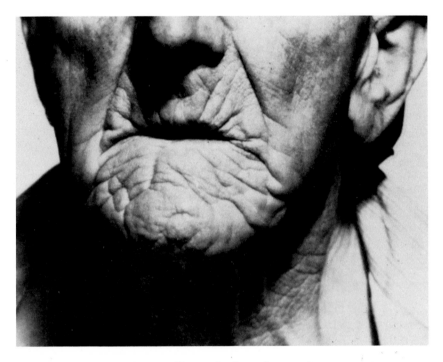

Fig. 5.   Wrinkles in the face of an aged man.

and later extend over the entire volar surfaces of the hands and feet. The width of the furrows increases at the same rate as the growth of the hands and feet (Hale, 1949). These early patterns remain unchanged during the lifetime of the individual and are altered only by permanent damage.

The palmar and plantar surfaces are filigreed by continuous and discontinuous alternating ridges and sulci (Fig. 7); the details of these markings and their configurations are collectively known as dermatoglyphics (Cummins and Midlow, 1961). Each area has unique regional and individual structural variations not matched elsewhere in the same or in any other individual. Nevertheless, the configuration patterns of the ridges and sulci can be grouped according to certain common characteristics; for example, the ridges are wider in men than in women. Although average measurements have been plotted, they mean very little because of the wide variations in the size of the digits. All ridges differ in length, ranging from "islands" which bear the one-pore orifice

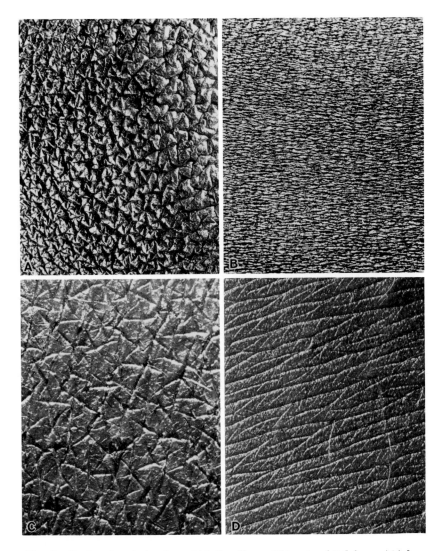

Fig. 6. Surface inscription from (A) the elbow, (B) antecubital fossa, (C) knee, and (D) popliteal fossa of the same subject. (Courtesy of Elizabeth Macpherson.)

of a sweat gland to long extended lines. Ridges branch and are so irregular that they rarely pursue a straight course. Instead they usually follow arched sweeps or form recognizable patterns; at the finger tips the principal designs are whorls, loops, arches, and combinations of

Fig. 7. Details of ridges and sulci from a toe. The white dots at the surface of the ridges are the openings of sweat glands. The white streak on the left is an acquired crease. (Courtesy of Elizabeth Macpherson.)

these (Fig. 8). Wherever three aggregations of ridge patterns with opposite directions come together, they form triangular patterns called triradia. This peculiar "signature," along with the permanence of dermatoglyphics, makes them unique factors for personal identification. Although only the fingertips are used for identification in dactyloscopy, as we shall see, other surface areas of the body could also be used. In addition to such variations, patterns can be individually abnormal. The simultaneous presence of ulnar loops on the digits, a simian crease (common in simian primates, it is a modified distal transverse flexion crease continuous from the radial to the ulnar margins of the palm, as in Fig. 9), and a triradium off the center would be rare in a normal person. The factors responsible for or predisposing to the conditions

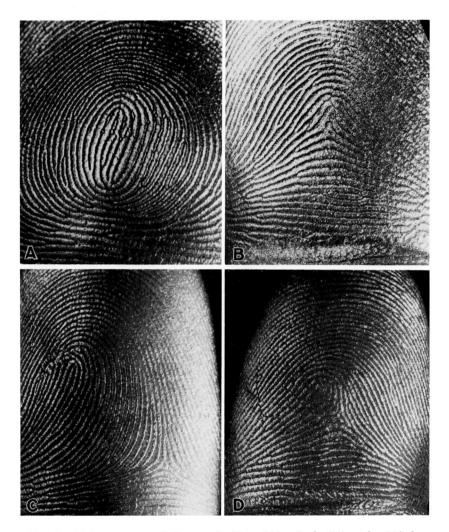

Fig. 8. Major patterns of dermatoglyphics: (A) whorl, (B) arch, (C) loop, (D) combined form. (Courtesy of Elizabeth Macpherson.)

mentioned above must have been at work in or before the fourth fetal month to alter the normal pattern of dermatoglyphics being laid down at that time. Perhaps dermatoglyphics are affected by the same agencies that impair the nervous system (Cummins, 1964). Some evidence that the development of dermatoglyphics is controlled by multiple genes can be gleaned in families that show unusual or rare configurations. Two

Fig. 9. The print of a palm, showing a well-defined simian crease. (Courtesy of Elizabeth Macpherson.)

applications have emerged from the various studies of heredity: dermatoglyphics can be used as supplementary criteria not only in diagnosing twin types but also in establishing paternity (Holt, 1968).

True dermatoglyphics are found in the friction surfaces only of primates and marsupials. Prehensile-tailed South American monkeys have them on the ventral surface of the tail, chimpanzees and gorillas on the knuckle pads. They obviously serve a common function in the animals that have them; they increase the surface and thereby provide room for more tactile sensory organs and give these surfaces a better grip in walking and in grasping. The function of such patterning is similar to the ridges on an automobile tire (Cummins, 1971).

Fingerprints, or the impressions of dermatoglyphics (Fig. 10), con-

Fig. 10. Fingerprints; the imprints of dermatoglyphics.

tain mostly water and traces of organic and inorganic compounds and are made visible with the ninhydrin reaction (Oden and von Hofsten, 1954) as well as with other agents. (These techniques are used to demonstrate prints on objects and could be useful in establishing the authenticity of manuscripts and documents.) Fingerprints can also be shown with reagents that have been used to visualize protein spots in paper electrophoresis.

In addition to the particularities of structure, function, and composition, skin has distinctive metabolic properties which will be elaborated in the chapters that follow. For example, a unique feature of steroid metabolism in skin is the steroid reduction of the Δ^4-double bond to form 5α-derivatives, in contrast to 5β-steroids which are the major part of the urinary metabolites. Human skin is capable of transforming testosterone into 5α-dihydrotestosterone, which is a more active androgen than testosterone itself. (See review by Strauss and Pochi, 1969.) Skin, then, participates not only in the catabolism of steroids but also in the formation of active hormones from inert steroid precursors supplied by the blood, and in this way partially regulates the activity of some steroid hormones. Furthermore, some skin appendages readily respond to certain steroids. Like the sebaceous glands, certain secondary sex characteristics such as beard, moustache, axillary and pubic hair, body hairs, male pattern alopecia, and acne vulgaris develop in response to androgenic stimulation.

Skin has been used for years to study allergic and immunological responses because such responses are easily produced and observable. It is also an ideal organ in which to demonstrate sensitization (Montagna et al., 1972). Furthermore, since certain reactions are produced predominantly or exclusively in skin, it is an excellent organ for investigating the biologic activities of antibodies.

Perhaps few pause to consider the vital role of skin in thermoregulation; yet, it is one of the skin's basic functions in mammals. The epidermis is a poikilothermic structure; that is, though being the true barrier between the external environment and the organism, it is separated from the deeper tissues by superficial cutaneous vascular beds which maintain the body temperature. Together with the role of thermoregulation, the importance of cutaneous thermoreceptors, blood vessels, and sweat glands and the integration of the central and vegetative nervous systems via the hypothalamic centers must also be considered. For that matter, human skin is so rich in blood that it can store as much as 4.5% of the total blood volume.

Thus, skin emerges as a highly complex organ possessing more than the discrete protective and adaptive qualities mentioned earlier. It is a vital organ whose properties *in toto* are directly involved in the biological economy of the whole organism.

References

Comel, M. 1953. "Fisiologia Normale e Patologica Della Cute Umana." Fratelli Treves Editori, Milan, Italy.

Cummins, H. 1964. Dermatoglyphics: A brief review. *In* "The Epidermis" (W. Montagna and W. C. Lobitz, eds.). Academic Press, New York.

Cummins, H. 1971. Dermatoglyphics. *In* "Dermatology in General Medicine" (T. B. Fitzpatrick, K. A. Arndt, W. H. Clark, A. Z. Eisen, E. J. Van Scott, and J. H. Vaughan, eds.). McGraw-Hill, New York.

Cummins, H., and C. Midlow. 1961. "Finger Prints, Palms and Soles. An Introduction to Dermatoglyphics." Dover, New York.

Gibson, T., and R. M. Kenedi. 1968. Factors affecting the mechanical characteristics of human skin. *In* "Proceedings of the Centennial Symposium on Repair and Regeneration." McGraw-Hill, New York.

Gibson, T., and R. M. Kenedi. 1970. The structural components of the dermis and their mechanical characteristics. *In* "Advances in Biology of Skin. The Dermis" (W. Montagna, J. P. Bentley, and R. L. Dobson, eds.), Vol. 10, pp. 19–38. Appleton, New York.

Hale, A. R. 1949. Breadth of epidermal ridges in the human fetus and its relation to the growth of the hand and foot. *Anat. Rec.* **105**: 763–776.

Hale, A. R. 1952. Morphogenesis of volar skin in the human fetus. *Amer. J. Anat.* **91**: 147–181.

Holt, S. B. 1968. "The Genetics of Dermal Ridges." Thomas, Springfield, Illinois.

Montagna, W., R. B. Stoughton, and E. J. Van Scott, eds. 1972. "Advances in Biology of Skin. Pharmacology and the Skin," Vol. 12. Appleton, New York.

Oden, S., and B. von Hofsten. 1954. Detection of fingerprints by the ninhydrin reaction. *Nature (London)* **173**: 449–450.

Strauss, J. S., and P. E. Pochi. 1969. Recent advances in androgen metabolism and their relation to the skin. *A.M.A. Arch. Dermatol.* **100**: 621–636.

Testut, L. 1945. "Anatomia Umana." Unione Tipografico-Torinese, Torino, Italy.

2

*The Epidermis**

I. Structural Features

Ever since Malpighi first described it, the epidermis has been divided into an inner layer of viable cells (stratum Malpighii) and an outer one of anucleated horny cells (stratum corneum) (Fig. 1). The stratum Malpighii is conventionally subdivided into a basal layer one-cell deep (stratum basale or stratum germinativum) in contact with the dermis, a prickle or spinous layer of variable thickness above it (stratum spinosum), and a granular layer (stratum granulosum) which contains various-sized keratohyalin granules easily stainable with many basic dyes (Fig. 2). In addition, in friction surfaces or in areas where the epidermis is very thick, there is a hyalin layer (stratum lucidum) of variable thickness which remains mostly unstained in histological preparations. The outer horny layer is composed of flattened cells, which when dissociated resemble squamae. In friction surfaces, the cells of the thick horny layer are firmly cemented together (Fig. 3).

Although epidermal cells generated in the basal layer seem to rise to the surface on a casual, first-come basis, Christophers (1971) and

* With contributions by Michael Im, Johns Hopkins Hospital, Baltimore.

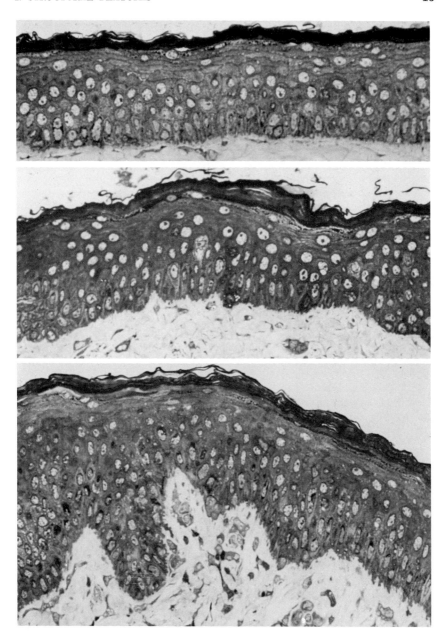

Fig. 1. The epidermis from the cheek (above), behind the ear (middle), and labia majora (below) showing the compact horny layer above and the Malpighian layer below. The horny layer is intact and compact. The differences in dermo–epidermal junction are characteristic of these areas. (Courtesy of Dr. M. Bell.)

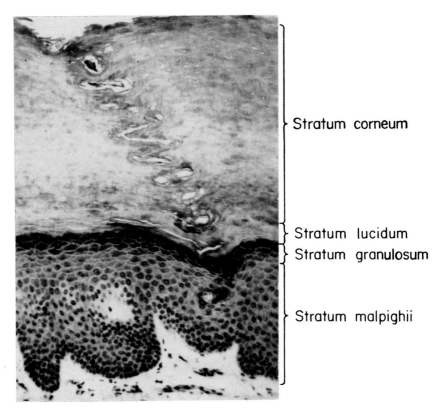

Stratum corneum

Stratum lucidum
Stratum granulosum

Stratum malpighii

Fig. 2. Epidermis from the palm showing all of its layers. The clefts in the horny layer are fragments of the spiralling duct of an eccrine sweat gland.

MacKenzie (1972) have shown that the horny layer, and to a degree the entire epidermis, is composed of precisely ordered structural units, rarely visible in histological preparations. This arrangement can be demonstrated in frozen sections of unfixed skin treated with dilute alkali or acids that swell the normally extremely thin horny cells. Seen after such expansion, the cell boundaries correspond to the acid- and alkali-resistant cell membranes.

The most precisely ordered stacking of squamae occurs in thin epidermis where, after expansion with alkali or acid, the stratum corneum is like a unified assemblage of individual cells stacked in orderly columns that extend from the stratum granulosum to the surface (Fig. 4). Along the greater part of their width, the cells in each column are in

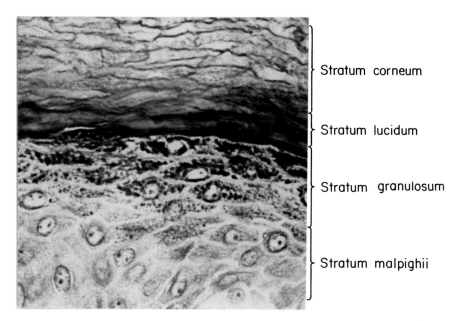

Fig. 3. The epidermis of the sole enlarged to make the layers more evident. The keratohyalin granules are of different sizes and shapes and the horny cells are firmly cemented. More than two-thirds of the upper stratum corneum has been trimmed from the figure.

contact with those above and below; laterally, where they meet the cells of adjacent columns, there is a slight overlap so that each squama interdigitates with its neighbors to produce a steplike pattern. Lateral interdigitations are usually aligned vertically throughout the thickness of the stratum corneum, although sudden breaks sometimes occur.

The position of individual squamae and the degree of their lateral interdigitation vary in regularity according to species and area. In man and other primates, where the squamae have a greater and more variable degree of overlap, the order is seldom as precise as that in rodents (Figs. 4 and 5).

How the regular columnar arrangement of keratinized cells is established and at what level in the Malpighian layer the alignment occurs are not known. In the thin epidermis of mouse ears, most of the Malpighian cells above the basal layer are flattened and aligned parallel to the overlying columns in the horny layer; the small basal cells show no precise alignment. In the thicker epidermis of these animals, flatten-

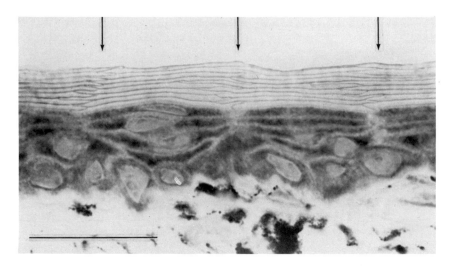

Fig. 4. Columns of stratum corneum cells from the ear skin of a hamster. Beneath these are aligned the 3 to 4 layers of the stratum granulosum and spinosum. The interdigitations between the columns of horny cells are indicated by arrows. Scale = 3 μm. (Courtesy of Dr. I. Mackenzie.)

ing and alignment under the columns occur only in the upper spinous and granular layers; no order can be seen in the lower spinous cells.

The maintenance of a columnar architecture in the horny layer seems to depend on a slow turnover of cells. The order of stacking, then, is best in those regions where the mitotic rate is normally low, and the patterns are less regular in tissues with a higher mitotic rate. Neither plantar nor palmar epidermis, characterized by great thickness and a high turnover rate, shows any evidence of columnar formation.

Although cell division in the basal cells of the epidermis is reportedly random, mitotic figures tend to occur more frequently beneath the regions that correspond to the junctions of overlying columns than beneath their central regions.

In man, some ordered structure can be seen only where the epidermis is thin. In general, however, human epidermis is thicker than that of most furry animals; hence the ordered structure is not evident. When the intact surface of the skin is viewed with the scanning electron microscope, the cells look remarkably loose (Figs. 6 and 7). At high magnifications they show imprints of irregular folds, convolutions, and occasionally villous projections (Fig. 8). Cell margins are sometimes difficult to see and adjacent cells appear to merge because of close

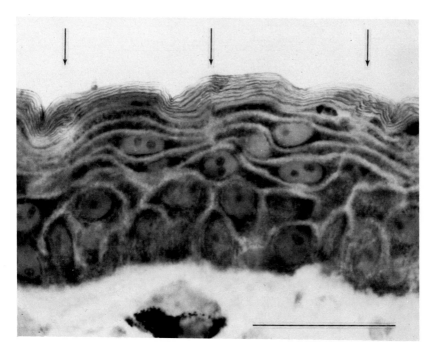

Fig. 5. Columns of cells in the stratum corneum of the skin of the arm of a rhesus monkey, with the interdigitations marked by arrows. The cells of the stratum spinosum show no alignment of cells. Scale = 30 μm. (Courtesy of Dr. I. Mackenzie.)

apposition, and desmosomal contacts are only occasionally visible at the cell margins (Dawber *et al.*, 1972). Mean cell size, measured as the horizontal diameter across the center of the cell, is 45 μm.

Basal epidermal cells are cuboidal or low columnar with the long axis aligned vertical to the skin surface (Figs. 9 and 10). Cell division occurs mainly in this layer; as the daughter cells are displaced toward the surface, they become larger and polyhedral. At higher levels, their polarity shifts as they become progressively flattened in a plane parallel to the surface of the skin (Fig. 9). The physical basis for such changes has been investigated by Branson (1968), who found in the stratum granulosum large numbers of microtubules which may be involved in the elongation and flattening of cells since they have been implicated in the maintenance of cell shape and projections in other tissues. But the property of the cytoplasmic filaments should not be discounted in the orientation of cells, many of which are oriented parallel to the surface of the skin in the stratum granulosum and corneum.

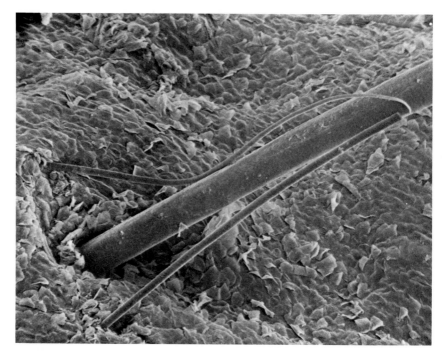

Fig. 6. Scanning electron micrograph of the intact surface of the scalp, showing the loose arrangement of the squamae and the emergence of three hairs of different sizes. (Courtesy of Dr. W. H. Fahrenbach.)

The integrity of the epidermis is maintained by attachments between adjacent cells at many small areas, first described by Bizzozero (1871). Once referred to as the nodes of Bizzozero or intercellular bridges, they are now called desmosomes or maculae adherentes (Fig. 11), one of the many types of cell junctions. (For detailed accounts of cell junctions in general, the reader is referred to Farquhar and Palade, 1963; and Kelly and Luft, 1966.) Despite reports on tight and intermediate junctions in human epidermis, little is known about their relative distribution in the different strata (Hashimoto, 1971a). Nonetheless, desmosomes are numerous in all layers of the epidermis (Fig. 9). Since new cells are constantly produced in the basal layer, desmosomes must also be continuously formed. However, because electron microscopic observations on basal cells in mitosis have been relatively few, little is known about their neogenesis in keratinizing epithelia. Perhaps they are formed

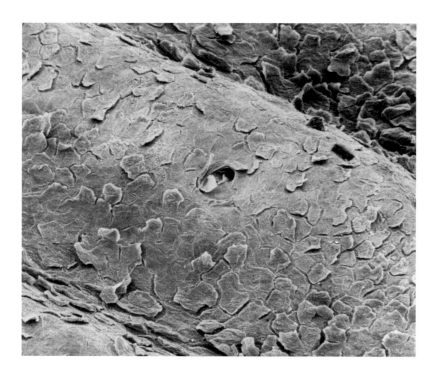

Fig. 7. Scanning electron micrograph of the intact surface of a toe showing loose surface squamae and one orifice of a sweat gland. Compare with Fig. 6. (Courtesy of Dr. W. H. Fahrenbach.)

so rapidly that their formative stages escape attention (Campbell and Campbell, 1971).

It is not clear how desmosomes affect the pressure and movement of cells. Presumably, if cells expand laterally, desmosomal contacts must be broken and then reestablished. Desmosomes disrupted after epidermal damage are re-formed within 72 hours, but how these attachments are lost and reestablished during the normal turnover of epidermal cells is not known.

Desmosomes, that is, discrete dense structures at irregular intervals along the cell membranes, have an ovoid shape and range from 5000 to 10,000 Å (Fig. 11). Despite their complex structure, they are remarkably symmetrical. Each half consists of a dense attachment plaque separated from the cell membrane by an electron-lucent zone of 150 Å. From the attachment plaques many tonofilaments radiate into the cytoplasm. In the area between contiguous cell membranes, there are

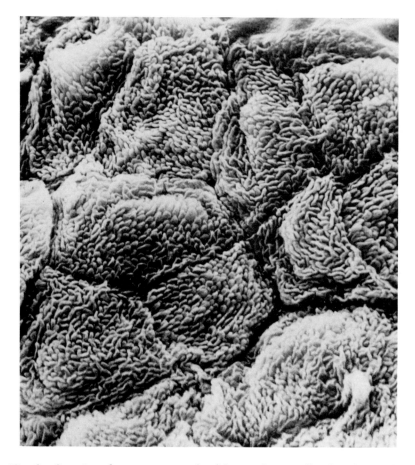

Fig. 8. Scanning electron micrograph of human horny cells after the top layer has been stripped with Scotch tape. The surfaces of the cells have villouslike projections. (Courtesy of Dr. D. Menton.)

Fig. 9. Low magnification electron micrograph of full-thickness epidermis showing basal, spinous, granular, and horny layers. Only part of the stratum corneum is seen on the upper left corner. The basal cells, at the bottom, are oriented perpendicular to the surface but those above them become parallel to the surface. There are more filament bundles in the upper layers. × 3600. (Courtesy of Dr. D. Knutson.)

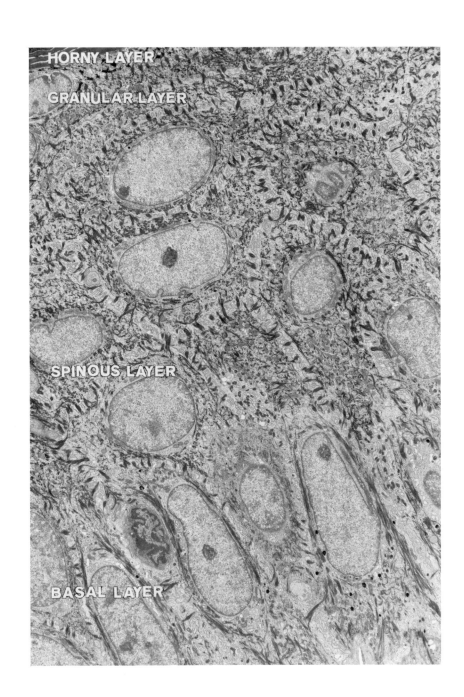

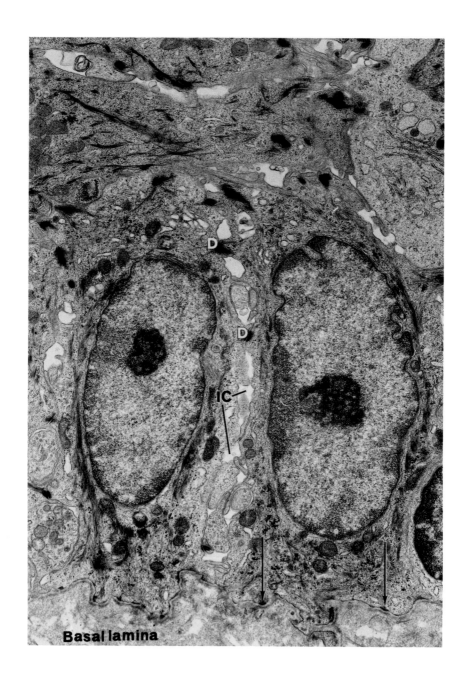

variable numbers of intercellular densities (Odland, 1958), probably mucopolysaccharides, which are stainable with ruthanium red.

What alterations occur in the desmosomes of the superficial cell layers of keratinizing epithelia is a matter of some conjecture. Listgarten (1964) found a gradual loss of contrast and definition, whereas Schroeder and Theilade (1966) found an abrupt disintegration. Snell (1965) reported that in the horny layers of guinea pig epidermis the desmosomal plaques lose their attachment to the tonofibrils and subsequently the extracellular material separates and rounds up to become a "fusiform body." Since desmosomes are thought to be important attachment areas between cells, it is vital to know what changes they undergo in the superficial layers because the exfoliation of the horny cells depends on the integrity of the cell attachments.

Hemidesmosomes, originally called "bobbins" by Weiss and Ferris (1954), form another type of attachment in the lower surface of basal cells. Like desmosomes, these half desmosomes (Fig. 10) have attachment plaques from which skeins of tonofilaments radiate into the cytoplasm.

The distinctive features of human epidermis are a relatively high total thickness, a well-developed stratum corneum and stratum granulosum, and an intimate and distinctive dermo–epidermal junction. The understructure of the epidermis in each part of the body is unique. In most domestic and laboratory animals, including nonhuman primates, the epidermis is relatively thin, with a thin stratum corneum, a poorly developed stratum granulosum, and an almost flat undersurface with shallow or no rete ridges. In the glabrous or sparsely haired areas of these animals, e.g., lips, rhinarium, muzzle, and friction surfaces, the undersurface of the epidermis is always variably complicated. For example, the underside of the thin epidermis in the hairy skin of monkeys has a nearly flat dermo–epidermal junction, but on their thick, glabrous ischial callosities it is complex (Figs. 12A,B). The epidermis of the glabrous Cetecea has an understructure composed of long, parallel laminae perforated by long, conical hollows that contain dermal papillae.

In histological sections the dermo–epidermal junction of human skin is variably irregular, with cones, ridges, and cords of different lengths

---

Fig. 10. Enlarged view of two basal cells at the dermo–epidermal junction. The basal lamina follows the contours of the plasma membrane on the dermal side. Hemidesmosomes (arrows) can be seen along the plasma membranes. Intercellular spaces (IC) appear where cells are not attached by desmosomes (D). The nucleus almost fills the cells. × 6700. (Courtesy of Dr. D. Knutson.)

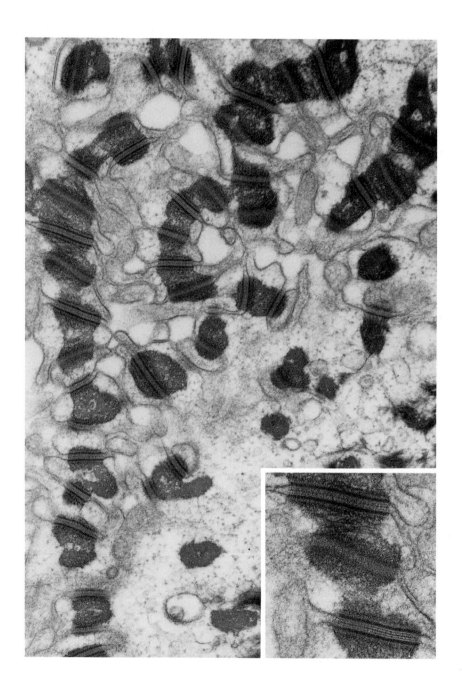

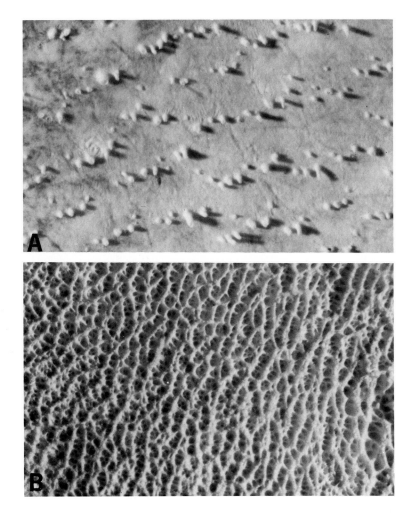

Fig. 12. The underside of the epidermis in (A) the hairy skin of a rhesus monkey, and (B) the ischial callosity of the same animal. Split skin preparation.

Fig. 11. Desmosomes at the apposing surfaces of keratinocytes in the spinous layer. Each desmosome, shown at higher magnification in the inset, consists of two highly electron-opaque attachment plaques continuous with the plasma membrane, into which are looped tonofilaments; other electron-opaque bands are extracellular between the plaques. × 54,000; Inset × 93,000. (Courtesy of Dr. M. Bell.)

extending at different depths into the dermis. In split-skin preparations, these cones, ridges, and cords resemble a series of branching ridges and mounds of different sizes which enclose channels, valleys, craters, and complex miniature systems of connecting caverns (compare Fig. 1 with Figs. 12–16). Through this uneven terrain, usually at the tip of the ridges, the ducts of glands and the pilary canals enter. Horstmann (1952, 1957) believed that the epidermal ridges form specific and predictable patterns around each of these structures, but our observations suggest that the ridges are molded at random around them. Where the epidermis characteristically has a relatively flat dermo–epidermal junction, as in the eyelids and scrotum, some crinkling of the epidermal ridges usually occurs around the attachment of the appendages (Fig. 13).

Fig. 13. The underside of the epidermis of an eyelid showing many small hair follicles and the relatively flat dermo–epidermal junction, except at the base of the hair follicle.

Fig. 14. Undersurface of the epidermis of the tip of the nipple (A), side of nipple (B), and areola (C).

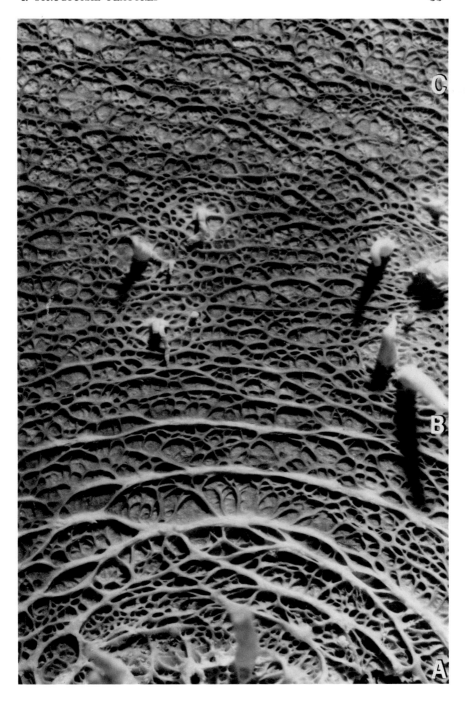

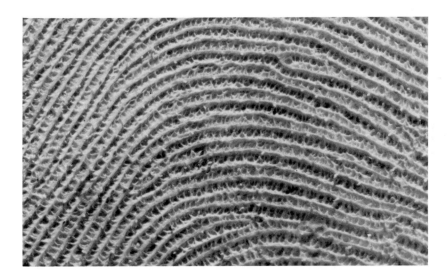

Fig. 15.   The complex undersurface of the epidermis of a thumb.

Well-formed characteristic ridge patterns can be found in the scalp, forehead, chin, jaw, ears, and neck, but not in the eyelids (except the palpebral borders) or malar areas of the cheeks. The epidermis of the breasts shows a characteristic complex system of ridges with spongelike patterns at the tip of the nipple, parallel horizontal ridges at the side of the nipple, mosaic patterns on the areola, and a low-relief pebbled pattern in the rest of the breast (Fig. 14). The most impressive undersurface is at the tip of the digits where parallel ridges form an intricate system of cavernouslike valleys and tunnels (Fig. 15). The valleys correspond to the sulci on the outer surface; this can be confirmed by the fact that the ducts of sweat glands enter the ridges on the underside and emerge on the ridges at the surface. The ridges of the underside of the labia minora form whorls and horseshoe patterns. Seen under the scanning electron microscope, the basal surface of the understructure is everywhere pitted and in the friction surfaces is actually thrown into solidly packed villous protrusions (Fig. 16). In old age, the systems of ridges tend to flatten out, particularly in the scalp and breasts.

Fig. 16.   Scanning electron micrographs of the underside of the epidermis of a toe at magnifications of (A) 80×, (B) 500×, and (C) 2000×. The villous projections of the basal cells interdigitate with the brushlike terminations of elastic fibers. (Courtesy of Dr. W. H. Fahrenbach.)

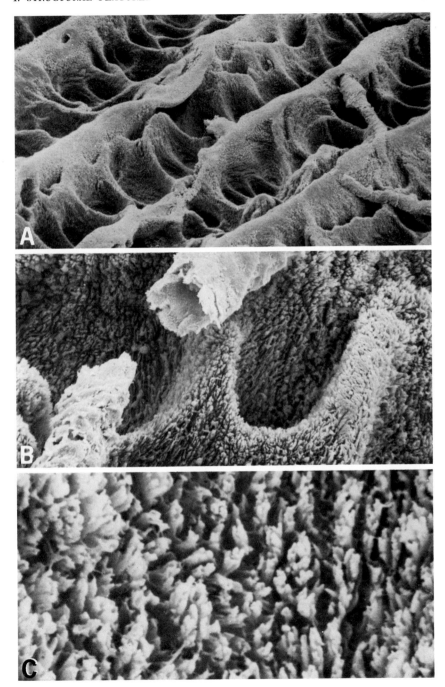

The total surface of the underside of such regions as the tip of the nipple or fingertips must be many times that of the outer surface but to what purpose is not known. Thus, the number of new cells that are formed in the basal layer of thick epidermis is much greater than that available to thin epidermis with a relatively uncomplicated dermo–epidermal junction.

At the dermo-epidermal junction a PAS-positive "basement membrane" follows the contours of the basal cells that face the dermis. Under the electron microscope, this structure, which resembles a continuous sheet of filamentous material about 500 Å thick, is separated from the plasma membrane of the basal cells by an equal distance (Fig. 10). Previously called "adepidermal membrane" or "dermal membrane," it is now called "basal lamina" (Fawcett, 1963).

Anchor fibers are attached at intervals along the basal lamina and extend about 250 nm into the dermis. These are often frayed into smaller units and are attached to the basal lamina and show banding patterns (Palade and Farquhar, 1965; Brody, 1960; Stern, 1965; Susi *et al.*, 1967). Anchor filaments are well developed in the skin of tadpoles and embryonic chicks, where they sometimes extend several micrometers into the dermis (Kallman *et al.*, 1967; Kischer and Keeter, 1971). The function and biochemical composition of these fibers are still unknown; Wessells (1965) suggested that they play an early role in the development of chick feathers. Because of their arrangement and position, they are thought to anchor the basal lamina in relationship to the overlying epidermis.

Just beneath the epidermis of aquatic or amphibious vertebrates is the basement lamella, a highly ordered array of collagen fibers (Nadol *et al.*, 1969; Rosin, 1946). The arrangement of these fibers in the lamella, i.e., successive piles oriented somewhat at right angles to one another, appears to be orthogonal. Mammalian skin does not have a well-ordered basement lamella, but the cornea has an orthogonal arrangement of collagen fiber layers (Jakus, 1961).

II. Histochemical and Microchemical Properties

A. *Carbohydrates*

Unlike the stratified squamous epithelium in mucous surfaces, which is rich in histochemically demonstrable glycogen, normal mammalian epidermis contains relatively small amounts of it scattered sporadically in restricted sites. It is often present in the epidermis of the scrotum and scalp, in the epidermal cells around the pilosebaceous and sweat gland orifices, in skin crevices, or in pilosebaceous orifices that enclose blocked cornified material. When present, glycogen is found mostly in the upper cells of the stratum spinosum.

Quantitative analyses of glycogen in microgram quantities of regenerating epidermis during wound healing reveal progressively increasing gradients of glycogen content from the marginal epidermis to the advancing epithelial tip. For example, in marginal epidermis, the content of glycogen is approximately doubled over the normal and in the distal halves of the actively migrating epidermis of wounds, it is a maximal nine times that of normal (Im and Hoopes, 1970a,b).

In the two- or three-layered epidermis of the human fetus less than 4 months old, basal as well as peridermal layers are laden with glycogen (Lombardo, 1907, 1934; Sasakawa, 1921; Achten, 1959). When in the fourth month the basal cells begin to form the anlagen of the cutaneous appendages, glycogen largely disappears from these cells, particularly at the sites of rapid cell division, and from 6 months to birth when keratinization takes place there is a sudden reduction in glycogen.

The epidermis of adult laboratory animals, like that of man (Lobitz and Holyoke, 1954), has demonstrable glycogen after injury (Argyris, 1952; Bradfield, 1951; Firket, 1951). Morphological aberrations in the epidermal cells are usually accompanied by an accumulation of glycogen. When the cells become hypertrophied and vacuolated and the cytoplasmic basophil substances become diluted, glycogen usually appears.

The epidermis also contains PAS-positive, nonglycogen substances that are hydrolyzed by saliva or diastase. Such reactive substances are usually found in the intercellular spaces, especially in the stratum granulosum.

When placed in culture media, skin slices can respire for several days. A supply of such carbohydrates as glucose, mannose, fructose, galactose, sorbitol of the monosaccharides, or maltose of the disaccharides maintains the respiration of skin cells *in vitro* (Carney et al., 1962). Of these nutrients, only glucose has been studied thoroughly (Decker, 1971).

The levels of free glucose in the epidermis depend on blood sugar levels. Since glucose diffuses freely into epidermal cells, its utilization is not limited, so that the glucose concentration in the epidermis is about 50% that in blood (Halprin and Ohkawara, 1967). Upon entering the cells, glucose becomes phosphorylated to glucose 6-phosphate through the action of hexokinase (Adachi and Yamasawa, 1966; Halprin and Ohkawara, 1966c).

The overall activities of each pathway of glucose metabolism have been studied with radioisotope-labeled substrates. Most of the glucose is converted to lactate through glycolysis, only a small portion being oxidized to carbon dioxide through the pentose phosphate and tricarboxylic acid cycles (Freinkel, 1960; Pomerantz and Asbornsen, 1961). A small amount of glucose may be utilized in the synthesis of glycogen (Adachi, 1961) and glucuronic acids (Jacobson and Davidson, 1962).

The intermediate steps of glucose metabolism are determined by measuring enzyme activities that catalyze each metabolic step. All the enzymes of the glycolytic pathway have been determined in human (Halprin and Ohkawara, 1966a,b; Hershey et al., 1960) and macaque (Adachi, 1967) epidermis. Glucose-6-phosphate dehydrogenase and 6-phosphogluconate dehydrogenase of the pentose phosphate shunt are extremely active in epidermis (Halprin and Ohkawara, 1966a,b; Hershey et al., 1960; Im and Adachi, 1966; Weber and Korting, 1964), which also exhibits the isocitrate dehydrogenase (Cruickshank et al., 1958), fumarase (Hershey et al., 1960), succinic dehydrogenase (Dushoff et al., 1965; Rosett et al., 1967), and malate dehydrogenase activities (Hershey et al., 1960) of the tricarboxylic acid cycle. Glycogen synthetase and phosphorylase activities of glycogen metabolism are low (Halprin and Ohkawara, 1966a,b), but the activities of uridinediphosphate glucose dehydrogenase, glucuronate reductase, xylulose reductase, and xylitol dehydrogenase of the uronic acid metabolism are also present (Fukui and Halprin, 1970). The activities of glutamate dehydrogenase (Adachi et al., 1967a) and of alanine and aspartate aminotransferase (Adachi et al., 1967b; Coffey et al., 1963), which acts as a bridge between carbohydrate and amino acid metabolism, are considerable.

Although 80 to 90% of epidermal energy, in the form of adenosine triphosphate (ATP), is generated through respiration, ATP produced from glucose catabolism is supplied mainly through glycolysis, only minimally by the tricarboxylic acid cycle (Im and Hoopes, 1970b; Freinkel, 1960; Pomerantz and Asbornsen, 1961). Conceivably the main supply of the substrate for the tricarboxylic acid cycle is from phospholipid through β-oxidation (Cruickshank et al., 1962; Yardley and Godfrey, 1961).

A specific feature of the epidermis—intense glycolytic activity in the presence of oxygen—is amplified by trauma. For example, during wound healing, glycolytic activity increases in the regenerating epidermis and a minor alteration in glucose catabolism occurs through the tricarboxylic acid cycle. Wounded skin also exhibits increased glucose utilization and lactate production. The activities of glycolytic enzymes increase 4 times over normal levels (Im and Hoopes, 1970a), and the rate of lactate production from uniformly labeled glucose-^{14}C is tripled in 3-day-old wounds (Im and Hoopes, 1970b). The activities of key glycolytic enzymes increase in some skin disorders and in experimental lesions (Halprin and Ohkawara, 1966b; Im, 1968).

Healing wounds show a fivefold increase in oxygen uptake (Paul et al., 1948). However, a small part of glucose metabolism through the tricarboxylic acid cycle decreases to half that of normal skin (Im and Hoopes, 1970b). Presumably, then, the fuel for increased respiratory oxidation in healing skin comes from other sources of nutrient.

Glucose metabolism through the pentose phosphate shunt supplies the pentose needed for nucleic acid formation and for reduced nicotinamide denucleotide phosphate (NADPH), a reducing equivalent needed for lipogenesis and indirectly for keratinization. Increased activity in this shunt in psoriasis, hyperplasia, tumors, and wound healing reflects the increased demands of the epidermis for synthetic activities. Glucose-6-phosphate dehydrogenase activity of the pentose phosphate shunt triples in psoriasis (Halprin and Ohkawara, 1966b), increases 5-fold in tumors and hyperplasia (Im, 1968), and 7-fold in the migrating epidermis during wound healing (Im and Hoopes, 1970a). There is a 50% increase in glucose flow along the pentose phosphate shunt during wound healing (Im and Hoopes, 1970b).

B. Enzymes

The many histochemically demonstrable enzymes found in the epidermis were listed by Montagna (1962) and need not be repeated. We call attention here to only a few hydrolytic enzymes.

Normal epidermis, whether in man or in other animals, contains no demonstrable alkaline phosphatase; however, it is always present in damaged epidermis and disappears when the damage is repaired.

An appreciable acid phosphatase reaction is obtained only in the granular layer and in the thin layer of cells immediately above it. Even the stratum corneum sometimes shows enzyme action.

With microanalytical methods and α-naphthyl phosphate as substrate, progressively increasing gradients of acid phosphatase activity can be seen from the lower layers to the upper ones in normal epithelium and in the regenerating epithelium of wounds from the marginal part to the advancing tip (Im et al., 1972). Low acid phosphatase activity in the actively migrating epithelial tip and the lower layers of epithelium indicates a lesser degree of differentiation.

There have been numerous studies of esterases in the epidermis (Montagna, 1962). The large amounts of these enzymes in the epidermis probably constitute an esterase spectrum of more or less specific enzymes that act on specific esters of carboxylic acid (Chessick, 1953). Heavy concentrations of nonspecific esterases have been found in the band of the epidermis that corresponds to its keratogenous zone (a thin layer of cells just above the granular layer, in which –SH groups are also concentrated).

There is appreciable reactivity for β-glucuronidase in all epidermal cells, from the basal layer upward. The most conspicuous reaction is in the stratum granulosum and the hyalin cells immediately above it. This reactive band corresponds to the keratogenous zone of the epidermis.

Yet microanalysis of β-glucuronidase activity with 4-methylumbelliferyl-β-D-glucuronide as substrate shows a distribution gradient of this enzyme with greater reactivity in the lower layers than in the upper ones. Regenerating epithelium shows markedly increased enzyme activity; progressively increasing gradients of β-glucuronidase activity have been observed from the marginal epithelium to the migrating epithelial tip (Im and Hoopes, 1972). The high activity of β-glucuronidase in the migrating epithelial tip may facilitate epithelial cell migration by the hydrolytic action of the enzyme on the ground substances in the dermis. Grillo et al. (1969) have suggested that collagenase liberates the epithelium for migration by breaking down collagen fibers.

III. Development

Human embryos 3 weeks old are covered with a single layer of cells, which by 3 months is two cells deep: the upper one represents the presumptive periderm, the lower the presumptive epidermis. The cells in each stratum divide rapidly and by the fourth fetal month both periderm and epidermis are stratified (Pinkus and Tanay, 1957). The periderm is two cells thick over most of the body except on the forehead, eyebrows, and around the openings of the mouth and nostrils where it is thicker. At the end of the fifth month, a stratum granulosum with keratohyalin granules appears around the orifices of hair follicles and where the epidermis is thickest (Breathnach, 1971; Achten, 1959);

later it develops over the entire body. When the epidermis forms a stratum corneum, periderm cells are shed into the amniotic fluid.

The epidermis develops asynchronously. For example, in the back and chest, it remains relatively undifferentiated until the fourth fetal month whereas in the eyebrows, lips, and nose it has undergone advanced differentiation even at 3 months.

A. The Periderm

The periderm is an embryonic or fetal tissue whose structure, function, and fate have been investigated with considerable thoroughness (Breathnach, 1971). Briefly, peridermal cells, initially cuboidal, become progressively flattened during later development and attain microvilli on the amniotic surface (Wolf, 1967a); on the underside, peridermal cells are attached to epidermal cells by desmosomes. Adjacent peridermal cells are attached by junctional complexes that consist of a tight junction (zonulae occludentes), intermediate junctions (zonulae adherentes), and desmosomes (maculae adherentes). Peridermal cells contain large amounts of glycogen and many filaments 80 Å thick (Wolf, 1967b). Before peridermal cells are shed, their cytoplasm becomes progressively clogged with filaments, and their nuclei and other organelles undergo partial or complete degeneration (Bonneville, 1965, 1968).

The function of the periderm is unknown. Since its ultrastructural properties resemble those of the amniotic epithelium (Sinha et al., 1970; Schmidt, 1963; Lyne and Hollis, 1971; Thomas, 1965; Tiedemann, 1972), it may do more than act as a protective covering for the developing epidermis. Hoyes (1968) and Wolf (1967c) suggested that the periderm participates in the exchange of material between the fetus and the amniotic fluid, the microvilli presumably increasing the surface area for such an exchange.

The periderm undergoes partial keratinization before it is shed into the amniotic cavity, and the cells of the stratum corneum and periderm are basically similar: both have large amounts of fibrillar protein and during the last stages of maturation neither has nuclei nor cytoplasmic organelles. Moreover, the increased density of the plasma membrane of mature periderm cells resembles the thickened cell envelopes of stratum corneum cells. On the other hand, periderm cells develop neither membrane-coating nor keratohyalin granules, both hallmarks of epidermal keratinization.

Chick periderm cells have large characteristic inclusions, several micrometers in diameter, called peridermal granules or cribiform bodies (Parakkal and Matoltsy, 1968; Mottet and Jensen, 1968) apparently formed by interlaced strands composed of about 80 Å-thick particles. These granules were once mistaken for keratohyalin or trichohyalin granules (Rosenstadt, 1897; Wessells, 1962; Kingsbury *et al.*, 1953).

B. The Epidermis

The early presumptive epidermis is one layer thick. Later, the flattened cells become cuboidal and their nuclei begin to occupy much of the cytoplasm. Glycogen gradually accumulates until the cytoplasm is replete with it. The prominent intercellular spaces that appear at this stage of development may be shrinkage artifacts (Breathnach and Robins, 1969). Under the comparatively smooth dermo–epidermal junction is a basal lamina about 500 Å thick. There are few desmosomes and hemidesmosomes.

By 12 weeks, the epidermis is 2 to 4 cells thick and contains much glycogen (Serri and Montagna, 1961; Brody and Larsson, 1965). Intercellular spaces become less prominent and desmosomes increase in number. During subsequent development, the number of cell layers increases and both membrane-coating and keratohyalin granules are formed. The first layers of the stratum corneum are formed by 20 weeks when the epidermis attains adult characteristics. The way glycogen is stored in the skin reflects the intimate relation between carbohydrate metabolism and cutaneous function. During the first half of fetal life, when the liver has not yet assumed this function, large amounts of glycogen are stored in the epidermis (Rothman, 1954).

C. Factors That Influence Skin Development

The epithelial and connective tissue elements of skin are derived from the ectoderm and mesoderm, respectively. The development of cutaneous structures in embryos depends on the interaction between these two primordial tissues (Billingham and Silvers, 1963; Wessells, 1967).

In some animals, the mesenchyme determines the type of differentiation that occurs in the overlying epithelium (McLoughlin, 1963). Epidermis from 6-day-old chick embryos grown *in vitro* on its own mesenchyme normally keratinizes; when grown on mesenchyme from the gizzard, it develops into a mucus-secreting epithelium and sometimes even becomes ciliated.

When epidermis from the ears of adult guinea pigs was combined with dermis from the foot pads, the regenerated epidermis resembled that of foot pads (Billingham and Silvers, 1967). Thus, even in adult animals, the dermis influences epidermal differentiation. This, however, is not universally true; foreign connective tissue has no effect on the normal development of differentiating epithelium from the tongue, cheek, and esophagus.

Furthermore, since reptilian epidermis differentiates normally regardless of extraneous factors (Flaxman et al., 1968), it must contain its own intrinsic developmental properties. When grown in vitro either with mesenchyme or by itself, it accumulates several "epidermal generations" one above the other. The complex layering of the epidermis in squamate reptiles (lizards and snakes), the "epidermal generation," consists of the oberhäutchen, a β-layer with the characteristics of feather keratin, mesos layer, and an α-layer similar to that in mammalian epidermis. Since the layers have different types of proteins (β- and α-keratin), patterns of differentiation alternate even though both develop from the same germinative layer (Maderson, 1965; Roth and Jones, 1967, 1970; Alexander and Parakkal, 1969). In this case, several epidermal generations can be grown in vitro, one above the other; thus the mechanism of controlling differentiation must be in the epidermis.

Epidermis develops normally even when it is separated from its mesenchyme by a Millipore filter which allows no direct contact between the cells of the two components. Moreover, epidermis can be grown alone on a collagen gel in 20% chick embryo extract (Wessells, 1964; Dodson, 1967). The most important element for histodifferentiation is the attachment of basal cells to a suitable substratum. When adult human skin is grown in culture, only peripheral epidermal cells wander onto the substratum, proliferate, and become organized into a squamous keratinizing epithelium (Flaxman et al., 1967). When they lose their attachment to the substratum, basal cells are incapable of differentiation.

Vitamin A can have profound effects on epidermal differentiation. When large amounts are added to cultures of embryonic chick skin, the epidermis differentiates into a mucus-secreting epithelium (Fell and Mellanby, 1953; Weiss and James, 1955; Fitton-Jackson and Fell, 1963) that eventually turns into goblet cells. Excess amounts of vitamin A produce similar but less pronounced effects on adult mammalian skin both in vitro and in vivo (Szabó, 1962). These changes are, however, reversible.

Epidermal specificities, then, are developed and maintained through the interactions between epidermis and dermis, between basal cells and substrate, and among the epidermal cells themselves.

IV. Life Cycle of Epidermal Cells (Keratinocytes)

The basal layer of the epidermis has a permanent population of germinal cells whose progeny undergo specific patterns of differentiation (keratinization) on their way up to the surface and are eventually shed. Thus, in the cycle of every epidermal cell three distinct phases can be discerned: (1) mitosis, (2) differentiation, and (3) exfoliation (Fig. 17).

A. *Mitosis*

New epidermal cells are continuously formed by the germinative layer to compensate for those that exfoliate at the surface. Thus, like hematopoietic tissues and intestinal mucosa, the epidermis belongs to a class of tissues whose cell production is commensurate with the constant loss.

As in all dividing cells, the mitotic cycle of epidermal cells can be divided into four arbitrary stages: (1) the S phase, characterized by nuclear DNA synthesis; (2) the G_2 phase, in which premitotic growth occurs; (3) the M phase, or actual division, and (4) the G_1 phase or postmitotic growth. In histological preparations, the synthetic phase can be recognized by means of autoradiographic techniques after the tissues have been exposed to tritiated thymidine. The different phases of mitosis, especially metaphase, are usually visualized, as in other biological preparations, by treating the tissues with colchicine *in vivo* or *in vitro*.

On the basis of histological analyses of mitosis (Pinkus, 1954), earlier reports on epidermal cell renewal contained several inherently erroneous statements, and even today the duration of the different phases of mitosis in epidermal cells has not been well established (Weinstein and Frost, 1969; Nicole and Cortese, 1972). Until the duration of mitosis in human skin is accurately determined, the data derived from converting mitotic counts to mitotic rates according to the following formula will have to remain problematic (Halprin, 1972):

$$\text{Turnover time of a germinative cell population} = \text{mitotic rate} = \frac{\text{mitotic count}}{\text{mitotic duration}}$$

In earlier studies, an important source of error in calculating mitotic rate was the practice of counting cells in whole epidermis instead of restricting the count to the germinative cells. Still another source of error was to count dividing cells in tangential sections.

The use of radioactive glycine to study epidermal turnover started with the work of Rothberg et al. (1961). After the administration of glycine-^{14}C, the upper layer of stratum corneum can be measured daily for radioactivity. This method presupposes that the epidermis is basically composed of two compartments: the viable cells of the Malpighian layer, which incorporate the radioactive amino acid, and the dead cells of the stratum corneum, which do not. Hence, the turnover time of stratum corneum cells is the time required for the isotope to appear in the top layers. According to this calculation, the average turnover time of normal stratum corneum is about 13 to 14 days (that of psoriatic stratum corneum is 2 days). Both the small number of subjects used in the study (2 patients each for normal and psoriatic) and the dispute regarding the uneven incorporation of glycine into the Malpighian layer have to be taken into account before any conclusions can be drawn on epidermal turnover time (Halprin, 1972). Tritiated thymidine has also been used to measure the transit time of cells through the Malpighian layer. After intradermal injections of ^{3}H-thymidine, biopsy specimens are removed at intervals from the area to see how far and how fast labeled cells migrate (Epstein and Maibach, 1965). All cells in the process of nuclear DNA synthesis incorporate the labeled thymidine; hence this method is valid only for measuring the transit time of viable cells because once they are completely keratinized they have no nucleus. Using this technique, Weinstein and Van Scott (1965) found that cells require 12 to 14 days (in psoriasis, 2 days) to move from the basal layer to the stratum corneum. The most important observation in these studies was that the labeled cells moved from the base to the surface in a random fashion either because of a differential migration rate or because they left the germinal compartment at different times. Therefore the transit time of 12 to 14 days should be regarded as the minimal, not the average, time (Halprin, 1972).

The diurnal cycle, the alternation of activity and rest, is one of the factors that affect mitotic rates in the epidermis. Mitosis is highest during periods of rest or sleep. Diurnal variations in mitotic activity have been reported in a number of animal tissues (Kahn et al., 1968). Bullough (1965) believes that the diurnal rhythm is related to epi-nephrine, a potent mitotic inhibitor, which has high concentrations during periods of activity and low ones during sleep. In spite of early

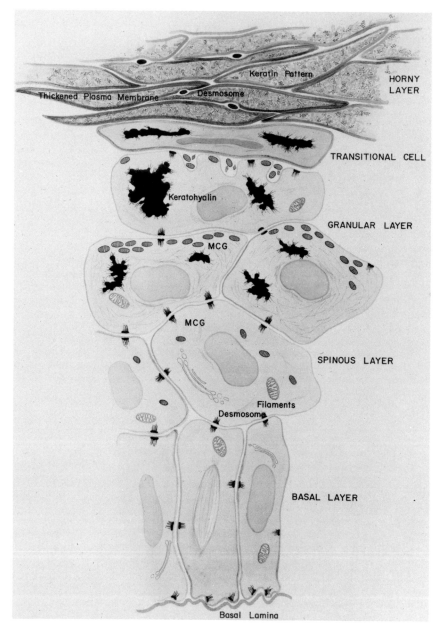

Fig. 17. Schematic diagram showing the layering of the epidermis. The basal cells are mitotically active. The differentiation products—filaments, membrane-coating

reports to the contrary, human epidermis shows only a slight increase in thymidine-labeled cells at midnight, an indication that it is less sensitive to diurnal variations than that of other animals.

In such cutaneous organs as hair follicles and sebaceous glands, mitotic activity is under the direct influence of androgens; in the stratified squamous epithelium of the vagina, it is under the influence of estrogen. However, there is no evidence that mitotic activity in the epidermis of the general body surface is directly affected by either hormone.

In normal epidermis, the production of cells at the base is balanced by exfoliation at the surface, but injuries such as radiation, tape stripping, scraping, and incisions upset this balance. Within 24 to 36 hours after trauma, there is a burst of mitotic activity in the lower layers of the affected epidermis (Pinkus, 1951, 1952; Weinstein and Frost, 1971), even in partially differentiated cells.

Bullough (1965) and Bullough and co-workers (1964, 1967) reported finding a tissue-specific "epidermal chalone" that inhibits mitosis. This substance, a heat-labile, basic glycoprotein with a molecular weight of about 25,000 which is obtained from mouse, pig, and human skin, is said to be effective in inhibiting mitosis *in vivo* and *in vitro*. For example, Elgjo and Hennings (1971) found that an intraperitoneal injection of a partially purified extract of pig skin, assumed to contain epidermal chalone, significantly depressed both mitotic rate and DNA synthesis in a transplantable squamous cell carcinoma in hamsters. However, adrenalin appears to be essential for epidermal chalone to exert its full antimitotic action (Bullough and Laurence, 1964). Moreover, adrenaline, alone is a mitotic inhibitor (Bullough *et al.*, 1967). Not all investigators have confirmed the inhibitory action of chalone. Baden and Sviokla (1968), for example, found no such substances in preparations tested on human skin *in vitro*. Because of these contradictory findings, unqualified acceptance of "epidermal chalones" is premature.

B. Differentiation (Keratinization)

As epidermal cells migrate upward from the basal layer, they lose their mitotic potential to a great extent and begin to synthesize such specific constituents as fibrillar and amorphous proteins, keratohyalin, and membrane-coating granules. Their surface becomes modified, and

---

granules (MCG), keratohyalin, and thickened plasma membrane—are shown in the different layers. The fully cornified cells are packed with a filament-matrix and show the "keratin pattern."

finally their nuclei and cytoplasmic organelles are lost. This sequence of differentiation will be dealt with under: (1) fibrillar protein, (2) amorphous proteins, (3) keratohyalin granules, (4) membrane-coating granules, (5) cell membrane modification, and (6) resorption of nuclei and organelles. After the cells have completed differentiation and become part of the protective system in the stratum corneum, they are exfoliated (Fig. 17).

1. FIBRILLAR PROTEIN

The "tonofibrillen" of Heidenhain, first described by Ranvier in 1879, are scattered throughout the cytoplasm of epidermal cells. Tonofibrils or fibrillar proteins are the major intracellular products of the epidermal cells. They were once erroneously believed to run from cell to cell across intercellular bridges, forming a continuous network of cytoplasmic connections between adjacent epidermal cells (Chambers and Renyi, 1925). In fresh or formalin-fixed frozen sections, tonofibrils are anistropic (Schmidt, 1937; Litvac, 1939); in paraffin sections, they can be stained with various methods, among them Heidenhain's hematoxylin and Mallory's phosphotungstic acid–hematoxylin.

In studies of x-ray diffraction patterns of the epidermis, Giroud and Champetier (1936) and Derksen *et al.* (1938) found that both the cornified and noncornified layers of the epidermis give an α-keratin pattern similar to that obtained by Astbury (1933) from wool. Giroud and Bulliard (1934, 1935) then demonstrated histochemically abundant thiol groups in noncornified cells and disulfide bonds in cornified cells. From these observations Giroud and Leblond (1951) concluded that the α-keratin patterns obtained from the epidermis are due to birefringent tonofibrils.

When histochemical preparations for –SH groups are viewed under the phase-contrast microscope, the reaction appears to be localized in the tonofibrils. Furthermore, Giroud and Leblond (1951) suggested that the –SH groups of the fibrous protein change into the disulfide bonds during the formation of the cornified cells and thus give the stratum corneum both strength and chemical inertness.

Most of the earlier attempts to study the fine structure of the epidermis with the electron microscope yielded meager results. There were assertions and denials of the syncytial nature of epidermal cells and of the presence and absence of tonofibrils. Porter (1954, 1956) made the first significant contribution to our knowledge of the structure of the epidermis; he found that the many fine filaments in undifferentiated basal epidermal cells in frog skin terminate at the desmosomal contact points. The diameter of these filaments in different keratinizing epithelia varies from 35 to 100 Å. Weiss and Ferris (1954) and Rhodin and Reith (1962) believed the filaments to be tubular. Differences in the size and shape of filaments are probably due to different methods of preparing material for electron microscopy and perhaps to species differences. All epidermal cells have a cytoplasm packed with filaments (Figs. 10 and 18), which when grouped into bundles form the tonofibrils seen in histological preparations.

Selby (1955, 1956, 1957) first demonstrated that filaments in the Malpighian cells of embryonic and adult human skin are about 100 Å in diameter. She believed that the cells of the stratum corneum are a "structureless gray mass" without filaments. Brody (1959, 1960), however, later demonstrated filaments in the stratum corneum of adult skin.

Filaments in the various layers of the epidermis have different staining properties (Brody, 1964). Those in the cells of the Malpighian layer are rendered opaque by uranyl acetate; those in cornified cells are unstained but the amorphous protein around them has a strong affinity for uranyl acetate (Figs. 10, 18, and 21). (This compact arrangement of fibril-matrix complex has been named "keratin pattern" by Brody.) Although such differences in staining characteristics have been attributed to chemical changes during differentiation, they could be due to the packing of the filaments. In the horny cells of the oral mucosa, for example, where the filaments are loosely packed, they stain intensely.

The main problem in analyzing epidermal fibrous proteins is their extreme insolubility. Using a 6 M urea solution, Rudall (1952) first isolated from the epidermis of bovine muzzles a fibrous protein which he called "epidermin," a substance that gave an α-keratin x-ray diffraction pattern. Matoltsy (1965), using 0.1 M citric acid-sodium citrate buffer at pH 2.6, also isolated a filamentous protein from the Malpighian layer of bovine muzzles. The isolated protein, called "prekeratin," gave an x-ray diffraction pattern like that of epidermin and had an estimated molecular weight of 640,000, a length of 1050 Å, and a width of 37 Å. Matoltsy (1965) also demonstrated *in vitro* that this protein has a strong tendency to aggregate into long fibers of various diameters. Crounse (1966) later prepared an alkali-soluble (0.02 M sodium hydroxide) fraction from human plantar callus that can be precipitated at pH 5.5 by the addition of dilute hydrochloric acid. The molecular weight of this purified protein was estimated to be 50,000, and its amino acid composition was similar to that of the protein extracted from cow muzzle.

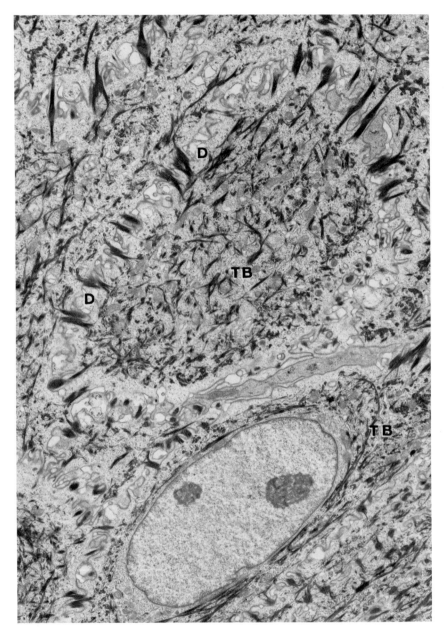

Fig. 18. Cytoplasmic details of cells in the stratum spinosum. Desmosomes (D) punctuate the plasma membrane. Clusters of tonofibrils (TB) are scattered throughout the cytoplasm. × 7000. (Courtesy of Dr. D. Knutson.)

Crounse suggested that the aggregation of the small units he had extracted forms subunits of Matoltsy's "prekeratin." We do not know whether these fibrous proteins are the substance of the visible intracellular filaments.

Analyses of amino acid from isolated fibrillar protein show it to be low in sulfur-containing amino acids like cystine and methionine and rich in proline (Matoltsy, 1965). Since labeled amino acids are incorporated into the proteins of all living epidermal cells, the filaments are probably synthesized in all of these cells. As the cells move upward, the number of filaments increases until finally the filaments in the cells of the stratum corneum account for almost half of the total protein; the amorphous proteins around the filaments constitute the rest of it. Brody (1960, 1964), however, denies that the number of filaments increases as the cells differentiate on their way to the horny layer.

We have learned much about fibrillar proteins by using both x-ray diffraction and the electron microscope. Astbury and Woods (1930) and Rudall (1947) showed that mammalian epidermis contains a protein that gives an α-diffraction pattern similar to that of unstretched wool, or myosin and fibrinogen. Hence, he called these proteins the KMEF group (keratin, myosin, epidermin, and fibrinogen). Comparative studies have shown that α-type keratin is present in the epidermis of all vertebrates (Fraser et al., 1972). On the other hand, both reptilian scales and bird feathers contain in addition another type of fibrous protein which gives a β-diffraction pattern. Since β-diffraction pattern is also obtained from stretched wool, the close relationship between α- and β-types of proteins is evident. Theoretically, the α-diffraction pattern is probably produced by the folded polypeptide chains of the fibrillar protein which when unfolded give a β-diffraction pattern.

Electron microscopic studies have shown that all keratinizing epithelia that show an α-diffraction pattern contain 80 Å filaments (Rudall, 1947; Parakkal and Alexander, 1972). On the other hand, feathers and reptilian scales, which give a β-diffraction pattern, are composed of 30 Å filaments. Attempts to visualize smaller subunits in the 80-Å filaments have not been entirely successful.

Crick (1953) postulated that the filaments are made up of two or three helical strands, each inclined toward the other in such a way that the amino acid of one helix fits into the gap between successive amino acids of the next helix and forms a compact ropelike structure. Pauling and Corey (1953) suggested that the number of intertwined α chains ranges between 6 and 8.

2. AMORPHOUS PROTEINS

The fibrillar component accounts for only about half the content of the horny cell. The other half, where the filaments are embedded, is thought to be the amorphous protein matrix. Rudall (1952) found that by using 6 M urea he could obtain an amorphous protein in addition to the fibrillar protein epidermin. Depending on the sulfur content, these two proteins have been called "low" and "high" sulfur fractions, a borrowing from studies of wool where "low" and "high" sulfur fractions are well established. Although not well characterized in the epidermis, amorphous proteins are thought to be heterogenous, a plausible theory since keratohyalin granules, which are synthesized in the granular layer, become part of the amorphous matrix of the horny cells (Matoltsy and Matoltsy, 1972).

Autoradiographic studies of wool have shown that labeled cystine is incorporated in the keratogenous zone and probably becomes part of the amorphous matrix (Downes *et al.*, 1963). Similar studies with radioactive cystine have shown that cystine is selectively incorporated high in the epidermis, but whether part of the cystine is incorporated into the amorphous matrix protein has not been determined.

3. MEMBRANE-COATING GRANULES

A characteristic differentiation product unique to mammals, the *membrane-coating granules* (keratinosome, Odland body) are ovoid or rod-shaped bodies that range from 0.1 to 0.5 μm long (Figs. 19 and 20). First identified by Selby (1957) in the cells of the well-developed granular layer in the epidermis of human foot pads, these granules were regarded as either degenerated or transformed mitochondria. When Odland (1960) later described their internal structure as consisting of parallel internal membranes resembling mitochondrial cristae, the hypothesis that they were attenuated mitochondria became more credible. At the same time, similar granules in both normal and pathological keratinizing epithelia led some investigators to regard them as unidentified virus particles (Frei and Sheldon, 1961). Matoltsy and Parakkal (1965, 1967) and Farbman (1964) further elucidated the

Fig. 19. Granular and horny layer cells. The keratohyalin granules (KH) appear as irregular amorphous masses. Note the thickened plasma membrane of the horny cells. Membrane-coating granules (MCG) are next to the cell membranes. × 15,000. (Courtesy of Dr. D. Knutson.)

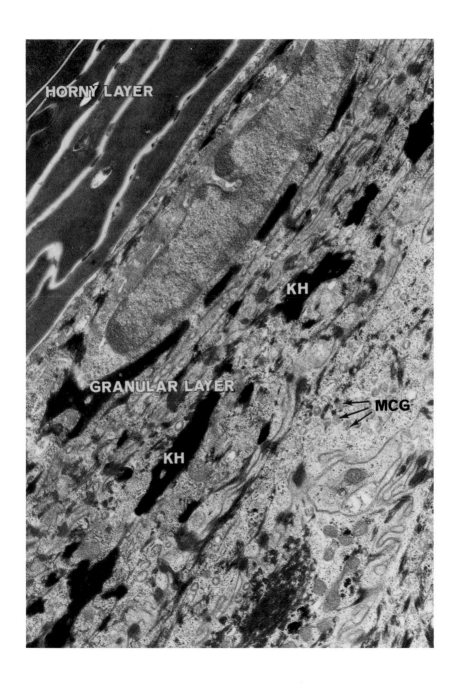

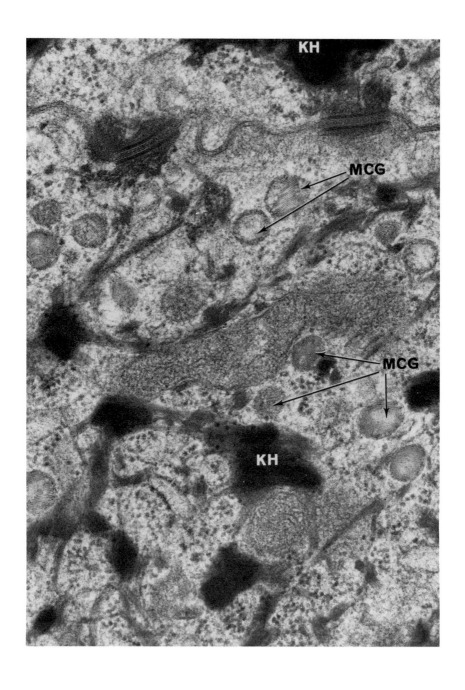

formation and fate of the granules, showing that they are formed in association with the Golgi complex and subsequently migrate and align themselves near the apical part of the cell membranes (Figs. 19 and 20). They later fuse with the latter and discharge their contents into the intercellular spaces in the granular layer (Fig. 21). Even after their discharge, their lamellar structure identifies them in the intercellular spaces where the contents of the granules probably become evenly spread around the cells (Fig. 21).

The fine structural details of the lamellar arrangement of the granules can be demonstrated with high resolution microscopy. Frithiof and Wersall (1965) and later Martinez and Peters (1971) showed the granules to have a complex internal structure, i.e., surrounded by an outer trilaminar membrane with alternating dense and light lamellae inside. These lamellae are about 60 to 75 Å thick, each consisting of a thick (30 Å) and a thin (20 Å) lamella.

Since membrane-coating granules have not been isolated, one can only speculate about their chemical nature. On the basis of histochemical studies, Matoltsy and Parakkal (1965) concluded that they contain mucopolysaccharides. The fact that the cell boundaries of the upper layers of stratified epithelia are periodic acid-Schiff reactive substances supports their hypothesis. On the other hand, because of the structural similarity of the granules to phospholipid, Frithiof and Wersall (1965) concluded that they contain phospholipids, a contention further strengthened by the fact that the granules are stainable with osmium iodide, which binds to phospholipids. Furthermore, phospholipase C digests the granules inside the cells and after their discharge into the intercellular spaces (Hashimoto, 1971b).

Although speculation abounds, the function of membrane-coating granules is still obscure. Because the plasma membrane of these cells becomes thicker as the cells become part of the horny layer, they were thought to function in the modification of cell membranes, especially since this thickening occurs simultaneously with the discharge of the membrane-coating granules into the intercellular spaces. However, Farbman (1966b) contends that the thickening of the plasma membrane occurs from inside the cell, not outside from the discharged membrane-

Fig. 20. Portions of granular cells showing both keratohyalin (KH) and membrane-coating granules (MCG) in the cytoplasm. × 78,000. (Courtesy of Dr. D. Knutson.)

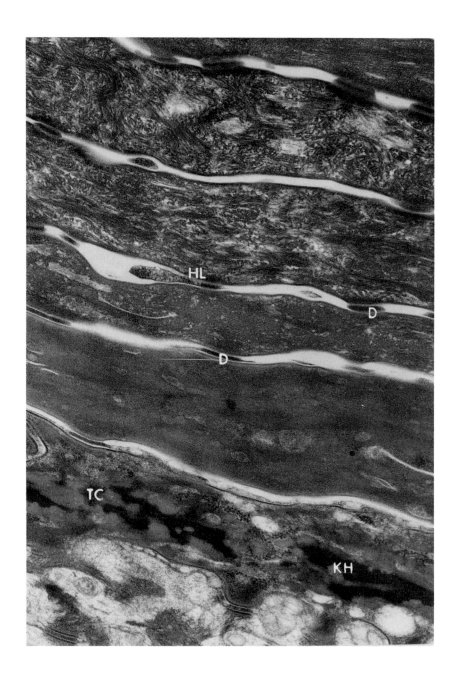

coating granules. This problem is still unresolved. Some believe that the granules contain lysosomal enzymes (Wolff and Holubar, 1967). Weinstock and Wilgram (1970) have demonstrated acid phosphatase and aryl sulfatase both inside the cells and after they are discharged and believe that the desquamation of horny cells at the surface is achieved by hydrolytic enzymes in the membrane-coating granules. But, this is not true of all keratinizing systems since large numbers of membrane-coating granules are present in the nails, which do not exfoliate.

Matoltsy and Parakkal (1965) also proposed that membrane-coating granules act as an intercellular cement substance; once the granules are discharged, they bind the cells together. Moreover, this amorphous substance acts as a barrier against the penetration of foreign material; lanthanum, for example, does not penetrate between the cells of the stratum corneum. Thus the connections between the cells of the stratum corneum may have the same qualities as the tight junction found in other epithelia. In addition, Hashimoto (1971b) has shown the amorphous substance in the intercellular spaces of the stratum corneum to be phospholipid similar to that found in tight junctions elsewhere.

4. KERATOHYALIN GRANULES

Auffhammer (1869) was probably the first to see clearly discrete cytoplasmic granules in the cells of the epidermis; four years later, Langerhans described the granular layer. Since the granules did not stain specifically for protein or lipid, their precise nature was long debated. Ranvier, who thought they contained lipid, named them "eleidine en graine," but others contended that they were composed of "hyalin"-like protein material. Waldeyer (1882) later called them "keratohyalin granules."

In mammalian epidermis, these granules are found in the cells immediately below the horny layer. In other vertebrates, they are so small that their existence has been questioned. They have been demonstrated in the epidermis of birds and reptiles with the electron microscope (Parakkal and Alexander, 1972).

Fig. 21. Transitional cells (TC) and horny layers (HL). The cells of the horny layers have filaments in various stages of aggregation. Remnants of desmosomes (D) are still seen between the horny cells. Just below the transitional layer the intercellular spaces contain discharged membrane-coating granules, and the transitional cell has intensely stained keratohyalin granules. × 49,000. (Courtesy of Dr. D. Knutson.)

The nature and function of keratohyalin granules are still somewhat enigmatic. Because they are located immediately below the horny layer, they have been regarded as precursors of "keratin." Giroud and Bulliard (1934) stressed that sulfhydryl groups are important in keratinization and many have attempted to demonstrate SH and S–S groups in them. Chevremont and Frederic (1943), who used the Prussian blue reaction to demonstrate –SH groups, found a delicate reaction in the Malpighian layer, an intense one in the keratohyalin granules, but none in the stratum corneum. With the method of Barrnett and Seligman (1952) for alcohol- and water-insoluble thiol groups, the Malpighian layer shows a homogenous distribution of –SH groups from the basal layer to the cells of the stratum granulosum, but the keratohyalin granules are not reactive.

It is also uncertain what other compounds are present in keratohyalin. When epidermis is microincinerated, the granules show the highest content of mineral ashes in the granular layer; these are believed to contain calcium (Opdyke, 1952). They are Feulgen-negative and do not contain deoxyribonucleoprotein, but are surrounded by a film or capsule of ribonucleoprotein, the basophilic staining of which is partially eliminated by treatment with ribonuclease. After proper fixation, the granules stain somewhat metachromatically with toluidine blue and sometimes contain some mucopolysaccharides. They readily take up stains for elastin and are gram positive; digestion with ribonuclease, hyaluronidase, or trypsin does not alter these staining properties. Not all keratohyalin granules have similar tinctorial characteristics. For example, none of those at the orifices of the pilosebaceous canals and sweat ducts stain with basic fuchsin, yet those in the rest of the epidermis stain weakly. The granules in mucous membranes have a stronger dye affinity at low pH than those in cutaneous membranes (Opdyke, 1952); and the stainability is destroyed after treatment with ribonuclease. However, in cutaneous surfaces, the enzyme does not alter their metachromatic property.

The best characterization of these granules has been obtained from autoradiographic methods. Fukuyama *et al.* (1965) and Fukuyama and Epstein (1966; 1968) obtained two major types of labeling patterns in newborn rats injected with tritiated amino acids. Labeled glycine, histidine, serine, arginine, and cystine are taken up preferentially by granular cells; tritiated phenylalanine, leucine, and methionine by the Malpighian layer. Labeled histidine is incorporated into the keratohyalin granules. Histidine has been demonstrated with histochemical methods preferentially in granules (Reaven and Cox, 1963, 1965; Nagy-Vezekenyi, 1969).

Biochemical findings support the presence of a histidine-rich protein in keratohyalin granules (Fukuyama and Epstein, 1967). The granular layer of newborn rats injected with tritiated histidine contained a protein that accounted for about 50% of the administered histidine. Amino acid analyses of this protein showed only trace amounts of sulfur-containing amino acids. Moreover, when keratohyalin was absent, no histidine-rich protein was isolated from human psoriatic skin.

Several techniques have been used to isolate keratohyalin granules for biochemical characterization, but it is difficult to do so because of their paucity. Matoltsy and Matoltsy (1970) used a mixture of 2 M urea and 1% trypsin to isolate them and subsequently dispersed the sticky and clumped keratohyalin granules with the detergent Brig 35. Amino acid analysis of such purified granules showed conspicuously low amounts of histidine, but relatively abundant proline and cystine. These results do not correspond with those of Hoober and Bernstein (1966) and Gumicio et al. (1967) who showed granular protein to be rich in histidine and low in cystine and proline. Since the extraction procedures used by these two groups are not the same, it is difficult to compare their results.

Ugel and co-workers (1969a,b, 1971) have recently isolated keratohyalin granules using 10 M phosphate buffer at pH 7 and dialysis of the extract against distilled water. The macroaggregates thus formed had histochemical and ultrastructural characteristics identical with in situ keratohyalin granules. Furthermore, Guss and Ugel (1972) claim that they produced antisera from rabbits immunized with macro-aggregates of keratohyalin. The antibodies against keratohyalin cross-react with those of goat, sheep, and rabbit but not those of man, rat, or guinea pig.

Numerous electron microscopic observations of keratohyalin granules, which vary from a few hundred angstroms to several micrometers (Figs. 19 and 20), show them avidly taking up osmium tetroxide and such heavy metal ions as uranyl acetate and lead citrate, which render them electron opaque. Small granules found in the lower cells of the granular layer coalesce in the upper cells and finally fill most of the cytoplasm (Fig. 21). These granules are as varied in shape as they are in size (Farbman, 1966a): spherical or ovoid in the epidermis of mice and rats, they are irregular in man, guinea pigs, and opossums.

Keratohyalin granules are not membrane bounded. However, ribosome-like particles are sometimes attached to their periphery or are present nearby since some granules have filaments abutting them. Keratohyalin granules have been reported as amorphous, granular, or fibrous masses; Brody (1960) even thinks that they are intensely stained regions of cytoplasmic fibrils.

Mercer (1961) has hypothesized that when these granules mysteriously disappear during the formation of horny cells, they are undergoing fibrous transformation. On the other hand, there is a growing belief that the amorphous material of keratohyalin granules becomes part of the interfilamentous matrix in the horny cell. But there is no definite evidence for any of these claims.

5. CHANGES IN PLASMA MEMBRANE

One of the most important changes during the terminal stages of keratinization is the almost universal thickening of the cell envelope of the horny cells. Unlike the plasma membranes of the basal, spinous, and granular cells, which range from 80 to 100 Å thick and have the typical triple-layer appearance of unit membranes, those in the horny layer become about 150 Å thick and the unit membrane structure is no longer discernible (Figs. 19 and 21).

Along with thickening, the cell membrane also undergoes qualitative changes that make it the toughest component of keratinized cells. Although the fibrous amorphous component of the cell can be dissolved with keratinolytic agents, the thickened plasma membranes cannot. The entire epidermis, except the membranes of horny cells, can be solubilized by 1 N NaOH, and cell envelopes can be isolated with this procedure (Matoltsy and Parakkal, 1965). When analyzed biochemically, the isolated membranes of horny cells contain large quantities of cystine and proline, an indication that they contain more sulfur than the low and high sulfur fraction of the epidermis (Matoltsy and Matoltsy, 1966). The high stability of cell membranes may be due to the disulfide bonds formed by the cystine.

How the cell membranes thicken is still uncertain. Farbman (1966b) believes that a peripheral band forms near the inner leaflet of the plasma membrane of the granular cells. As this band thickens, it becomes indistinguishable from the inner leaflet, whereas the middle and outer leaflets remain unchanged. Martinez and Peters (1971), however, believe that the outer leaflet also thickens when an amorphous intercellular material derived from the membrane-coating granules is added to it.

6. THE RESORPTION OF NUCLEI AND ORGANELLES

After the differentiation of fibrillar and amorphous proteins, membrane-coating and keratohyalin granules, and thickened plasma membranes, epidermal cells become transformed into horny cells. During this process, the nucleus and other cell organelles disintegrate and are eliminated from the cells. Just before they are exfoliated, fully differentiated horny cells contain a fibrous-amorphous complex surrounded by a thickened, tough cell envelope.

The mechanism of resorption of the nuclei and organelles is not completely understood, although increasing evidence points to the participation of lysosomal hydrolytic enzymes. All viable epidermal cells contain histochemically demonstrable acid phosphatase. These enzymes, however, are much more stable in the basal layers: as the cells differentiate, acid phosphatase becomes more labile. Most of the breakdown of nuclei and organelles takes place in the differentiating cells. Biochemical and histochemical studies have shown a wide variety of hydrolytic enzymes— acid phosphatase, cathepsin, β-glucuronidase, and aryl sulfatase—in the epidermis.

Although the early attempts to localize and demonstrate lysosomes in epidermal cells with combined histochemical and electron microscopic techniques were not successful, small numbers of lysosomes were subsequently shown in the epidermal cells of man and other animals (Olson et al., 1968; Rowden, 1967; Wolff and Schreiner, 1970). Since many of the cellular components have to be broken down during the last phases of keratinization, the scarcity of lysosomal-like structures is puzzling. However, pathological epidermis, like that in eczema and psoriasis or after exposure to ultraviolet radiation or irritation by foreign substances, contains many autophagic vacuoles (Prose et al., 1965; Johnson and Daniels, 1969).

C. Exfoliation

At the end of differentiation, the cornified cells are cemented together to form the stratum corneum, which varies in thickness in different vertebrates and in different parts of the same animal. The horny cells at the surface of the stratum corneum are continuously shed. If cups are fastened to the skin for a month or more, an arrangement that traps all exfoliated materials (Goldschmidt and Kligman, 1964), a loose, fat-soaked, horny mass accumulates which can be easily scraped down to a firm surface below where the coherent stratum corneum is normal. Therefore, the horny layer desquamates at a normal rate even when not affected by external forces. Even after six months, the exfoliated mass consists of intact cells: single cells, small aggregates of cells, and macroscopic clusters several millimeters in size.

How exfoliation occurs is not fully understood. Membrane-coating granules are thought to be somehow involved in the process (see Section IV,B,3). Most probably, the material of membrane-coating granules, once they are spread in the intercellular spaces, acts as a banding substance. The acid hydrolytic activity demonstrated in them is necessary for an even spreading of the contents in the intercellular spaces. As the cells reach the surface, the cementing material becomes less effective and desquamation ensues.

V. Properties and Function of Stratum Corneum

At the end of differentiation, epidermal cells become constituents of the horny layer, which shields the animal against damage from the environment and maintains the "internal milieu." Any mechanical or chemical alteration in this layer impairs its "barrier" function and makes the skin permeable to water and soluble substances (Winsor and Burch, 1944).

In histological preparations, the horny layer resembles a loose, desquamating mass of scales so separated by spaces and cracks as to raise serious doubts about how such a porous mass can function as an efficient protective barrier to both water loss and penetration. After special treatments and fixation, however, the stratum corneum is seen to be composed of highly organized units of horny cells stacked in vertical columns one above the other (Mackenzie, 1972; Christophers, 1971; Menton and Eisen, 1971).

Intact sheets of horny layer can be separated from the epidermis of living subjects by cantharidin blisters (Kligman and Christophers, 1964). These thin, transparent sheets can be stored, dried indefinitely, and used for studies in permeability and physical properties. The siltlike orifices of the sweat ducts and the funnel-shaped openings of the hair follicles, once thought to render the membrane leaky, swell shut when the membrane is immersed in aqueous media.

To determine the amount of stratum corneum lost per day, Gold-schmidt and Kligman (1964) trapped the flaked horny cells within cups fastened to various skin areas for one month. After this material had been defatted, washed, dried, and weighed, they calculated that 70% of the weight consisted of the keratinous content of the horny cells, the rest of cell membranes and bound lipids. With a body surface some 2 m², the amount of horny material produced is only 0.5 to 1.0 gm daily. Freedberg and Baden (1962) observed daily losses of only 9 to 17 gm even in generalized exfoliative diseases. Thus skin functions economically, making only insignificant demands on nitrogen.

Resorting to the now well-known Scotch tape stripping technique, Wolf (1939) examined the broad flat surface of the cells and saw details not visible in cross sections. He observed the topographic patterns of cell surfaces and described numerous nipplelike elevations and foldings. Under the scanning electron microscope, the surface of the horny cells from the human palmar and plantar surfaces is covered with evenly distributed long foliate villi (Fig. 8). The horny cells from trunk and limb have fewer and smaller villous processes.

Depending on site and specific function, two types of horny layers can be distinguished: that of the palms and soles, adapted for weight-bearing and friction (horny pads) and that over the rest of the body which is membranous and adapted for flexibility and impermeability. Although the difference in thickness between the horny pads and the membranous horny layer is at least 40-fold (about 600 μm versus 15 μm, respectively), there are other even more remarkable differences. The cells of the pads are more easily dissociated by mechanical trauma or strong chemicals and the cell membranes are more easily dissolved in alkali than those in the membranous horny layer. They contain only about half the amount of water-soluble substances, are extremely brittle when dry, and are many times more permeable to water and chemicals. What protection the pads give is almost entirely due to their great thickness. Much of the published data on the chemical and physical properties of the stratum corneum is derived from the callus and does not apply to the membranous horny layer.

One of the major functions of the stratum corneum is associated with its low permeability properties. It not only effectively retards water loss from the inner hydrated layers but also prevents the entrance of most toxic agents. The question of where the "barrier" mechanism is located still remains. Since most substances penetrate the skin only after injury or the removal of the stratum corneum, the barrier function would seem to reside in this layer. Rein (1924) postulated that ions are prevented from entering by a membrane located in the transitional zone between living and dead epidermis. Years later Rothman (1955), Blank (1953), and Szakall (1955) speculated that a special impenetrable membrane resided at the base of the horny layer. Szakall asserted that horny cells come off individually on the first 8 to 12 strippings with Scotch tape (stratum disjunction). Below this he obtained a coherent, thin sheet known as the stratum conjunctum, which could be floated free of the tape with petroleum ether. He analyzed its physical and chemical properties and concluded that this was the isolated "barrier" (Stupel and Szakall, 1957). Unfortunately, Szakall's barrier or stratum conjunction is not an anatomical entity but only a modest portion of the stratum corneum (Kligman, 1964), and it is now generally accepted that the bulk of the stratum corneum is a uniformly good diffusion barrier. Thus if an area of skin is denuded of stratum corneum, the barrier function is retarded only for 2 to 3 days, during which time a part of a new stratum corneum develops (Matoltsy *et al.*, 1962; Monash and Blank, 1958).

Although all the strata of the horny layer contribute to its barrier function, a gradient of increasing resistance from above to below probably corresponds to the increasing cohesiveness of the cells. Aside from this gradient, there are no well-defined structural "barrier" layers within the stratum corneum.

A certain amount of body water is lost by sweating and diffusion. In man, however, both diffusion and water loss are remarkably low and inversely related to the thickness of the horny layer; both are dependent on air flow, temperature, and the humidity of the surface (Blank, 1952).

Laboratory animals, with a thinner and less well-consolidated horny layer than that of man, have higher diffusion rates. Yet the shell membrane of hen's eggs, a much thicker keratinized sheet, transpires water almost as freely as an uncovered chamber of water, and frog skin offers little resistance to the passage of water.

References

Achten, G. 1959. Recherches sur la keratinization de la cellular epidermique chez l'Homme et le Rat. *Arch. Biol.* **70**: 1–119.

Adachi, K. 1961. Metabolism of glycogen in the skin and the effect of x-rays. *J. Invest. Dermatol.* **37**: 381–395.

Adachi, K. 1967. Enzyme activities in mammalian pigment cells. *In* "Advances in Biology of Skin. The Pigmentary System" (W. Montagna and F. Hu, eds.), Vol. 8, pp. 223–240, Pergamon, Oxford.

Adachi, K., and S. Yamasawa. 1966. Quantitative histochemistry of the primate skin. I. Hexokinase. *J. Invest. Dermatol.* **46**: 473–476.

Adachi, K., C. Lewis, and F. B. Hershey. 1967a. Enzymes of amino acid metabolism in normal human skin. I. Glutamate dehydrogenase. *J. Invest. Dermatol.* **48**: 226–229.

Adachi, K., C. Lewis, and F. B. Hershey. 1967b. Enzymes of amino acid metabolism in normal human skin. II. Alanine and asparate transaminases. *J. Invest. Dermatol.* **49**: 240–245.

Alexander, N. J., and P. F. Parakkal. 1969. Formation of α- and β-type keratin in lizard epidermis during the molting cycle. *Z. Zellforsch. Mikrosk. Anat.* **101**: 72–87.

Argyris, T. S. 1952. Glycogen in the epidermis of mice painted with methylcholanthrene. *J. Nat. Cancer Inst.* **12**: 1159–1165.

Astbury, W. T., and H. J. Woods. 1930. The x-ray interpretation of structure and elastic properties of hair keratin. *Nature (London)* **126**: 913–914.

Auffhammer, H. 1869. Kritische Bemerkungen zu Schroen' Satz: "Lo strato corneo trae la sua origine dalle ghiandole sudorifere." *Verh. Phys-med. Ges. Würzb.*, n.F., **1**: 192–209. (Quoted by Waldeyer, 1882.)

Baden, H. P., and S. Sviokla. 1968. The effect of chalone on epidermal DNA synthesis. *Exp. Cell Res.* **50**: 644–646.

Barrnett, R. J., and A. M. Seligman. 1952. Histochemical demonstration of protein-bound sulfhydryl groups. *Science* **116**: 323–327.

Billingham, R. E., and W. K. Silvers. 1963. The origin and conservation of epidermal specificities. *N. Engl. J. Med.* **268**: 477–480; 539–545.

Billingham, R. E., and W. K. Silvers. 1967. Studies on the conservation of epidermal specificities of skin and certain mucoses in adult mammals. *J. Exp. Med.* **125**: 429–446.

Bizzozero, G. 1871. Sulla struttura degli epiteli pavimemtosi stratificati. *Zentral. med. Wochschr.* **9**: 482–483. (Abstr.)

Blank, I. H. 1952. Factors which influence the water content of the stratum corneum. *J. Invest. Dermatol.* **18**: 433–440.

Blank, I. H. 1953. Further observations on factors which influence the water content of the stratum corneum. *J. Invest. Dermatol.* **21**: 259–271.

Bonneville, M. A. 1965. The periderm of human fetal skin. *In* "Electron Microscopy, 1964" (M. Titlbach, ed.), pp. 565–569. Proc. 3rd European Regional Conf., Prague, Vol. B., London Royal Microscopical Society.

Bonneville, M. 1968. Observations on epidermal differentiation in the fetal rat. *Amer. J. Anat.* **123**: 147–164.

Bradfield, J. R. G. 1951. Glycogen of vertebrate epidermis. *Nature (London)* **167**: 40–42.

Branson, R. J. 1968. Orthogonal arrays of microtubules in flattening cells of the epidermis. *Anat. Rec.* **160**: 109–122.

Breathnach, A. S. 1971. Embryology of human skin. A review of ultrastructural studies. *J. Invest. Dermatol.* **57**: 133–143.

Breathnach, A. S., and E. J. Robins. 1969. Ultrastructural feature of epidermis of a 14 mm (6 weeks) human embryo. *Brit. J. Dermatol.* **81**: 504–516.

Brody, I. 1959. The keratinization of epidermal cells of normal guinea pig skin as revealed by electron microscopy. *J. Ultrastruct. Res.* **2**: 482.

Brody, I. 1960. The ultrastructure of the tonofibrils in the keratinization process of normal human epidermis. *J. Ultrastruct. Rec.* **4**: 264–267.

Brody, I. 1964. Observations on the fine structure of the horny layer in the normal human epidermis. *J. Invest. Dermatol.* **42**: 27–31.

Brody, I., and K. S. Larsson. 1965. Morphology of mammalian skin: Embryonic development of the epidermal sub-layers. *In* "Biology of the Skin and Hair Growth" (A. G. Lyne and B. F. Short, eds.), pp. 267–290. Amer. Elsevier, New York.

Bullough, W. S. 1965. Mitotic and functional homeostasis: a speculative review. *Cancer Res.* **25**: 1683–1727.

Bullough, W. S., and E. B. Laurence. 1964. Mitotic control by internal secretion: the role of the chalone-adrenalin complex. *Exp. Cell Res.* **33**: 176–194.

Bullough, W. S., E. B. Laurence, O. H. Iversen, and K. Elgjo. 1967. The vertebrate epidermal chalone. *Nature (London)* **214**: 578–580.

Campbell, R. D., and J. H. Campbell. 1971. Origin and continuity of desmosomes. *In* "Origin and Continuity of Cell Organelles" (J. Reinert and H. Ursprung, eds.), pp. 251–293, Springer-Verlag, Berlin and New York.

Carney, S. A., J. C. Lawrence, and C. R. Ricketts. 1962. Carbohydrate substrates for skin cells. *Biochem. J.* **83**: 533–539.

Chambers, R., and G. S. Renyi. 1925. The structure of the cells in tissues as revealed by microdissection. I. Are physical relationships of the cells in epithelia. *Amer. J. Anat.* **35**: 385–402.

Chessick, R. D. 1953. Histochemical study of the distribution of esterases. *J. Histochem. Cytochem.* **1**: 471–485.

Chevremont, M., and J. Frederic. 1943. Une nouvelle méthode histochimique de mise en évidence des substances à fonction sulfhydrile. Application à l'épiderme, au poil et à la levure. *Arch. Biol.* **54**: 589–605.

Christophers, E. 1971. Cellular architecture of the stratum corneum. *J. Invest. Dermatol.* **56**: 165–169.

Coffey, W., P. Finkelstein, and K. Laden. 1963. The effect of U.V. irradiation on enzyme systems in the epidermis. *J. Soc. Cosmet. Chem.* **14**: 55–61.

Crick, F. H. C. 1953. The packing of α-helixes: simple coiled-coils. *Acta Crystallogr. Sect. B* **6**: 689–697.

Crounse, R. G. 1966. Epidermal keratin and epidermal prekeratin. *Nature (London)* **211**: 1301–1302.

Cruickshank, C. N. D., F. B. Hershey, and C. Lewis. 1958. Inociturate dehydrogenase activity of human epidermis. *J. Invest. Dermatol.* **30**: 33–37.

Cruickshank, C. N. D., M. D. Trotter, and J. R. Cooper, 1962. A study of available substrates for the endogenous respiration of skin *in vitro*. *J. Invest. Dermatol.* **39**: 175–178.

Dawber, R. P. R., R. Marks, and J. A. Swift. 1972. Scanning electron microscopy of the stratum corneum. *Brit. J. Dermatol.* **86**: 272–281.

Decker, R. H. 1971. Nature and regulation of energy metabolism in the epidermis. *J. Invest. Dermatol.* **57**: 351–363.

Derksen, J. C., G. C. Heringa, and A. Weidinger. 1938. On keratin and cornification. *Acta Neerl. Morphol. Norm. Pathol.* **1**: 31–37.

Dodson, J. W. 1967. The differentiation of epidermis. 1. The interrelationship of epidermis and dermis in embryonic chicken skin. *J. Embryol. Exp. Morphol.* **17**: 83–105.

Downes, A. M., L. F. Sharry, and G. E. Rogers. 1963. Separate synthesis of fibrillar and matrix protein in the formation of keratin. *Nature (London)* **199**: 1059–1061.

Dushoff, I. M., J. Payne, F. B. Hershey, and R. C. Donaldson. 1965. Oxygen uptake and tetrazolium reduction during skin cycle of the mouse. *Amer. J. Physiol.* **209**: 321–325.

Elgjo, E., and H. Hennings. 1971. Epidermal chalone and cell proliferation in a transplantable squamous cell carcinoma in hamsters. I. *In vivo* results. *Virchows Arch. B*, **7**: 1–7.

Epstein, W. L., and H. L. Maibach. 1965. Cell renewal in human epidermis. *Arch. Dermatol.* **92**: 462–468.

Farbman, A. I. 1964. Electron microscope study of a small cytoplasmic structure in rat oral epithelium. *J. Cell Biol.* **21**: 491–495.

Farbman, A. I. 1966a. Morphological variability of keratohyalin. *Anat. Rec.* **154**: 275–286.

Farbman, A. I. 1966b. Plasma membrane changes during keratinization. *Anat. Rec.* **156**: 269–282.

Farquhar, M. G., and G. E. Palade. 1963. Junctional complexes in various epithelia. *J. Cell. Biol.* **17**: 375–412.

Fawcett, D. W. 1963. Comparative observations on the fine structure of blood capillaries. *In* "The Peripheral Blood Vessels" (J. L. Orbison and D. E. Smith, eds.), pp. 17–44. Williams & Wilkins, Baltimore, Maryland.

Fell, H. B., and E. Mellanby. 1953. Metaplasia produced in cultures of chick ectoderm by high vitamin A. *J. Physiol. (London)* **119**: 470–488.

Firket, H. 1951. Recherches sur la régénération de la peau de mammifère. Deuxième partie: etude histochimique. *Arch. Biol.* **62**: 335–351.

Fitton-Jackson, S., and H. B. Fell. 1963. Epidermal fine structure in embryonic chicken skin during atypical differentiation induced by vitamin A in culture. *Develop. Biol.* **7**: 394–419.

Flaxman, B. L., M. A. Lutzner, and E. J. Van Scott. 1967. Cell maturation and tissue organization in epithelial outgrowths from skin and buccal mucosa *in vitro*. *J. Invest. Dermatol.* **49**: 322–332.

Flaxman, A. B., P. F. A. Maderson, G. Szabo, and S. I. Roth. 1968. Control of cell differentiation in lizard epidermis *in vitro*. *Develop. Biol.* **18**: 354–374.

Fraser, R. D. B., T. P. MacRae, and G. E. Rogers. 1972. "Keratins. Their Composition, Structure and Biosynthesis." Thomas, Springfield, Illinois.

Freedberg, I. M., and H. P. Baden. 1962. The metabolic response to exfoliation. *J. Invest. Dermatol.* **38**: 277–284.

Frei, J. V., and H. Sheldon. 1961. A small granular component of the cytoplasm of keratinizing epithelia. *J. Biophys. Biochem. Cytol.* **11**: 719–724.

Freinkel, R. K. 1960. Metabolism of glucose-C-14 by human skin *in vitro*. *J. Invest. Dermatol.* **34**: 37–42.

Frithiof, L., and J. A. Wersall. 1965. A highly ordered structure in keratinizing human oral epithelium. *J. Ultrastruct. Res.* **12**: 371–379.

Fukui, K., and K. M. Halprin. 1970. Enzymes of the uronic acid pathway in the human epidermis. *J. Invest. Dermatol.* **55**: 179–183.

Fukuyama, K., and W. L. Epstein. 1966. Epidermal keratinization: Localization of isotopically labelled amino acids. *J. Invest. Dermatol.* **47**: 551–560.

Fukuyama, K., and W. L. Epstein. 1967. Ultrastructural autoradiographic studies of keratohyalin granule formation. *J. Invest. Dermatol.* **49**: 595–604.

Fukuyama, K., and W. L. Epstein. 1968. Protein synthesis studied by autoradiography in the epidermis of different species. *Amer. J. Anat.* **122**: 269–274.

Fukuyama, K., T. Nakamura, and I. A. Bernstein. 1965. Differentially localized incorporation of amino acids in relation to epidermal keratinization in the newborn rat. *Anat. Rec.* **152**: 525–535.

Giroud, A., and H. Bulliard. 1934. Mise en evidence des substances a fonction sulfhydryle. *Bull. Histol. Appl. Physiol. Pathol. et Tech. Microsc.* **11**: 169–172.

Giroud, A., and H. Bulliard. 1935. Les substances a fonction sulfhydryle dans l'epiderme. *Arch. Anat. Microscop. Morphol. Exp.* **31**: 271–290.

Giroud, H., and C. Champetier. 1936. Recherche sur les roentgenogrammes des keratines. *Bull. Soc. Chim. Biol.* **18**: 656–664.

Giroud, A., and C. P. Leblond. 1951. The keratinization of epidermis and its derivatives, especially the hair, as shown by x-ray diffraction and histochemical studies. *Ann. N. Y. Acad. Sci.* **53**: 613–626.

Goldschmidt, H., and A. M. Kligman. 1964. Quantitative estimation of keratin production by the epidermis. *Arch. Dermatol.* **88**: 709–712.

Grillo, H. C., J. E. McLennan, and F. G. Wolfort. 1969. Activity and properties of collagenase from healing wounds in mammals. *In* "Repair and Regeneration, the Scientific Basis for Surgical Practice" (J. E. Dunphy and W. VanWinkle, Jr., eds.), pp. 185–197. McGraw-Hill, New York.

Gumucio, J., C., Feldkamp, and I. A. Bernstein. 1967. Studies on localization of "histidine-rich" peptide material present in epidermis of the newborn rat. *J. Invest. Dermatol.* **49**: 545–551.

Guss, S. B., and A. R. Ugel. 1972. Immunofluorescent antibodies to bovine keratohyalin and immunologic confirmation of homology. *J. Histochem. Cytochem.* **20**: 97–106.

Halprin, K. M. 1972. Epidermal "turnover time"—a reexamination. *J. Invest. Dermatol.* **86**: 14–19.

Halprin, K. M., and A. Ohkawara. 1966a. Glucose and glycogen metabolism in the human epidermis. *J. Invest. Dermatol.* **46**: 43–50.

Halprin, K. M., and A. Ohkawara. 1966b. Carbohydrate metabolism in psoriasis: an enzymeatric study. *J. Invest. Dermatol.* **46**: 51–68.

Halprin, K. M., and A. Ohkawara. 1966c. Glucose utilization in the human epidermis: its control by hexokinase. *J. Invest. Dermatol.* **46**: 278–282.

Halprin, K. M., and A. Ohkawara. 1967. Glucose entry into the human epidermis. II. The penetration of glucose into the human epidermis *in vitro*. *J. Invest. Dermatol.* **49**: 561–568.

Hashimoto, K. 1971a. Intercellular spaces of the human epidermis as demonstrated with lanthanum. *J. Invest. Dermatol.* **57**: 17–31.

Hashimoto, K. 1971b. Cementosome; a new interpretation of the membrane-coating granule. *Arch. Dermatol.* **240**: 349–364.

Hershey, F. B., C. Lewis, J. Murphy, and T. Schiff. 1960. Quantitative histochemistry of human skin. *J. Histochem. Cytochem.* **8**: 41–49.

Hoober, J. K., and I. A. Bernstein. 1966. Protein synthesis related to epidermal differentiation. *Proc. Nat. Acad. Sci. U.S.* **56**: 594–601.

Horstmann, E. 1952. Über den Papillarkörper der menschlichen Haut und seine regionalen Unterschiede. *Acta Anat.* **14**: 23–42.

Horstmann, E. 1957. Die Haut. *In* "Handbuch der Mikroskopischen Anatomie des Menschen." (W. V. Möllendorff, ed.), Vol. 3, Part 3, pp. 1–488. Springer-Verlag, Berlin and New York.

Hoyes, A. D. 1968. Fine structure of human amniotic epithelium in early pregnancy. *J. Obstet. Gynaecol. Brit. Commonw.* **75**: 949–962.

Im, M. J. C. 1968. Glucose metabolism in skin tumor and hyperplasia induced in the rhesus monkey by 7,12-dimethylbenz[a]anthracene. *J. Nat. Cancer Inst.* **41**: 73–79.

Im, M. J. C., and K. Adachi. 1966. Quantitative histochemistry of the primate skin. V. Glucose-6-phosphate dehydrogenase and 6-phosphogluconate dehydrogenase. *J. Invest. Dermatol.* **47**: 121–124.

Im, M. J. C., and J. E. Hoopes. 1970a. Enzyme activities in the repairing epithelium during wound healing. *J. Surg. Res.* **10**: 173–179.

Im, M. J. C., and J. E. Hoopes. 1970b. Energy metabolism in healing skin wounds. *J. Surg. Res.* **10**: 459–464.

Im, M. J. C., and J. E. Hoopes. 1972. Enzyme activities in regenerating epithelium during wound healing. II. β-glucuronidase. *J. Surg. Res.* **12**: 406–410.

Im, M. J. C., J. E. Hoopes, and Y. T. Sohn. 1972. Enzyme activities in regenerating epithelium during wound healing. I. Acid phosphatase. *J. Surg. Res.* **12**: 402–405.

Jacobson, B., and E. Davidson. 1962. Biosynthesis of uronic acid by skin enzymes. *J. Biol. Chem.* **237**: 635–637.

Jakus, M. A. 1961. The fine structure of the human cornea. *In* "The Structure of the Eye" (G. K. Smelser, ed.), pp. 343–366, Academic Press, New York.

Johnson, B. E., and F. Daniels. 1969. Lysosomes and the reactions of skin to ultraviolet radiation. *J. Invest. Dermatol.* **53**: 85–94.

Kahn, G., G. D. Weinstein, and P. Frost. 1968. Kinetics of human epidermal cell proliferation. *J. Invest. Dermatol.* **50**: 459–462.

Kallman, F., J. Evans, and N. M. Wessells. 1967. Anchor filament bundles in embryonic feather germ and skin. *J. Cell Biol.* **32**: 236–240.

Kelly, D. E., and J. H. Luft. 1966. Fine structure, development and classification of desmosomes and related attachment mechanisms. 6th Int. Congr. Electron Microsc., Kyoto, pp. 401–402.

Kingsbury, J. W., V. G. Allen, and B. A. Rotheram. 1953. The histological structure of the beak in the chick. *Anat. Rec.* **116**: 95–110.

Kischer, C. W., and J. S. Keeter. 1971. Anchor filament bundles in embryonic skin: origin and termination. *Amer. J. Anat.* **130**: 179–194.

Kligman, A. M. 1964. The biology of the stratum corneum. *In* "The Epidermis" (W. Montagna and W. C. Lobitz, Jr., eds.), pp. 387–433. Academic Press, New York.

Kligman, A. M., and E. Christophers. 1964. Preparations of isolated sheets of human stratum corneum. *Arch. Dermatol.* **88**: 702–705.

Listgarten, M. A. 1964. The ultrastructure of human gingival epithelium. *Amer. J. Anat.* **114**: 49–69.

Litvac, A. 1939. Sur la keratinisation epitheliale *in vitro*. *Arch. Anat. Microsc.* **35**: 55–63.

Lobitz, W. C., Jr., and J. B. Holyoke. 1954. The histochemical response of the human epidermis to controlled injury; glycogen. *J. Invest. Dermatol.* **22**: 189–198.

Lombardo, C. 1907. Il glicogeno della cute. *Giorn. Ital. Mal. Vener.* **42**: 448–464.

Lombardo, C. 1934. Il glicogeno in alcuni derivati epidermici della cute umana. *G. Ital. Dermatol. Sifilol.* **75**: 185–186.

Lyne, A. G., and D. E. Hollis. 1971. Ultrastructural changes in Merino sheep epidermis during foetal development. *J. Anat. (London)* **108**: 211.

Mackenzie, I. C. 1972. The ordered structure of mammalian epidermis. *In* "Epidermal Wound Healing" (H. I. Maibach and D. T. Rovee, eds.), pp. 5–25. Yearbook Publ., Chicago, Illinois.

Maderson, P. F. 1965. The structure and development of the squamate epidermis. *In* "Biology of the Skin and Hair Growth" (A. G. Lyne and B. F. Short, eds.), pp. 129–153. Amer. Elsevier, New York.

Martinez, I. R., and A. Peters. 1971. Membrane-coating granules and membrane modifications in keratinizing epithelia. *Amer. J. Anat.* **130**: 93–120.

Matoltsy, A. G. 1965. Soluble prekeratin. *In* "Biology of the Skin and Hair Growth" (A. G. Lyne and B. F. Short, eds.), pp. 291–305. Amer. Elsevier, New York.

Matoltsy, A. G., and M. Matoltsy. 1966. The membrane protein of horny cells. *J. Invest. Dermatol.* **46**: 127–129.

Matoltsy, A. G., and M. Matoltsy. 1970. The chemical nature of keratohyalin granules of the epidermis. *J. Cell Biol.* **47**: 593–603.

Matoltsy, A. G., and M. Matoltsy. 1972. The amorphous component of keratohyalin granules. *J. Ultrastruct. Res.* **41**: 550–560.

Matoltsy, A. G., and P. F. Parakkal. 1965. Membrane-coating granules of keratinizing epithelia. *J. Cell Biol.* **24**: 297–307.

Matoltsy, A. G., and P. F. Parakkal. 1967. Keratinization. *In* "Ultrastructure of Normal and Abnormal Skin" (A. Zelickson, ed.), pp. 76–104. Lea & Febiger, Philadelphia, Pennsylvania.

Matoltsy, A. G., A. Schragger, and M. N. Matoltsy. 1962. Observations on regeneration of the skin barrier. *J. Invest. Dermatol.* **38**: 251–253.

McLoughlin, C. B. 1963. Mesenchymal influences on epithelial differentiation. *Symp. Soc. Exp. Biol.* **17**: 359–388.

Menton, D. N., and A. Z. Eisen. 1971. Structure and organization of mammalian stratum corneum. *J. Ultrastruct. Res.* **35**: 247–264.

Mercer, E. H. 1961. "Keratin and Keratinization. An Essay on Molecular Biology," Vol. 12. Pergamon, Oxford.

Monash, S., and H. Blank. 1958. Location and reformation of the epithelial barrier to water vapor. *Arch. Dermatol.* **78**: 584–590.

Montagna, W. 1962. "The Structure and Function of Skin." 2nd edition. Academic Press, New York.

Mottet, N. K., and H. M. Jensen. 1968. The differentiation of chick embryonic skin. *Exp. Cell Res.* **52**: 261–283.

Nadol, J. B., J. R. Gibbins, and K. R. Porter. 1969. A reinterpretation of the structure and development of the basement lamella: An ordered array of collagen in fish skin. *Develop. Biol.* **20**: 304–331.

Nagy-Vezekenyi. 1969. On the histidine content of human epidermis. *Brit. J. Dermatol.* **81**: 685–691.

Nicole, P. A., and T. A. Cortese, Jr. 1972. The physiology of skin. *Annu. Rev. Physiol.* **36**: 177–203.

Odland, G. F. 1958. The fine structure of the interrelationship of cells in the human epidermis. *J. Biophys. Biochem. Cytol.* **4**: 529–538.

Odland, G. F. 1960. A submicroscopic granular component in human epidermis. *J. Invest. Dermatol.* **34**: 11–15.

Olson, R. L., R. G. Nordquist, and M. A. Everett. 1968. Ultrastructural localization of aryl sulfatase in human epidermis. *Acta Derm. Venereol.* **48**: 556–562.

Opdyke, D. L. 1952. Observations on the chemical morphology of the keratohyalin granules. Ph.D. thesis, Washington University, St. Louis, Missouri.

Palade, G. E., and M. G. Farquhar. 1965. A special fibril of the dermis. *J. Cell Biol.* **27**: 215–224.

Parakkal, P. F., and N. J. Alexander. 1972. "Keratinization: A Survey of Vertebrate Epithelia." Academic Press, New York.

Parakkal, P. F., and A. G. Matoltsy. 1968. An electron microscopic study of developing chick skin. *J. Ultrastruct. Res.* **23**: 403–416.

Paul, H. E., M. F. Paul, J. D. Taylor, and R. W. Marsters. 1948. Biochemistry of wound healing. IV. Oxygen uptake of healing tissue of skin wounds. *Arch. Biochem.* **17**: 429–434.

Pauling, L., and R. B. Corey. 1953. Compound helical configurations of polypeptide chains: Structure of proteins of the α keratin type. *Nature (London)* **171**: 59–61.

Pinkus, H. 1951. Examination of epidermis by strip method of remaining horny layers. I. Observations on thickness of the horny layer and on mitotic activity after stripping. *J. Invest. Dermatol.* **16**: 383–386.

Pinkus, H. 1952. Examination of the epidermis by strip method. II. Biometric data on regeneration of the human epidermis. *J. Invest. Dermatol.* **19**: 431–447.

Pinkus, H. 1954. Biology of epidermal cells. *In* "Physiology and Biochemistry of the Skin (S. Rothman, ed.), Chap. 25, pp. 584–600. The Univ. of Chicago Press, Chicago.

Pinkus, H., and A. Tanay. 1957. Die Embryologie der Haut. Pt. 1. *In* "Handbuch der Haut- und Geschlechtskrankheiten." (O. Gans and G. K. Steigleder, eds.), Vol. 1. Springer-Verlag, Berlin and New York.

Pomerantz, S. H., and M. T. Asbornsen. 1961. Glucose metabolism in young rat skin. *Arch. Biochem. Biophys.* **93**: 147–152.

Porter, K. R. 1954. Observations on the submicroscopic structure of animal epidermis. *Anat. Rec.* **118**: 433 (Abstract).

Porter, K. R. 1956. Observations on the fine structure of animal epidermis. *Proc. 3rd Int. Conf. Electron Microscopy, London, 1954*, pp. 539–546. Royal Microscopical Society, London.

Prose, P. H., Ed. Sedlis, and M. Bigelow. 1965. The demonstration of lysosomes in the diseased skin of infants with eczema. *J. Invest. Dermatol.* **45**: 448–457.

Reaven, E. P., and A. J. Cox. 1963. The histochemical localization of histidine in the human epidermis and its relationship to zinc binding. *J. Histochem. Cytochem.* **11**: 782–790.

Reaven, E. P., and A. J. Cox. 1965. Histidine and keratinization. *J. Invest. Dermatol.* **45**: 422–431.

Rein, H. 1924. Experimentale Studien über Elektroendosmose an überlebender menschlicher Haut. *Z. Biol. (Munich)* **81**: 125–140.

Rhodin, J. A. G., and E. J. Reith. 1962. Ultrastructure of keratin in oral mucosa, skin, esophagus, claw and hair. *In* "Fundamentals of Keratinization (E. O. Butcher and R. F. Sognnaes, eds.), Pub. No. 77. Amer. Assoc. Advance. Sci., Washington, D. C.

Rosenstadt, B. 1897. Über das Epitrichium des Hühnchens. *Arch. Mikr. Anat. Entwicklungsmech.* **49**: 561–585.

Rosett, T., M. Ohkido, J. G. Smith, Jr., and H. Yardley. 1967. Studies in the biochemistry of skin. IV. Some properties of mitochondria isolated from the epidermis of the adult rat. *J. Invest. Dermatol.* **48**: 67–78.

Rosin, S. 1946. Über Bau und Wachstum der Grenzlamelle der Epidermis bei Amphibienlarven: Analyse einer orthogonalen Fibrillarstruktur. *Rev. Suisse Zool.* **53**: 133–210.

Roth, S. I., and W. A. Jones. 1967. The ultrastructure and enzymatic activity of the boa constrictor *(Constrictor constrictor)* skin during the resting phase. *J. Ultrastruct. Res.* **18**: 304–323.

Roth, S. I., and W. A. Jones. 1970. The ultrastructure of epidermal maturation in the skin of the boa constrictor *(Constrictor constrictor).* *J. Ultrastruct. Res.* **32**: 69–93.

Rothberg, S., R. G. Crounse, and J. C. Lee. 1961. Glycine ^{14}C incorporation into the proteins of normal stratum corneum and the abnormal stratum corneum of psoriasis. *J. Invest. Dermatol.* **37**: 497–505.

Rothman, S. 1954. "Physiology and Histochemistry of Skin." Univ. of Chicago Press, Chicago, Illinois.

Rothman, S. 1955. The mechanism of percutaneous penetration and absorption. *J. Soc. Cosmet. Chem.* **6**: 193.

Rowden, G. 1967. Ultrastructural studies of keratinized epithelia of the mouse. 1. Combined electron microscope and cytochemical study of lysosomes in mouse epidermis and esophageal epithelium. *J. Invest. Dermatol.* **49**: 181–197.

Rudall, K. M. 1947. X-ray studies of the distribution of protein chain types in the vertebrate epidermis. *Biochem. Biophys. Acta* **1**: 549–562.

Rudall, K. M. 1952. The proteins of the mammalian epidermis. *Advan. Protein Chem.* **7**: 253–290.

Sasakawa, M. 1921. Beiträge zur Glykogenverteilung in der Haut unter normalen und pathologischen Zuständen. *Arch. Dermatol. Syph.* **134**: 418–443.

Schmidt, W. J. 1937. Neuere polarisationsoptische Arbeiten auf dem Gebiete der Biologie. I. *Teil Protoplasma.* **29**: 300–312.

Schmidt, W. 1963. Struktur und Funktion des Amnionepithels Von Mansa und Huhn. *Z. Zellforsch. Mikrosk. Anat.* **61**: 642–660.

Schroeder, H. E., and J. Theilade. 1966. Electron microscopy of normal human gingival epithelium. *J. Periodont. Res.* 1: 95–119.

Selby, C. C. 1955. An electron microscope study of the epidermis of mammalian skin in thin sections. 1. Dermoepidermal junction and basal cell layer. *J. Biophys. Biochem. Cytol.* 1: 429–444.

Selby, C. C. 1956. Fine structure of human epidermis. *J. Soc. Chem.* 7: 584–599.

Selby, C. C. 1957. An electron microscope study of thin sections of human skin. II. Superficial cell layers of footpad epidermis. *J. Invest. Dermatol.* 29: 131–149.

Serri, F., and W. Montagna. 1961. The structure and function of the epidermis. *Pediatr. Clin. North Am.* 8: 917–941.

Sinha, A. A., U. S. Seal, and A. W. Erickson. 1970. Ultrastructure of the amnion and amniotic plaques of the white-tailed deer. *Amer. J. Anat.* 127: 369–396.

Snell, R. 1965. The fate of epidermal desmosomes in mammalian skin. *Z. Zellforsch. Mikrosk. Anat.* 66: 471–487.

Stern, I. B. 1965. Electron microscopic observations of oral epithelium. 1. Basal cells and the basement membrane. *Periodontics* 3: 224–238.

Stupel, H., and A. Szakall. 1957. "Die Wirkung von Waschmitteln auf die Haut." Huthig, Heidelberg.

Susi, F. R., W. D. Belt, and J. W. Kelly. 1967. Fine structure of fibrillar complexes associated with basement membrane in human oral mucosa. *J. Cell Biol.* 34: 686–690.

Szabo, G. 1962. Cultivation of skin, pure epidermal sheets and tooth germs *in vitro*. *In* "Fundamentals of Keratinization" (E. O. Butcher and R. F. Sognnaes, eds.), Publ. No. 70. Amer. Assoc. Advance. Sci., Washington, D. C.

Szakall, A. 1955. Über die Eigenschaften, Herkunft und physiologischen Funktionen der die H-Ionenkonzentration bestimmenden Wirkstoffe in der verhornten Epidermis. *Arch. Klin. Exp. Dermatol.* 201: 331–360.

Thomas, C. E. 1965. The ultrastructure of human amnion epithelium. *J. Ultrastruct. Res.* 13: 65–84.

Tiedemann, K. 1972. Die Ultrastruktur des Amnion-Nabelstrang-und Hautepithels beim Sihafembryo verschiedener Entwicklungsstadien. *Z. Zellforsch. Mikrosk. Anat.* 125: 252–276.

Ugel, A. R. 1969a. The isolation of keratohyalin-like granules by *in vitro* aggregation of solubilized keratohyalin. *J. Cell Biol.* 43: 148 (abstract).

Ugel, A. R. 1969b. Keratohyalin: extraction and *in vitro* aggregation. *Science* 166: 250.

Ugel, A. R. 1971. Studies on isolated aggregating oligoribonucleoproteins of the epidermis with histochemical and morphological characteristics of keratohyalin. *J. Cell Biol.* 49: 405–422.

Waldeyer, W. 1882. Untersuchungen über die Histogenese der Horngebilde, besonders Haare und Federn. *Beitr. Anat. Embryol. Henle Testagebe.* p. 141.

Weber, G., and G. W. Korting. 1964. Glucose-6-phosphate dehydrogenase in human skin. *J. Invest. Dermatol.* 42: 167–169.

Weinstein, G. D., and P. Frost. 1969. Cell proliferation kinetics in benign and malignant skin diseases in humans. *Nat. Cancer. Inst. Monogr.* 30: 225–246.

Weinstein, G. D., and P. Frost. 1971. Replacement kinetics. *In* "Dermatology in General Medicine" (T. B. Fitzpatrick, K. A. Arndt, W. H. Clark, A. Z. Eisen, E. J. Van Scott, and J. H. Vaughan, eds.), pp. 78–87. McGraw-Hill, New York.

Weinstein, G. D., and E. J. Van Scott. 1965. Turnover times of normal and psoriatic epidermis. *J. Invest. Dermatol.* **45**: 257–262.

Weinstock, M., and G. F. Wilgram. 1970. Fine structural observations on the formation and enzymatic activity of keratinosomes in mouse tongue filiform papillae. *J. Ultrastruct. Res.* **30**: 262–274.

Weiss, P., and W. Ferris. 1954. Electron microscopic study of the texture of the basement membrane of larval amphibian skin. *Proc. Nat. Acad. Sci. U.S.* **40**: 528–540.

Weiss, P., and R. James. 1955. Skin metaplasia *in vitro* induced by brief exposure to vitamin A. *Exp. Cell Res. Suppl.* **3**: 381–394.

Wessells, N. K. 1962. Tissue interactions during skin histodifferentiation. *Develop. Biol.* **4**: 87–107.

Wessells, N. K. 1964. Substrate and nutrient effects upon epidermal basal cell orientation and proliferation. *Proc. Nat. Acad. Sci. U.S.* **52**: 252–269.

Wessells, N. K. 1965. Morphology and proliferation during early feather development. *Develop. Biol.* **12**: 131–153.

Wessells, N. K. 1967. Differentiation of epidermis and epidermal derivatives. *N. Engl. J. Med.* **277**: 21–33.

Winsor, T., and G. E. Burch. 1944. Differential roles of layers of human epigastric skin on diffusion rate of water. *Arch. Intern. Med.* **74**: 428.

Wolf, J. 1939. Die innere Struktur der Zellen des Stratum Desquamans der menschlichen Epidermis. *Z. Mikroskop. Anat. Forsch.* **46**: 170–202.

Wolf, J. 1967a. Structure and function of periderm. I. Superficial structure of the peridermal epithelium. *Folia Morphol.* **15**: 296–305.

Wolf, J. 1967b. Structure and function of periderm. II. Inner structure of peridermal cell. *Folia Morphol.* **15**: 306–317.

Wolf, J. 1967c. The relationship of the periderm to the amniotic epithelium. *Folia Morphol.* **15**: 384–391.

Wolff, K., and K. Holubar. 1967. Odland-Korper (Membrane-coating granules, keratinosomen) als epidermale Lysosomen. *Arch. Klin. Exp. Dermatol.* **231**: 1–19.

Wolff, K., and E. Schreiner. 1970. Epidermal lysosomes. Electron microscopic-cytochemical studies. *Arch. Dermatol.* **101**: 276–286.

Yardley, H. J., and G. Godfrey. 1961. Incorporation of [^{32}p]orthophosphate into the phosphate esters of skin maintained *in vitro* in the absence of added substrate. *Biochem. J.* **81**: 37P.

3

Other Cells in the Epidermis

I. Melanocytes

A. Introduction

Sandwiched between tightly packed keratinocytes in the germinative layer, melanocytes are important residents of the epidermis and hair follicle. Streamers of cytoplasm radiating from them distinguish them from the surrounding cells (Fig. 1). In mammalian skin as well as in that of other vertebrates, melanocytes derive from the neural crest during early embryonic development (Rawles, 1947). Melanoblasts, for example, are recognizable in the dermis of Negro fetuses during the tenth week of development. Between the twelfth and fourteenth weeks, they invade the epidermis and become numerous. During the fourth, fifth, and sixth months, very few melanocytes remain in the dermis; the so-called Mongolian spot is one of the few exceptions (Zimmerman and Becker, 1959).

The only known function of melanocytes is the synthesis of pigment granules (melanosomes), the main component of which is a light yellow to nearly black biochrome called melanin. Eventually melanosomes are transferred to the surrounding keratinocytes which interact with melanocytes to determine the amount of pigmentation in skin and hair.

75

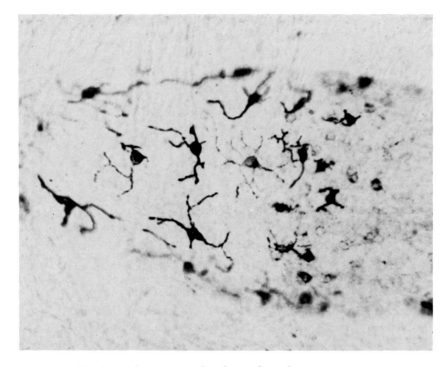

Fig. 1. Melanocyte in a lactiferous duct of a young woman.

Epidermal melanin acts as a screen against solar ultraviolet radiation. However, the functional significance of melanocytes at other locations, such as the viscera and around blood and nerves, is not yet known. In these remote areas, melanocytes do not transfer their pigment to receptor cells as they do in the epidermis and hair follicles.

B. Structure of Melanocytes and Melanogenesis

Grossly, melanocytes are arachnoid in shape, bipolar, or even amorphous, depending on their location and state of activity. In electron micrographs of epidermis, they can be distinguished from the surrounding keratinocytes by the scarcity of tonofilaments and the absence of desmosomal attachments. Their unique characteristic, however, is the presence of melanosomes (Fig. 2), distinctive organelles that represent stages in the formation of pigment granules. The cytoplasm, like that of other cells, contains ribosomes, mitochondria, Golgi membranes and

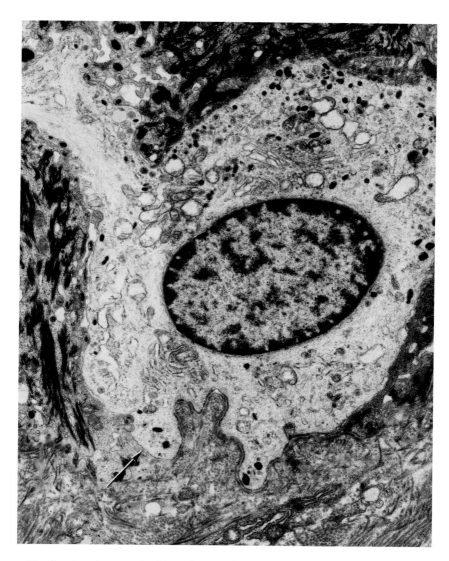

Fig. 2. A melanocyte in the epidermis of a young man. There are no desmosomal attachments between its cell membrane and that of adjacent keratinocytes. A basal lamina (arrow) separates the lower border of the cell from the dermis. The premelanosomes in the cytoplasm are in various stages of development. Numerous mitochondria, fine filaments, large Golgi zones, and a few profiles of granular endoplasmic reticulum can be seen in the cytoplasm. × 3200. (Courtesy of Dr. M. Bell.)

vesicles, centrioles, and rough-surfaced endoplasmic reticulum. Seiji and Fitzpatrick (1961) and Seiji *et al.* (1963) studied the biochemical properties of melanosomes and other cytoplasmic components isolated with ultracentrifugal density gradient techniques. A number of other investigators have correlated the findings of these authors with electron microscopic, histochemical, and radioautographic observations. It is generally agreed that melanogenesis or the formation of melanosomes proceeds in a predictable fashion. The enzyme tyrosinase is synthesized in ribosomes and conveyed via the endoplasmic reticulum to the Golgi region where it accumulates in small, round, membrane-limited vesicles. These enlarge and become oval and develop a characteristically patterned internal structure consisting of an ordered arrangement of tyrosinase molecules on a protein matrix with a periodicity of 100 Å. Melanin is deposited on the inner structural units only when they become fully developed. When melanin is deposited on the protein matrix of premelanosomes, the periodicity of the internal membrane becomes progressively obscured until the melanosomes resemble uniformly dense granules with no discernible internal structure.

Melanin is a biochrome of high molecular weight produced by the oxidation of the amino acid tyrosine with the copper-containing enzyme tyrosinase. The first step in the reaction is the formation of dopa (3, 4-dihydroxyphenylalanine), oxidized to dopa-quinone. Further oxidation and polymerization of dopa-quinone gives rise to tyrosine-melanin (eumelanin). When frozen sections are incubated with buffered solutions of either tyrosine or dopa, the tyrosinase in the melanocytes converts them into the easily recognized melanin. Melanogenesis, then, is the synthesis of melanin pigment and its deposition on the protein matrix within the unique membrane-bounded granules. Melanosomes that contain blackish-brown melanoproteins are called eumelanins.

The brownish-yellow pigment commonly found in blonde and red hair is phaeomelanin. Eumelanin and phaeomelanin have common pathways; at some stage in the formation of eumelanin, some intermediate products react with cysteine to form phaeomelanin.

Basic structural differences distinguish eumelanin from phaeomelanin. The melanosomes of red and blonde hair follicles are spherical whereas those of brown or black follicles are ellipsoidal. Both kinds of pigments have been produced *in vitro* (Cleffmann, 1963). *In vitro* follicles of agouti hairs, which normally produce eumelanin, produce yellow pigment when glutathione is added to the medium. Whether eumelanin and phaeomelanin are produced by two types of melanocytes or whether the same melanocyte changes from the synthesis of one to the other is not known.

C. Epidermal Melanin Units

In mammalian epidermis melanin pigmentation depends on the inter-action between melanocytes and keratinocytes, the rate of melano-genesis in the melanocytes, the rate of transfer of melanosomes from melanocytes into keratinocytes, and the disposition of melanosomes in the keratinocytes (Fitzpatrick and Breathnach, 1963; Quevedo, 1969, Fitzpatrick and Quevedo, 1971; Quevedo, 1971). Melanocytes are not distributed among keratinocytes at random. An orderly structural and functional association between the two makes up what has come to be called the epidermal-melanin unit, that is, a melanocyte and a constant population of satellite keratinocytes. In man the ratio of active ("dopa-positive") melanocytes to keratinocytes in the Malpighian layer is about 1 to 36 (Frenk and Schellborn, 1969), but this varies in different parts of the body (Szabo, 1967a,b).

Like keratinocytes, melanocytes must be self-perpetuating to maintain the melanocyte–keratinocyte ratio. Bullough and Laurence (1964, 1968) and Bullough (1971) asserted that the mitotic activity of melanocytes is also regulated by tissue-specific inhibitory chalones, but they have not clarified the role of epidermal and melanocyte chalones and their inter-action in maintaining epidermal melanin units.

Some authors have challenged the chalone concept of mitotic regula-tion (Powell et al., 1971; Voorhees and Duell, 1971), holding that mitotic activity in the epidermis is controlled by levels of cyclic AMP. Further-more, instead of strengthening the action of chalones, adrenaline appears to bind to cell surfaces at β-adrenergic receptor sites and to stimulate the production of cyclic AMP through the action of adenyl cyclase. If this is correct, chalones may be part of the cyclic AMP system. Iversen (1969) even proposed that the regulation of mitosis involves the inter-action of the chalone and cyclic AMP.

We are still not absolutely certain how melanosomes are transferred to keratinocytes. Melanocytes were once thought to inject their pigment into keratinocytes by a process that Masson (1948) called "cytocrine" activity. Now, however, evidence shows that keratinocytes actively phagocytize the melanin-laden type of melanocyte dendrites (Fitz-patrick and Breathnach, 1963; Mottaz and Zelickson, 1967). Perhaps there is even a feedback mechanism between melanocytes and keratinocytes. Thus when the number of keratinocytes increases, e.g., after exposure to sunlight or UV light, melanogenesis and tanning increase accordingly. Neither ultraviolet nor x-rays, however, stimulate melanocytes to produce pigment *in vitro* (Hu, 1968; Kitano and Hu, 1969).

In human skin the distribution of melanosomes within keratinocytes is determined by the size of the individual melanosomes (Wolff and Konrad, 1971; Toda *et al.*, 1972). Those larger than 0.8 μm are found singly (Fig. 3) within membrane-limited vesicles of the keratinocytes (Toda *et al.*, 1972) whereas smaller ones tend to aggregate in groups of two or more in membrane-limited vesicles called "melanosome complexes" (Fig. 3).

Szabo (1959, 1967b) showed that neither sex nor race affects the number of melanocytes. But racial color differences are affected by the

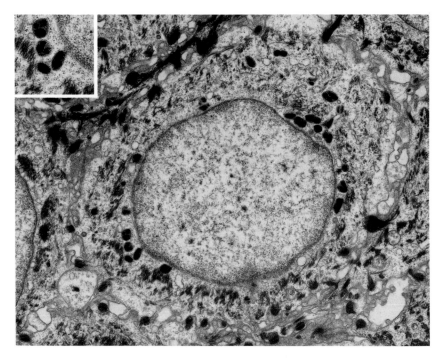

Fig. 3. Melanosomes in an epidermal keratinocyte from the face of a young man. The melanosomes are in complexes, typical of the skin of Caucasians; each complex consists of clusters of two to four melanosomes enclosed within a limiting membrane. $\times$ 12,600; Inset (melanosome complex at higher magnification) $\times$ 17,100. (Courtesy of Dr. M. Bell.)

number and size of mature melanosomes and the way they are distributed in the keratinocytes. In Negroids and Australoids, melanosomes are larger than 0.8 μm at their greatest diameter and are numerous in the melanocytes. In Caucasoids and Mongoloids, they are smaller than 0.8 μm. Furthermore, for the most part, melanosomes occur singly in the keratinocytes of Negroids and Australian aborigines but as complexes in Caucasoids and Mongoloids (Szabo, 1967a; Mitchell, 1968).

Inside the keratinocytes, melanosomes are associated with lysosomal enzymes (Hori et al., 1968; Olson et al., 1970; Ohtaki and Seiji, 1971) and apparently undergo gradual degradation; most become fragmented into small dense particles. In heavily pigmented skin, however, melanosomes remain intact inside the keratinocytes despite the lysosomal activity.

Reflectance spectrophotometric studies suggest that in protected human integument the melanogenic activity of epidermal melanin units varies from region to region (Edwards and Duntley, 1939). Skin is darkest in the axillae, other "fold regions," and perineum. The skin of adult human beings appears to be a mosaic of many unit areas, whose properties are maintained by the nervous system. Whimster (1971) found a "unit nature" and symmetry in such cutaneous diseases as vitiligo, in which the loss of melanocytes from the epidermis can be so precise that cutaneous depigmentation occurs in symmetrical patterns (Breathnach, 1969; Whimster, 1971). Such regional differentiation of the integument into a mosaic must occur during development. Lerner (1971) suggests that the symmetrical loss of function in epidermal melanin units in particular sites of the body is the result of neurocytotoxic agents.

Quevedo (1972) believes that region-specific programs could determine the responsiveness of epidermal melanin units to endogenous (e.g., hormonal) or exogenous (e.g., UV radiation) factors. The patterns of depigmenting cutaneous diseases indicate that some epidermal melanin units are controlled by a given program and that melanocyte function is specified from within the epidermal melanin units by cues from the dermis. Billingham and Silvers (1967) showed that the regional specificity of the epidermis depends largely on regulatory mechanisms in the dermis.

A clear example of melanocyte–keratinocyte interrelationship is seen in the pigmentation of hair, where the activity of melanocytes is totally related to the hair growth cycle. Melanocytes synthesize and transfer pigment into the cortical and medullary cells of the hair shaft only during the growth phase (anagen) of the growth cycle. Just before the end of the cycle, follicular melanocytes stop producing pigment and mysteriously "disappear." Although very small melanocytes are seen occasionally inside the hair germ of quiescent follicles or in the diurnal papilla, the absolute location of melanocytes in the quiescent follicle is not known. When quiescent follicles become active, melanocytes reappear and resume pigment production.

D. Control of Skin Pigmentation

It is not our intention to enumerate all of the factors that control skin pigmentation but to classify them all under two general categories, genetic and hormonal.

1. GENETIC

In laboratory mice, more than 70 genes, singly or in combination, at 40 loci control the origin and differentiation of melanoblasts, the number and shape of melanocytes, the different steps in the production of melanin polymer and melanosomes, and finally the transfer to and deposition of melanosomes in keratinocytes. In albinism melanocytes are unable to synthesize tyrosinase, a step in the production of melanin polymer that arises from a mutant gene operating at the C locus. This is comparable to oculocutaneous albinism in man. Two kinds of oculocutaneous albinism have been distinguished on the basis of their phenotypic expression, tyrosinase positive and negative. The skin of the albino mouse possesses a normal number of melanocytes which produce the membrane-bounded protein matrices but without the melanin polymer. The nonmelanized melanosomes of albinos are even transferred to the keratinocytes (Parakkal, 1967). Genes at two loci control the shape of melanocytes: dilute (d) and leaden (ln). Any defect in these genes produces stubby melanocytes with few short dendritic processes and reduces the intensity of coat coloration. In piebaldism, melanocytes are absent in the white spots, specific genes having influenced the migration and primary differentiation of neural crest cells into melanoblasts with certain properties.

2. HORMONAL

Hormones profoundly influence mammalian pigmentation. When α-MSH (melanocyte stimulating hormone) and β-MSH are administered to human subjects, marked hyperpigmentation occurs (Lerner and McGuire, 1961) because of a greater dispersion of melanosomes into the dendritic process and an increase in the number of melanosomes in the keratinocytes (Snell, 1967). Estrogen, progesterone, and MSH markedly increase the pigmentation of nipples, areolae, linea nigra (Figs. 4 and 5), and genitalia during pregnancy, but the mode of action of these hormones on melanocytes is not known.

Pigmentary changes characterize certain disease states that involve endocrine imbalances. In hyperpituitarism, increased amounts of MSH cause the skin to darken; the reverse is true in hypopituitarism. In Addison's disease, adrenal steroids decrease and fail to inhibit the production of large amounts of ACTH, which also causes increased pigmentation.

In frogs, α-, β-MSH catecholamines and ACTH cause dramatic color changes (McGuire, 1967); melanosomes are dispersed to the long dendritic process of the melanophores and the skin becomes visibly

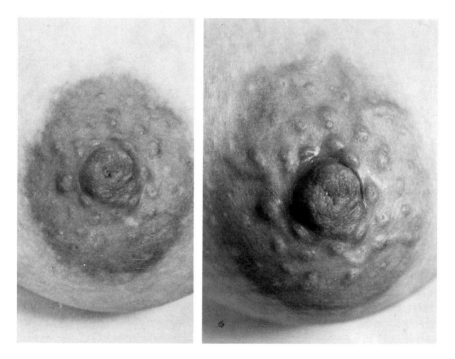

Fig. 4. On the left is the nipple and areola of a woman after four months of pregnancy; on the right is the same breast four months later. Both natural size.

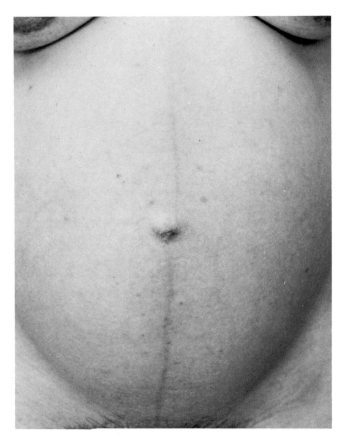

Fig. 5. A linea nigra on the abdomen of a woman pregnant eight months. The line largely disappears after pregnancy.

darker. In lighter colored frogs, the melanosomes are withdrawn into the melanophore perikaryon. Similar changes, produced by placing frogs in a dark and light environment, are also believed to be mediated by hormones (Hadley and Quevedo, 1967).

The action of MSH is mediated through α-adrenoceptor, that of catecholamines through the β-adrenoceptors; α-adrenoceptor blocking agents inhibit MSH-induced darkening, β-adrenoceptor blocking agents inhibit catecholamine-induced darkening. Both increase cyclic AMP levels resulting in a dispersion of melanoma (Goldman and Hadley, 1970; McGuire, 1970).

E. Comparative Pigmentary Systems

Ultimately, melanin pigmentation in man, which often differs from area to area, will have to be clarified by studies on experimental animals. Since it is impossible to find an experimental model whose pigmentary system is identical with that of man, any information that will make the pigmentary complex more intelligible, whether in man or in other animals, must be sought. Studies in melanin pigmentation have been confined to some fishes and amphibians, some domestic fowl, rodents, and man himself. Some of the key theories about the biology of melanocytes have emerged from studies of melanoma cells. Yet the epidermis of some laboratory mammals, regardless of hair color, has at best only a sparse population of mostly amelanotic melanocytes. The polar bear, for example, with a yellowish-white coat has a black skin. (It is of some interest that polar bears often suffer from malignant melanomas.) Although nonhuman primates offer a rich source of experimental material for the study of melanin pigmentation, that source remains almost untapped except for a few descriptive reports (Yun and Montagna, 1965; Hu and Montagna, 1971; F. Hu, unpublished data). A few brief observations should suffice to illustrate the potential value of these animals.

Although adult galagos, whether brown or black, have a nonpigmented epidermis, near-term fetuses and postnatal infants up to two weeks have an extremely heavily pigmented epidermis. When the skin of adults is damaged with either mechanical or chemical agents, pigmentation occurs only in or around the damaged epidermis. Much the same situation obtains in adult rhesus monkeys, whose epidermis, heavily pigmented in the second and early trimester of gestation (Bell, 1969), contains practically no melanotic dopa-positive melanocytes. After irradiation with ultraviolet light, however, the normally dopa-negative melanocytes become positive and melanotic and have two peculiarities: (1) they continue to enlarge and synthesize melanosomes up to 30 days of irradiation; and (2) the melanosomes remain inside the cytoplasm and only occasionally are found inside keratinocytes. Furthermore, if irradiation is continued beyond 30 days, the melanocytes gradually become amelanotic (Yun and Montagna, 1966).

Melanocytes are common in the dermis of nonhuman primates. Celebes apes are born with a white skin but become black-skinned by three months. Some chimpanzees have white epidermis and others black, even as adults. Stump-tailed macaques develop lentigines after puberty, more copious in males than in females. In rhesus monkeys melanocytes are aggregated in confluent patches that give a piebald appearance and in spider monkeys they are widely disseminated throughout the dermis. Normally melanocytes are found around the blood vessels and nerves of the scalp and face of many primates (Montagna, 1972) and in the sebaceous glands of seals and several primates, including man.

When melanocytes are found far removed from the surface and from exposure to ultraviolet light, either they are misplaced through inborn developmental errors or they perform other functions than that of synthesizing a sun screen. For example, many mammals have active melanocytes in the spleen, and some fish, reptiles, and birds have them throughout the viscera and even skeletal muscles. The predictable presence of active melanocytes in the substantia nigra of the brain and in the meninges supports the suggestion that melanocytes have some still unknown function. Even granting that nature sometimes blunders, it seems unrealistic to conclude that melanocytes have no biological significance in these areas. Whether or not epidermal melanocytes, which protect the skin from ultraviolet light, are a case of serendipity and have as their real function something other than the synthesis of melanin remains a moot question.

II. The Cells of Langerhans

Dendritic cells were first seen in the upper layers of the epidermis by Langerhans (1868) and were later named after him. Notwithstanding their long history and much recent investigation, their origin remains somewhat obscure and their function is completely unknown. Most of the earlier, and some of the recent difficulties in demonstrating and identifying these cells with the light microscope have been due to capricious or nonselective techniques. For example, the gold impregnation techniques (Breathnach, 1965; Ferreira-Marques, 1951) give uncertain and unreliable results, and neither the osmium iodide (Niebauer, 1956; Mishima and Miller-Milinska, 1961) nor the ATPase methods (Jarrett and Riley, 1963; Mustakallig, 1962; Bradshaw et al., 1963) distinguish between melanocytes and Langerhans cells. Quevedo and Montagna (1962), however, were able to demonstrate alkaline phosphatase activity in the Langerhans cells in the skin of African Lorisidae.

Because of their characteristic ultrastructural features, Langerhans cells can be identified with certainty only with the electron microscope. They have a markedly indented nucleus and a comparatively clear cytoplasm (Fig. 6) but no tonofilaments in their cytoplasm and no desmosomes in the plasma membrane (Bell, 1966). Furthermore, in addition to a Golgi complex these cells, like melanocytes, contain limited amounts of smooth and rough-surfaced endoplasmic reticulum (Breathnach, 1965); however, they have racquet-shaped organelles or granules in their cytoplasm and no formative stages of melanosomes (Fig. 6). Three-dimensional models of the granules reconstructed from serial sections (Sagebiel and Reed, 1968) show a flattened or curved orthogonal net of particles bound externally by a limiting membrane which is shaped like a disc, a cup, or a combination of both. Many of these granules, ranging from 140 to 500 μm by 50 μm, are close to and directly continuous with cell membranes. With lanthanum as a tracer, many are seen to contain lanthanum and to be continuous with the intracellular space (Cancilla, 1968). Whether the granules originate from the plasma membrane and move into the cell or develop within the cell along with the Golgi complex and then migrate to the cell periphery is still controversial (Tarnowski and Hashimoto, 1967; Zelick-son, 1965, 1966; Niebauer et al., 1969).

Langerhans cells have also been found in such other stratified squamous epithelia as oral mucosa, gingiva, and vagina as well as in the pilosebaceous system, the dermis, and lamina propria of the trachea. In pathological conditions, they abound in the urinary bladder of vitamin A-deficient rats and in lesions of histiocytosis X. Despite the similarity of the Langerhans granules in histocytosis and in normal epidermis, it is debatable whether they have the same lineage.

Functionally, Langerhans cells were considered by Masson (1948, 1951) to be effete melanocytes. However, Breathnach et al. (1968) showed them to be present in the epidermis of animals deprived of the neural crest. Since these animals have no epidermal melanocytes, Langerhans cells and melanocytes cannot belong to the same lineage.

Langerhans (1868) himself considered the cells to be related to the peripheral nervous system but had little experimental proof for his belief. Studies with the electron microscope have not demonstrated any specific connection between nerve fibers and Langerhans cells. Again, they were thought to function as phagocytes, but experiments with peroxidase and ferritin showed no phagocytic activity (Wolff and Schreiner, 1970; Sagebiel, 1972).

Despite the continuing uncertainty about their function, Langerhans cells are undoubtedly an independent cell population in the epidermis. After irradiation and other injury, they incorporate labeled thymidine (Giacometti and Montagna, 1967) and mitotic figures have been seen.

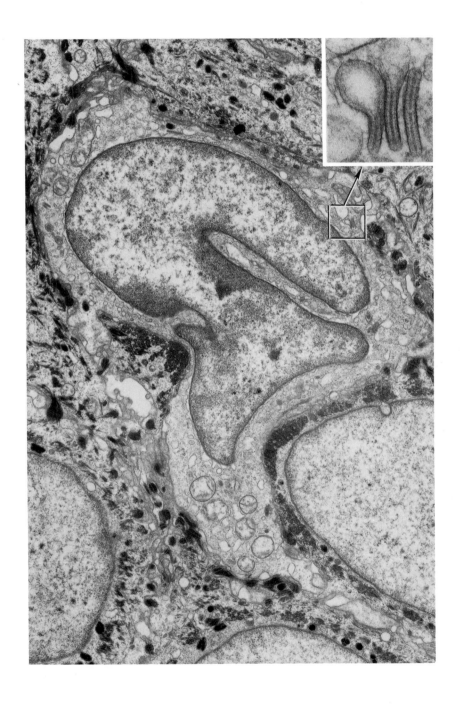

III. Merkel Cells

Throughout the mammalian epidermis and adjacent to certain hair follicles are discrete thickened regions called epidermal pads, tactile discs, or Häar-scheiben (Pinkus, 1927; Lyne and Hollis, 1971; Breathnach and Robins, 1970; Smith, 1970; Hashimoto, 1972a; Winkelmann and Breathnach, 1973). Scattered among the basal keratinocytes in these areas are certain apparently specialized cells, called Merkel cells, which because of their association with neural terminations are thought to be sensory receptors or transducers. The Merkel cells found elsewhere than in embryonic epidermal pads are not associated with neurites and probably have nothing to do with sensory mechanisms.

Markel cells have been reported in a wide variety of mammals (Lyne and Hollis, 1971) and in other vertebrates (Nafstad, 1971). In mammals they are found in general hairy surfaces, in the nail matrix, oral mucosa, the palms, and in the mesenchyme immediately underneath the epidermis (McGavran, 1968). The cells have lobated, irregular nuclei with less electron-dense cytoplasms than adjacent keratinocytes. They also contain numerous dense osmiophilic granules in the cytoplasm that vary in size from 700 to 1800 Å (Fig. 7). These granules, which seem to be formed in the prominent Golgi complex, are identical with those in the adrenal medulla that contain noradrenalin or in the chromaffin cells in other locations. They are more concentrated on the basal side of the cells where the neurites from the dermis come to rest against them. The intercellular space between their respective plasma membranes is extremely small. There are numerous mitochondria in the cytoplasm of the neurites next to Merkel cells.

Merkel cells are attached to adjacent epidermal cells by numerous desmosomes. The tonofilaments found in Merkel cells, unlike those in the keratinocytes, are not grouped into bundles.

The origin of these cells is not known. Because of their resemblance to keratinocytes, they are believed by some to have differentiated from epidermal keratinocytes (Munger, 1965; Smith, 1967). Breathnach

---

Fig. 6. A Langerhans cell from the epidermis of a young man. There are no desmosomal attachments between the apposing surfaces of this cell and adjacent keratinocytes. The cell has a typically deeply infolded nucleus. Fine filaments, mitochondria, Golgi elements, and Langerhans granules are distributed in the cytoplasm. The inset shows Langerhans granules at higher magnification. The granules consist of vesicles associated with rod-shaped structures with electron-opaque cores (periodicity ~ 90 Å). × 12,600. Inset × 151,000. (Courtesy of Dr. M. Bell.)

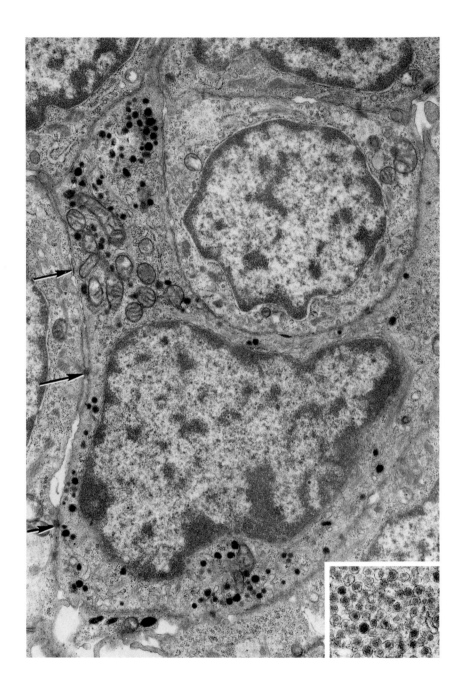

(1971), however, believes that they originate from neuroectoderm and along with peripheral nerves migrate into the skin and come to rest in the basal layer of the epidermis. Breathnach and Robins (1970) have seen Merkel cells and Schwann cells closely associated underneath the epidermis. Hashimoto (1972b) has further shown thàt Merkel cells are often surrounded by a Schwann sheath and that the basal lamina of the epidermis and of the Schwann cells becomes continuous.

References

Bell, M. 1966. Electron microscopy and histochemistry of the Langerhans cells of subhuman primates. *In* "Advances in Biology of Skin. The Pigmentary System" (W. Montagna and F. Hu, eds.), Vol. 8, pp. 115–134. Pergamon, Oxford.

Bell, M. 1969. The ultrastructure of differentiating hair follicles in fetal rhesus monkeys *(Macaca mulatta)*. *In* "Advances in Biology of Skin. Hair Growth" (W. Montagna and R. L. Dobson, eds.), Vol. 9, pp. 61–81. Pergamon, Oxford.

Billingham, R. E., and W. K. Silver. 1967. Studies on the conservation of epidermal specificities of skin and certain mucosas in adult mammals. *J. Exp. Med.* **125**: 429–446.

Bradshaw, M., M. Wachstein, J. Spence, and J. M. Elias. 1963. Adenonine triphosphatase activity in melanocytes and epidermal cells of human skin. *J. Histochem. Cytochem.* **11**: 465–473.

Breathnach, A. S. 1965. The cell of Langerhans. *Int. Rev. Cytol.* **18**: 1–27.

Breathnach, A. S. 1969. Normal and abnormal melanin pigmentation of the skin. *In* "Pigments in Pathology" (M. Wolman, ed.), pp. 353–394. Academic Press, New York.

Breathnach, A. S. 1971. "An Atlas of the Ultrastructure of Human Skin. Development, Differentiation, and Postnatal Features." Churchill, London.

Breathnach, A. S., and E. J. Robins. 1970. Ultrastructural observations on Merkel cells in human fetal skin. *J. Anat.* **106**: 4110.

Breathnach, A. S., W. K. Silver, T. Smith, and S. Heyner. 1968. Langerhans cells in mouse skin experimentally deprived of its neural crest component. *J. Invest. Dermatol.* **50**: 147–160.

Bullough, W. S. 1971. The actions of the chalones. *Agents and Actions* **2**: 1–7.

Bullough, W. S., and E. B. Laurence. 1964. Mitotic control by internal secretion: the role of the chalone-adrenalin complex. *Exp. Cell Res.* **33**: 176–194.

Bullough, W. S., and E. B. Laurence. 1968. Control of mitosis in mouse and hamster melanoma by means of the melanocyte chalone. *Eur. J. Cancer* **4**: 607–615.

Fig. 7. An electron micrograph of a Merkel cell in the epithelium of a fetal rhesus monkey. The cell contains characteristic membrane-bounded granules, which probably contain catecholamines with electron-opaque cores. Desmosomal attachments (arrows on the left) are found at the surfaces of the cells and adjacent keratinocytes. The inset shows granules from a Merkel cell in human fetal epidermis. × 11,400. Inset × 37,100. (Courtesy of Dr. M. Bell.)

Cancilla, P. A. 1968. Demonstration of the Langerhans granule by lanthanum. *J. Cell Biol.* **38**: 248–252.

Cleffmann, G. 1963. Agouti pigment cells *in situ* and *in vitro. Ann. N. Y. Acad. Sci.* **100**: 749–761.

Edwards, E. A., and S. Q. Duntley. 1939. The pigments and color of living human skin. *Amer. J. Anat.* **65**: 1–33.

Ferreira-Marques, J. 1951. Systema sensitioum intra-epidermicum. Die Langer-hannschen Zellen als Rezeptoren des hellen Schmerzes: Doloriceptors. *Arch. Dermatol. Syph.* **193**: 191–250.

Fitzpatrick, T. B., and A. S. Breathnach. 1963. Das epidermale Melanin-Einheit-System. *Dermatol. Wochenschr.* **147**: 481–489.

Fitzpatrick, T. B., and W. C. Quevedo, Jr. 1971. Biological process underlying melanin pigmentation and pigmentary disorders. *Mod. Trends Dermatol.* **4**: 122–149.

Frenk, E., and J. P. Schellborn. 1969. Zur Morphologie der epidermalen Melanin-linheit. *Dermatol.* **21**: 339–348.

Giacometti, L., and W. Montagna. 1967. Langerhans cells: Uptake of tritiated thy-midine. *Science* **157**: 439–440.

Goldman, J. M., and M. E. Hadley. 1970. Evidence for separate receptors for melanophore stimulating hormone and catecholamine regulation of cyclic AMP in the control of melanophore responses. *Brit. J. Pharmacol.* **39**: 160–166.

Hadley, M. E., and W. C. Quevedo, Jr. 1967. The role of epidermal melanogenesis in adaptive color changes in amphibians. *In* "Advances in Biology of Skin. The Pigmentary System" (W. Montagna and F. Hu, eds.), Vol. 8, pp. 337–359, Pergamon, Oxford.

Hashimoto, K. 1972a. The ultrastructure of human embryos. X. Merkel tactile cells in the finger and nail. *J. Anat.* **111**: 99–120.

Hashimoto, K. 1972b. Fine structure of Merkel cells in human oral mucosa. *J. Invest. Dermatol.* **58**: 381–387.

Hori, Y., K. Toda, M. Pathak, W. Clark, and T. Fitzpatrick. 1968. A fine-structure study of the human epidermal melanosome complex and its acid phosphatase activity. *J. Ultrastruct. Res.* **25**: 109–120.

Hu, F. 1968. Melanocytes and melanin production. *J. Soc. Cosmet. Chem.* **19**: 565–580.

Hu, F. Primate pigment cells in culture. *Excerpta Medica* (In press).

Hu, F., and W. Montagna. 1971. The development of pigment cells in the eyes of rhesus monkey. *Amer. J. Anat.* **132**: 119–132.

Iversen, O. H. 1969. Chalones of the skin. *In* "Ciba Foundation Symposium on Homeostatic Regulators" (G. E. W. Wolstenholme and J. Knight, eds.), pp. 29–53. Churchill, London.

Jarrett, A. J., and P. A. Riley. 1963. Esterase activity in dendritic cells. *Brit. J. Dermatol.* **75**: 79–81.

Kitano, Y., and F. Hu. 1969. The effects of ultraviolet light on mammalian pigment cells *in vitro. J. Invest. Dermatol.* **52**: 25–31.

Langerhans, D. 1868. Uber die Nerven der Menschlichen Haut. *Virchows Arch. Pathol. Anat. Physiol.* **44**: 326–337.

Lerner, A. B. 1971. On the etiology of vitiligo and gray hair. *Amer. J. Med.* **51**: 141–147.

Lerner, A. B., and J. S. McGuire. 1961. Effect of alpha- and beta-melanocyte stimulating hormones on the skin colour of man. *Nature (London)* **189**: 176–179.

Lyne, A. G., and D. E. Hollis. 1971. Merkel cells in sheep epidermis during fetal development. *J. Ultrastruct. Res.* **34**: 464–472.

Masson, P. 1948. Pigment cells in man. *In* "The Biology of Melanosomes" (R. W. Miner and M. Gordon, eds.), Vol. 4, pp. 15–37. Special Publication of the N. Y. Acad. of Sci.

Masson, P. 1951. My conception of cellular naevi. *Cancer* **4**: 9–38.

McGavran, M. H. 1968. *In* "The Skin Senses" (D. R. Kenshalo, ed.), pp. 61–62. Thomas, Springfield, Illinois.

McGuire, J. S. 1967. The epidermal melanocytes of the frog. *In* "Advances in Biology of Skin. Vol. 8. The Pigmentary System" (W. Montagna and F. Hu, eds.), pp. 329–336. Pergamon, Oxford.

McGuire, J. 1970. Adrenergic control of melanocytes. *Arch. Dermatol.* **101**: 173–179.

Mishima, Y., and A. Miller-Milinska. 1961. Junctional and high-level dendritic cells revealed with osmium iodide reaction in human and animal epidermis under conditions of hyperpigmentation and depigmentation. *J. Invest. Dermatol.* **37**: 107–120.

Mitchell, R. E. 1968. The skin of the Australian aborigine; a light and electron microscopical study. *Aust. J. Dermatol.* **9**: 314–328.

Montagna, W. 1972. The skin of nonhuman primates. *Amer. Zool.* **12**: 109–124.

Mottaz, J. H., and A. S. Zelickson. 1967. Melanin transfer: a possible phagocytic process. *J. Invest. Dermatol.* **49**: 605–610.

Munger, B. L. 1965. The intraepidermal innervation of the snout skin of the opossum. *J. Cell Biol.* **26**: 78–97.

Mustakallig, K. 1962. Adenisone triphosphatase activity in neural elements of human epidermis. *Exp. Cell Res.* **28**: 449–451.

Nafstad, D. H. J. 1971. Comparative ultrastructural study on Merkel cells and dermal basal cells in poultry *(Gallus domesticus). Z. Zellforsch. Mikrosk. Anat.* **116**: 342–348.

Niebauer, G. 1956. Über die interstitiellen Zellen der Haut. *Hautarzt* **7**: 123–126.

Niebauer, G., W. S. Krawczyk, R. L. Kidd, and G. F. Wilgram. 1969. Osmium zinc iodide reactive sites in the epidermal Langerhans cell. *J. Cell Biol.* **43**: 80–89.

Ohtaki, N., and M. Seiji. 1971. Degradation of melanosomes by lysosomes. *J. Invest. Dermatol.* **57**: 1–5.

Olson, R. L., J. Nordquist, and M. A. Everett. 1970. The role of epidermal lysosomes in melanin physiology. *Brit. J. Dermatol.* **83**: 188–199.

Parakkal, P. F. 1967. Transfer of premelanosomes into the keratinizing cells of albino hair follicle. *J. Cell Biol.* **35**: 473–477.

Pinkus, F. 1927. Normale anatomie der Haut. *In* "Handbuch der Haut- und Geschlechtskrankheiten" (J. Jadassohn, ed.), Vol. 1/1, pp. 344. Springer-Verlag, Berlin and New York.

Powell, J. A., E. A. Duell, and J. J. Voorhees. 1971. Beta adrenergic stimulation of endogenous epidermal cyclic AMP formation. *Arch. Dermatol.* **104**: 359–365.

Quevedo, W. C., Jr. 1969. The control of color in mammals. *Amer. Zool.* **9**: 531–540.

Quevedo, W. C., Jr. 1971. Genetic regulation of pigmentation in mammals. *In* "Biology of Normal and Abnormal Melanocytes" (T. Kawamura, T. B. Fitzpatrick, and M. Seiji, eds.), pp. 99–115. Univ. of Tokyo Press, Tokyo.

Quevedo, W. C., Jr. 1972. Epidermal melanin units. Melanocyte–keratinocyte interactions. *Amer. Zool.* **12**: 35–41.

Quevedo, W. C., Jr., and W. Montagna. 1962. A new system of melanocytes in the skin of the potto *(Perodicticus potto). Anat. Rec.* **144**: 279–286.

Rawles, M. E. 1947. Origin of pigment cells from the neural crest in the mouse embryo. *Physiol. Zool.* **20**: 248–266.

Sagebiel, R. W. 1972. *In vivo* and *in vitro* uptake of ferritin by Langerhans cell of the epidermis. *J. Invest. Dermatol.* **58**: 47–54.

Sagebiel, R. W., and T. A. Reed. 1968. Serial reconstruction of the characteristic granule of the Langerhans cell. *J. Cell Biol.* **36**: 595–602.

Seiji, M., and T. B. Fitzpatrick. 1961. The reciprocal relationship between melanization and tyrosinase activity in melanosomes (melanin granules). *J. Biochem.* **49**: 700–706.

Seiji, M., K. Shimao, M. S. C. Birbeck, and T. B. Fitzpatrick. 1963. Subcellular localization of melanin biosynthesis. *Ann. N. Y. Acad. Sci.* **100**: 497–533.

Smith, K. R. 1967. The structure and function of the Haarscheibe. *J. Comp. Neurol.* **131**: 459.

Smith, K. R. 1970. The ultrastructure of the human Haarschiebe and Merkel cell. *J. Invest. Dermatol.* **54**: 150–159.

Snell, R. S. 1967. Hormonal control of pigmentation in man and other mammals. *In* "Advances in Biology of Skin. Vol. 8. The Pigmentary System" (W. Montagna and F. Hu, eds.), pp. 447–466. Pergamon, Oxford.

Szabo, G. 1959. Quantitative histological investigations on the melanocyte system of the human epidermis. *In* "Pigment Cell Biology" (M. Gordon, ed.), pp. 99–125. Academic Press, New York.

Szabo, G. 1967a. Photobiology of melanogenesis: Cytological aspects with reference to differences in racial coloration. *In* "Advances in Biology of Skin. The Pigmentary System" (W. Montagna and F. Hu, eds.), Vol. 8, pp. 379–396. Pergamon, Oxford.

Szabo, G. 1967b. The regional anatomy of the human integument with apical reference to the distribution of hair follicles, sweat glands and melanocytes. *Phil. Trans. Roy. Soc. London, Ser. B.* **252**: 447–485.

Tarnowski, W. M., and K. Hashimoto. 1967. Langerhan's cell granules in histiocytosis. X. The epidermal Langerhans cell as a macrophage. *Arch. Dermatol.* **96**: 298–304.

Toda, K., M. A. Pathak, J. A. Parrish, T. B. Fitzpatrick, and W. C. Quevedo, Jr. 1972. Alterations of racial differences in melanosome distribution in human epidermis after exposure to ultraviolet light. *Nature (London)* **236**: 143–145.

Voorhees, J. J., and E. A. Duell. 1971. Psoriasis as a possible defect of the adenyl cyclase-cyclic AMP cascade. *Arch. Dermatol.* **104**: 352–358.

Whimster, I. W. 1971. Symmetry in dermatology. *Mod. Trends Dermatol.* **4**: 1–30.

Winkelmann, R. K., and A. S. Breathnach. 1973. The Merkel cell. *J. Invest. Dermatol.* **60**: 2–15.

Wolff, K., and K. Konrad. 1971. Melanin pigmentation; an *in vivo* model for studies of melanosome kinetics within keratinocytes. *Science* **174**: 1034–1035.

Wolff, K., and E. Schreiner. 1970. Uptake, intracellular transport and degradation of exogenous protein by Langerhans cells. *J. Invest. Dermatol.* **54**: 37–47.

Yun, J. S., and W. Montagna. 1965. The skin of primates. XXV. Melanogenesis in the skin of bushbabies. *Amer. J. Physiol. Anthropol.* **23**: 143–148.

Yun, J. S., and W. Montagna. 1966. The melanocytes in the epidermis of the rhesus monkey *(Macaca mulatta)*. *Anat. Rec.* **154**: 161–174.

Zelickson, A. S. 1965. The Langerhans cell. *J. Invest. Dermatol.* **44**: 201–212.

Zelickson, A. S. 1966. Granule formation in the Langerhans cell. *J. Invest. Dermatol.* **47**: 498–502.

Zimmerman, A. A., and S. W. Becker, Jr. 1959. Melanoblasts and melanocytes in fetal Negro skin. *Ill. Mong. Med. Sci.* **6**: 1–59.

4

*The Dermis**

I. General Considerations

The dermis, or corium, lying between the epidermis and the sub-cutaneous panniculus, consists basically of a matrix of loose connective tissue composed of the fibrous proteins—collagen, elastin, and reticulin—embedded in an amorphous ground substance. The matrix is traversed by blood vessels, nerves, and lymphatics, and into it penetrate the epidermal appendages: eccrine sweat glands, apocrine glands, and the pilosebaceous units.

The dermal matrix contains few cells (more in the upper papillary layer than in the lower reticular), which are predominantly fibroblasts with the potential to produce most, if not all, the components of the extracellular matrix. More abundant are the mast cells; in addition, histiocytes or macrophages, melanocytes, and extravasated leukocytes are often found. During dermal remodeling, macrophages participate in collagen degradation (see p. 124). In man, the whole mass of dermis constitutes from 15 to 20% of total body weight. The versatility of the dermis is seen in its range of functions, from ion exchange to protection

---

* Written largely by Jeffrey Pinto, University of Oregon Medical School, Portland.

from mechanical injury. It provides nourishment to the epidermis and interacts with it during embryogenesis and morphogenesis and during repair and remodeling. Its various properties stem primarily from the matrix of extracellular connective tissue, the ground substance, and the fibrous proteins.

The specific nature and function of the dermis in any given area of the body are closely related; that is, what it *is* depends primarily on what it *does*. The corneal stroma, for example, contains a highly structured orthogonal arrangement of exceedingly thin collagen "fibers" (mean diameter, about 250 to 300 Å) which provide minimal optical distortion. Quantitatively and qualitatively, the glycosaminoglycan (mucopolysaccharide) content of the dermis also differs regionally. In the swollen "sex skin" of some female baboons and macaques (Fig. 1), for example, the content of hyaluronic acid increases many fold in response to hormonal stimulation. In the skin, sulfated glycosaminoglycan is predominantly dermatan sulfate (chondroitin sulfate B), whereas in the cornea it is keratin sulfate. Except for arrectores pilorum muscles, the dermis is generally free of muscle. However, skin from the areola and nipple of the breast, the penis, scrotum, and perineum contains variable numbers of smooth muscle fibers which when contracted produce wrinkling. Dermal vascularity also varies regionally according to the degree of heat exchange in any given area. Dermal attachments, too, vary from one region to the next; e.g., skin is loose over the joints but firmly attached over the friction surfaces.

Grossly, the dermis is a tough and resilient tissue with viscoelastic properties. Under the light microscope, the collagen fibers resemble an irregular meshwork, oriented roughly parallel to the epidermis. At this level, the dermis can be arbitrarily divided into an upper papillary layer or body of thin fibers immediately underneath the epidermis and a lower reticular layer composed of a network of thick collagen fibers. The human papillary layer has thinner fibers than the reticular layer and its interfibrillar spaces, which contain the ground substance, are greater. When the epidermis is removed, the upper surface of the papillary layer forms a negative image of the lower surface of the epidermis (Figs. 2 and 3).

The three-dimensional arrangement of the collagen meshwork can be seen with the scanning electron microscope at relatively low power. Collagen fibrils 10 to 15 μm in diameter are woven into a mat of remarkable structural integrity and flexibility. At the level of the transmitting electron microscope, the collagen in this network is seen to be

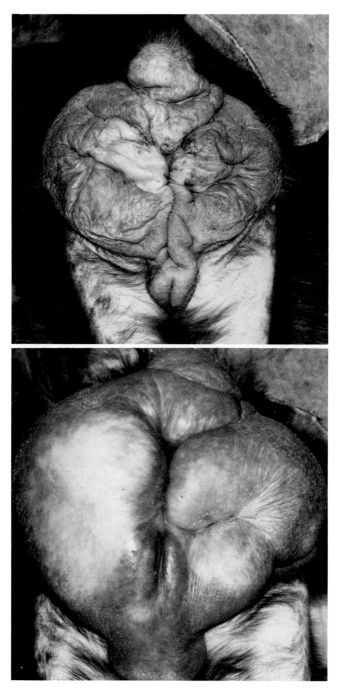

arranged hierarchically. Although there is no universal nomenclature for these various organizational levels of collagen fiber, we have adopted the terminology proposed by Grant and Prockop (1972). Collagen bundles that can be seen with the naked eye are referred to as "fibers," which under the light microscope consist of numerous smaller "fibrils." Fibrils, by definition the smallest units that can be seen under the light microscope, are about 10 to 15 μm in diameter. Under the electron microscope, the fibrils themselves are seen to consist of bundles of smaller, identical and parallel "microfibrils" 300 to 600 Å in diameter. On unfixed material viewed at very high magnifications, Bouteille and Pease (1971) have shown that the microfibrils consist of bundles of ultrathin "filaments" only 30 to 50 Å in diameter. Since this is too large to be a single collagen molecule (which is about 14 Å in diameter), it probably represents an aggregation of 3 to 5 molecules in cross section.

Structurally, collagen shows a helical twisting at various levels of organization. As we shall see later, the collagen molecule itself, referred to as tropocollagen, consists of 3 polypeptide chains wound into a triple helix. The filaments just described are helically wound within the microfibril. Whether this helical winding of the collagen molecules takes place in the filament or in the microfibrils within the fibrils has not been determined.

Within each microfibril, the filaments are surrounded by material containing carbohydrate, which may stabilize the collagen structure. Although the precise nature of the interaction between collagen and proteoglycans is unknown, it is probably ionic and occurs between the negatively charged sulfate groups on the proteoglycans and the positively charged groups on collagen. (For a review of this subject, see Jackson and Bentley, 1968.) The structure of collagen at the molecular level will be discussed later.

Although the electron microscope shows the ground substance as a homogenous mass, ultrastructural studies of highly purified proteoglycan (the complex formed from protein core and glycosaminoglycan side chains) show that these substances reaggregate in a specific manner in the test tube (Rosenberg *et al.*, 1970). Whether the aggregates so visualized represent an artifact or a true *in vivo* macromolecule configuration remains to be determined. The elastic fibers, on the other hand (see Section II, B), are seen ultrastructurally as component fibers,

---

Fig. 1. The buttocks of a postovulatory (above) and of a preovulatory (below) pigtail macaque in estrus.

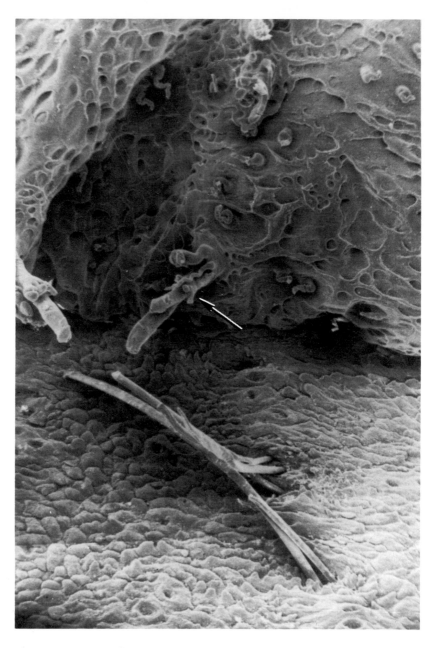

Fig. 2. Scanning electron micrograph of a partially split skin showing the epidermis above and the dermis below. Some epidermal appendages are still adherent to the epidermis (arrow). A tuft of hairs is implanted in the dermis. Though not very clear in this photograph because of opposite reliefs, the dermis forms a negative image of the underside of the epidermis. (Courtesy of Dr. W. H. Fahrenbach.)

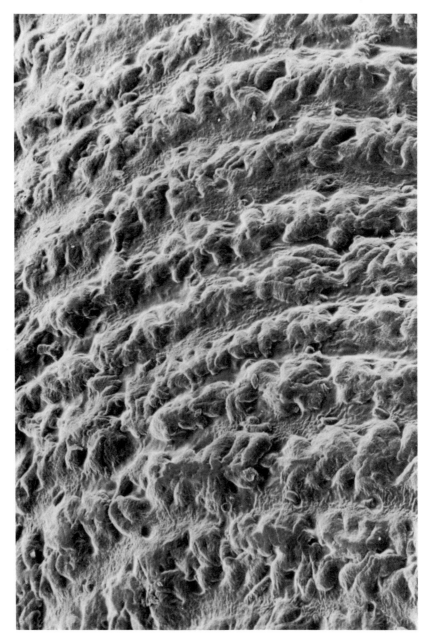

Fig. 3. Scanning electron micrograph of the surface of the papillary body of the dermis after the epidermis has been split off. (Courtesy of Dr. W. H. Fahrenbach.)

only the amorphous core of which corresponds to the protein elastin. The outer component, a casing of microfibrils, is thought to play a role in elastogenesis (Ross and Bornstein, 1969).

II. Dermal Components

A comprehensive review of the molecular biology of connective tissue is beyond the scope of this book; however, the short discussion that follows will provide the reader with some basic facts about the dynamics of connective tissue and their relation to dermal function.

A. Collagen

Since the substructure of collagen filaments cannot be seen with the electron microscope, the exact packing of collagen molecules within them is not known. At the level of the filament and at each successive level of organization, there are cross striations with a periodicity of 640 to 780 Å (Figs. 4 and 5). The individual collagen molecule, tropocollagen, is a rod-shaped triple helix about 2900 Å long and 14 Å in diameter which has five charged regions 680 Å apart with an affinity for stains. The periodicity in collagen filaments and higher aggregates is explained by the registration of these charged regions on laterally adjacent molecules. When collagen is stained with phosphotungstic acid, which binds to the basic amino acids, the banding corresponds to local concentrations of lysine and arginine. When collagen is stained with uranyl salts, the uranyl cation combines with the acidic amino acids, and the banding reflects local concentrations of glutamic and aspartic acids. Laterally adjacent molecules are aligned in the same direction with an approximate quarter stagger between their like ends and an approximate 300 Å overlap of their unlike ends (Fig. 5). Between the ends of longitudinally adjacent molecules is a so-called "hole" region of about 400 Å. A dense overlap region and a loose hole region constitute a single 700 Å axial period and explain the pattern of striation found in electron micrographs of negatively stained collagen fibrils (Fig. 5).

At each end of the tropocollagen molecule, which is a triple helix, the molecule appears to have a nonhelical pleated region. The three polypeptide chains making up the triple helix are called α chains. The two types of α chains (α_1 and α_2) have a molecular weight of about 100,000 each but differ slightly in amino acid composition. They are

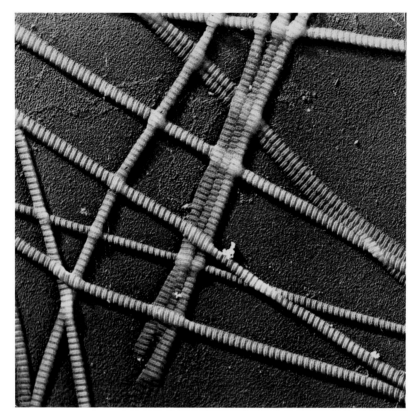

Fig. 4. High magnification of calf-skin collagen fibrils showing their distinctive periodicty. × 26,100. (Courtesy of Dr. J. Gross.)

present in mammalian collagen in the ratio $(\alpha_1)_2:(\alpha_2)_1$. However, the fact that some invertebrate collagens possess three α_1 chains per molecule may have evolutionary significance.

The helix of collagen differs from the α helix in other proteins in that the axial distance between adjacent amino acid residues is 2.9 Å rather than the 1.5 Å of the α helix. This results from the high content of the amino acids, proline and hydroxyproline, the latter being a relatively specific marker for collagen. The ring structure of these amino acids, which together constitute about 20% of the total residues, prevents α helix formation. However, the heterocyclic ring does not prevent all helix formation, and the bond angles it imposes on the chain, as well as the fact that glycine forms every third residue, permit the formation

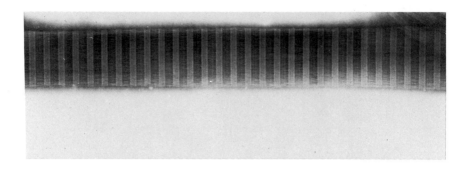

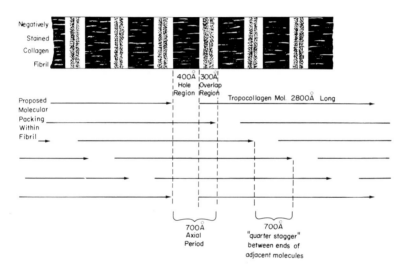

Fig. 5. Above, electron micrograph of negatively stained collagen fibril. (Courtesy of Dr. Romaine Bruns.) Below is a schematic representation of the quarter stagger arrangement of the tropocollagen molecules.

of a stable triple helix in which the small glycyl residues are tucked into the center; the bulky proline and hydroxyproline rings as well as the side chains of the heavier amino acids are on the outside. Unlike the tropocollagen helix itself, which has a right-handed twist, each individual chain twists to the left. These opposing twists resist lengthwise deformation and thereby contribute to the tensile strength of the molecule. Covalent bonding between laterally and longitudinally adjacent molecules within the filament increases its tensile strength. For detailed discussions of the molecular structure of collagen, see Rich and Crick (1961), Ramachandran and Sasisekharan (1961), Hodge (1967), and Kühn (1969).

The biosynthesis of collagen has been investigated by various means. Apparently collagen is first assembled in the fibroblasts or their near equivalents—osteoblasts, chondrocytes, odontocytes, and possibly also smooth muscle cells and epidermal cells. During its formation, portions of the proline and lysine residues contained in the nascent collagen chain are acted upon intracellularly by the enzymes, protocollagen proline hydroxylase and protocollagen lysine hydroxylase, to form residues of hydroxyproline and hydroxylysine, respectively, within the molecule. This pathway is obligatory in that neither hydroxyproline nor hydroxylysine can be incorporated directly into the molecule. The hydroxylase enzymes require ferrous iron atmospheric oxygen, α-ketoglutarate, and ascorbate. Subsequently, when still within the fibroblast, the collagen molecule undergoes glycosylation, which adds glucose or glucose-galactose units to hydroxylysine residues. The cell then releases the molecule. Both hydroxylation and glycosylation appear to be necessary for this extrusion process.

An intracellular biosynthetic precursor of the polypeptide chains, procollagen, has recently been described (Bellamy and Bornstein, 1971). Procollagen contains an α chain precursor with a molecular weight of about 120,000 (compared with 100,000 for the α chain); the extra peptide sequence differs from the remainder of the collagen molecule in that it contains cysteine (Dehm et al., 1972).

Speakman (1971) has proposed that during the biological assembly of the tropocollagen triple helix, each new α chain leaves the polyribosomal complex with a registration peptide attached to one end. Registration peptides from $1\alpha_2$ and $2\alpha_1$ chains would then interact laterally perhaps through disulfide bonding to form a stable aggregate. This registration peptide complex would thus bring the three new α chains in the right proportions into close proximity and in register and would then permit them to coil into a tropocollagen triple helix. Once the triple helix is formed, the registration peptide would be enzymatically cleaved off. Recent evidence suggests the existence of such an enzyme (Bornstein et al., 1972). This appended peptide may also facilitate molecular transport or prevent intracellular fibrillogenesis.

Radioautography has shown that during biosynthesis and within 15 minutes after injection a tritiated proline label administered intraperitoneally appears in the endoplasmic reticulum of fibroblasts, then successively in the Golgi apparatus, the peripheral cytoplasmic vesicles, and the extracellular space, becoming extracellular less than 1 hour after injection (Ross and Benditt, 1965). The fully hydroxylated and glycosylated collagen is released from the cell in the soluble molecular form, tropocollagen, which lacks covalent cross-links (tropocollagen cannot be seen with the electron microscope). This soluble tropocollagen then aggregates with other tropocollagen molecules to initiate fibrillo-genesis, after which it stabilizes by means of intra- and intermolecular covalent cross-linking. The most easily extractable collagen is the most recently synthesized; progressively older collagen is solubilized only by increasing the ionic strength or decreasing the pH of the extracting medium, the oldest collagen being insoluble in all extractants (Jackson and Bentley, 1960). That solubility is related to cross-linking has been shown by chromatography of the denaturation components from the various extracts. The most recently synthesized and extractable material has the lowest content of covalent cross-links, most of which are intra-molecular links. The degree of covalent cross-linking and the number of intermolecular links increase as the collagen ages and becomes less extractable. Covalent cross-linking stabilizes the molecule with time, not only to extracting solutions, as seen above, but also to enzymatic degrada-tion (Kohn and Rollerson, 1960). The result is that general body collagen in a mature individual is relatively inert, with an estimated half-life of as long as 5 years. However, under certain conditions, e.g., postpartum uterine involution, the half-life is as short as 1 to 2 days.

Collagen cross-linking begins by the oxidative deamination of the ε-amino groups of lysine and hydroxylysine residues to form allysine, an aldehyde, which combines either with another residue of lysine to form a Schiff-base type of cross-link or with another allysine residue to form a cross-link through aldol condensation (Fig. 6). The enzyme for the oxidative deamination of lysine and hydroxylysine, lysyl oxidase, differs from the serum amine oxidase and requires copper. It is inhibited by β-aminopropionitrile (βAPN), the lathyrogenic principle from the seeds of the sweet pea (Lathyrus odoratus). Experimental lathyrism, produced by the oral administration of βAPN, is associated with an abnormal fragility of collagen-containing tissues. Since copper deprivation also inactivates lysyl oxidase, copper-deficient animals show most of the signs of lathyrism. Collagen cross-linking can also be blocked by the administration of penicillamine which combines with the aldehyde group of allysine and thus prevents it from further interaction. The covalent cross-links in collagen are intramolecular and intermolecular, with only one intramolecular cross-link per tropocollagen molecule; this, an aldol

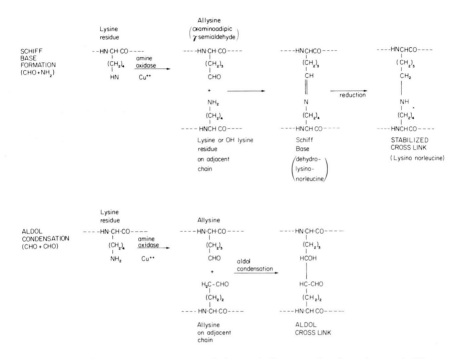

Fig. 6. Scheme summarizing cross-linking of allysine molecule to form Schiff-base and aldol condensation products.

type, is located in the amino-terminal nonhelical region. Several inter-molecular cross-links of the Schiff-base type are scattered throughout the helical region of the molecule.

Whereas collagen is relatively stable, as reflected by its long half-life *in vivo,* under certain circumstances it can turn over at a rapid rate. In the tail fin of tadpoles, the rapid resorption of dermal collagen during metamorphosis is associated with the elaboration or activation of a collagenolytic factor (Gross and Lapiere, 1962).

Since the demonstration of this collagenolytic principle, appropriately named collagenase, collagen has been resorbed during various remodeling phenomena; in each case tissue collagenase has appeared. This enzyme is heat labile, acts at physiological pH and temperature, and diffuses through Millipore filters that exclude cells; it is inactivated by a serum inhibitor (Eisen *et al.,* 1971). Collagenase is the only known mammalian enzyme that splits the collagen helix. A bacterial collagenase from *Clostridium histolyticum* also degrades collagen but in a different way. Tissue collagenase specifically cleaves the tropocollagen molecule into one-quarter- and three-quarter-length pieces, whereas the bacterial enzyme completely degrades the molecule. Attempts to extract collagenase from tissues during collagen resorption have been largely unsuccessful. Therefore, either the collagenase is present in tissues in an inactive zymogen form (Harper *et al.*, 1971) or it is inactivated during extraction.

Another equally important but less thoroughly studied aspect of collagen degradation is its ingestion by tissue macrophages. During the catagen stage of the hair follicle, when much collagen is resorbed, macrophages cluster around follicles and ingest the perifollicular collagen (Parakkal, 1969) exactly as in the postpartum uterus and in the tadpole tail fin during metamorphosis. In each of these tissues, typical collagen bundles can be demonstrated inside the macrophages throughout the remodeling process. The intracellular collagen bundles and many of the digestion vacuoles are positive for acid phosphatase and are probably related to the terminal intracellular breakdown of collagen by lysosomal cathepsins. Whether or not collagen bundles have to be acted upon by tissue collagenase before being ingested by macrophages is not known.

B. Elastin

Elastic tissue, like reticulin, has until recently been recognized primarily in histological and histochemical preparations. Elastic fibers are stained semiselectively with orcein (dark brown), resorcin fuchsin (dark blue), and the Verhoeff Van Giesen technique (dark purplish-brown). An often neglected peculiarity of human skin is the relative abundance of elastic fibers. However, despite their abundance, their insolubility remains a major barrier to investigators. Here, even at a gross level, morphological studies give significant insight into this obscure but important tissue component. In thick, well-prepared histological sections, coarse elastic fibers are entwined with the collagenous fibers of the pars reticularis dermis. Their abundance varies from area to area; elastic fibers are always more abundant in the scalp and face than elsewhere, even in newborn infants. The really fine architectural distribution of elastic fibers, however, can be appreciated best in the papillaris dermis and around the cutaneous adnexa and blood vessels.

In the papillary dermis, elastic fibers that originate in the surface of the reticular layer form a palisade, the individual strands of which run perpendicular to the epidermis (Fig. 7). As each fiber rises toward the epidermis, it sprays out into penicillate formations often called inverted candelabra. Each of these fine fibers may spray again into barely resolvable fiberlets with a globose end that pushes its way against the base of the membranes of the basal layer and makes a dimple on it. Most likely, the function of these fine fibers is to anchor the epidermis to the dermis. Around the upper half of hair follicles, very fine elastic fibers arranged into longitudinal and circular patterns attach the arrectores pilorum muscles to the bulge of hair follicles, maintain the muscle in a bundle, and anchor its origin to the surface of the reticular layer. These fibers form a skeleton around the secretory segments of eccrine and apocrine glands and anchor blood vessels to their environ-ment by way of a loose reticulum. It must be emphasized, and will be again, that in most animals there is an inverse relationship between the abundance of terminal hairs in an area and the quantity of elastic fibers. In man all dermis is rich in elastic tissue, regardless of hair cover.

Like rubber, teased elastic fibers stretch easily with little loss of energy and resume their original shape when the stress is relieved. When fully extended, they have moderate tensile strength. Thus, the fibers behave like an elastomer: they have a low coefficient of elasticity, i.e., they can be deformed by a small force, and they recover their original dimensions even after considerable deformation.

Unlike the ultrastructure of collagen and reticulin, the protein elastin is nonfibrillar and homogenous. Elastic fibers consist of two components: an inner amorphous "medulla," the elastin, and an outer "cortex" con-sisting of nonelastin proteinous microfibrils 110 Å in diameter (Fig. 8). During early elastogenesis, the elastic fibers consist predominantly of the microfibrils. The fibrillar component has been found in grooves in the surface of fibroblasts where it appears to be a tubular mold into which the amorphous inner material is secreted (Ross and Bornstein, 1969). The sequence of appearance of the fibrillar and the amorphous com-ponents suggests that one is the precursor of the other. Ross and Bornstein (1969) separated chemically the fibrillar and amorphous components from the ligamentum nuchae of the ox, which contains about 70% elastin by weight. The two proteins differed markedly in amino acid composition, and neither resembled collagen. The amorphous component had essentially the same amino acid composition as a soluble

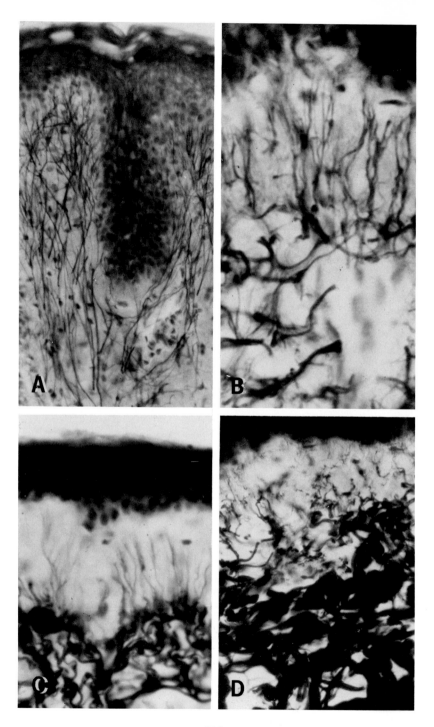

elastin precursor, tropoelastin, isolated from copper-deficient pigs. So, the elastin of the elastic fiber is the central amorphous material and is chemically unrelated to the outer microfibrillar component.

As mentioned, elastin is insoluble, a property often used as a criterion of its purity during isolation. Unlike collagen, elastin has extraordinary hydrothermal stability. Whereas insoluble collagen is completely solubilized to gelatin in boiling water, elastin is not. To be solubilized, elastin must be autoclaved with acid and alkalis, a harsh procedure which results in its degradation. A low content of polar amino acids may explain the insolubility of elastin.

Acid hydrolysates of purified elastin contain the unique amino acids, desmosine and isodesmosine, which probably form a special kind of cross-link within the molecule. The empirical formula of these compounds suggests that each forms from the condensation of 4 lysyl residues. The first step in elastin cross-linking is apparently the same as in collagen, that is, the oxidative deamination of lysyl residues by a copper-dependent lysyl oxidase. In copper-deficient animals, the elastin formed is rich in lysine but lacking in desmosine or isodesmosine. β-Aminopropionitrile (βAPN), which blocks the enzyme, also gives rise to an increased ratio of lysine to desmosine and isodesmosine. If an organ culture of chick aorta is given a pulse of radioactive lysine and one day after that lysine is replaced with cold lysine, the lysine radioactivity in the aortic elastin drops precipitously during the next 10 days, whereas that of the desmosines rises sharply (Miller et al., 1965). This indicates that the desmosine and isodesmosine of elastin are derived from lysine. About one-third of the lysine residues of tropoelastin are converted to desmosine cross-links, whereas much of what remains participates in other cross-links. If one desmosine cross-link united two, three, or four peptide chains, it could confer upon elastin a structure similar to that of rubber, where long-chain molecules are united into a loose, three-dimensional network by widely spaced covalent cross-links. Unlike rubber, dehydrated elastin loses its elastic properties when dehydrated and becomes hard, brittle, and inextensible. Various hypothetical molecular models have been constructed to summarize our knowledge of elastin structure (Partridge, 1970; Sandberg et al., 1971).

Soon after its synthesis, newly formed tropoelastin, with a molecular weight of about 70,000, begins the process of cross-linking and rapidly

Fig. 7.   Elastic fibers form (A) areola of a young breast; (B) auricula; (C) areola of a senile breast showing elastotic changes; (D) extensive elastotic changes in the areola of a senile breast.

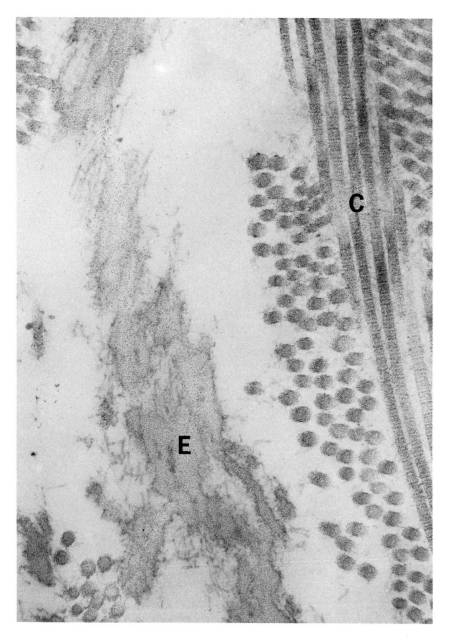

Fig. 8. Electron micrograph showing both elastic fibers (E) and collagen fibrils (C). The elastic fibers consist of both amorphous and fibrillar components. × 52,250. (Courtesy of Dr. M. Bell.)

becomes insoluble. This insolubility makes elastin a highly stable product; chronic labeling has shown that adult animals retain their elastin for life. Aging elastin forms cross-links, and its content of polar amino acids appears to increase. However, this is probably the result of increased association with and contamination by other proteins. Aging elastin also gradually becomes calcified. Despite its insolubility, elastin can be broken down by some proteolytic enzymes such as the digestive enzyme, pancreatic elastase. However, no tissue elastase has ever been demonstrated. Like collagenase, elastase is first adsorbed onto the elastin and then begins to split peptide bonds. Unlike collagen, elastin can also be degraded by a number of other proteolytic enzymes.

C. Reticulin

Reticulin can be identified in the dermis as fine-branching fibers blackened by ammoniacal silver nitrate, whereas collagen fibers are non-branching, coarse bundles and are not argyrophilic. Because reticulin is present in minute quantities and difficult to separate from collagen, some have theorized that they are the same substance. However, an analysis of the midcortical layer of human kidneys, which in histological preparations contains mostly reticulin fibers, has shown a smaller content of proline, larger amounts of hydroxyproline and hydroxylysine, and more neutral sugar than collagen (Windrum et al., 1955). Reticulin also contains large quantities (10.9%) of firmly bound fatty acid and can thus be considered a lipoglycoprotein. Electron microscopy shows reticulin composed of sparsely distributed fibrils, with the typical 640 Å axial periodicity of collagen, and surrounded by an amorphous matrix. Reticulin may provide a template for the extracellular aggregation of collagen fibrils.

D. Ground Substance

The structureless portion of connective tissue, the ground substance, lies outside the cells, fibers, blood vessels, and nerves; all other dermal components are embedded in this amorphous matrix. The ground substance is a sol or gel which does not leak out of the dermis even under pressure. Normally impermeable to microorganisms or injected dyes, its permeability can be altered by injections of testicular or bacterial hyaluronidase. The ground substance appears to be a multicomponent system of: (1) substances derived from the blood such as water, inorganic ions, blood sugars, blood proteins, and urea; (2) metabolic products of parenchymal cells; and (3) metabolic products of connective tissue cells: the soluble precursors of the fibrous proteins, and the proteoglycans, glycoproteins, and complexes formed from these (Dorfman, 1953). There is no discernible free fluid in the ground substance, but the glycosaminoglycan components of the proteoglycans are hydrophilic and bind water. The metabolites of all cells, mesenchymal or epithelial, which are separated from their blood supply by connective tissue, must pass through this bound water. The ground substance, then, is the internal milieu, the immediate environment of dermal and epidermal cells, and an extension of the vascular system.

The basic molecular building block of ground substance proteoglycan is a polymeric sugar, the glycosaminoglycan, consisting of approximately 100 up to 5000 monosaccharide residues in glycosidic linkage. Several different species of GAG, hyaluronate (HA), chondroitin sulfate (ChS), dermatan sulfate (DS), keratin sulfate (KS), and heparin sulfate (hep SO_4) are found in connective tissue. All of these compounds are (1) straight chain molecules; (2) composed of repeating disaccharide units with one member, hexosamine, with the amino group generally N-acetylated, and the other either a hexuronate (glucuronate or iduronate) or galactose; and (3) carry fixed negative charges along the entire length of the chains in the form of carboxyl, or in some ester sulfate groups usually attached to the number 4 or number 6 carbon of the hexosamine moiety. These negative charges are associated with sodium cations *in vivo* and the glycosaminoglycans behave like polyanionic electrolytes. For this reason, they are easily stainable with such cationic dyes as alcian blue, Hale's colloidal iron, and various dyes that give a metachromatic color. They precipitate from aqueous solution with cationic detergents such as cetylpyridinium or cetyltrimethylammonium.

The three principal glycosaminoglycans in skin are hyaluronate, dermatan sulfate (chondroitin sulfate B), and chondroitin sulfate A and C. The repeating disaccharide unit of hyaluronate consists of glucuronate and N-acetyl glucosamine, whereas that of dermatan sulfate contains the epimers of these sugars, iduronate and N-acetyl galactosamine, the latter possessing a sulfate in the C-4 position (Fig. 9). Chondroitin sulfate A and C contains repeating disaccharide units of glucuronate and N-acetyl galactosamine, the latter being sulfated in the 4 position (CS-A) or the G position (CS-C). Hyaluronate is a huge molecule with a molecular weight of several million, far larger than the 20,000 MW of dermatan sulfate and the chondroitin sulfates. All are linear polymers and in solution are randomly coiled.

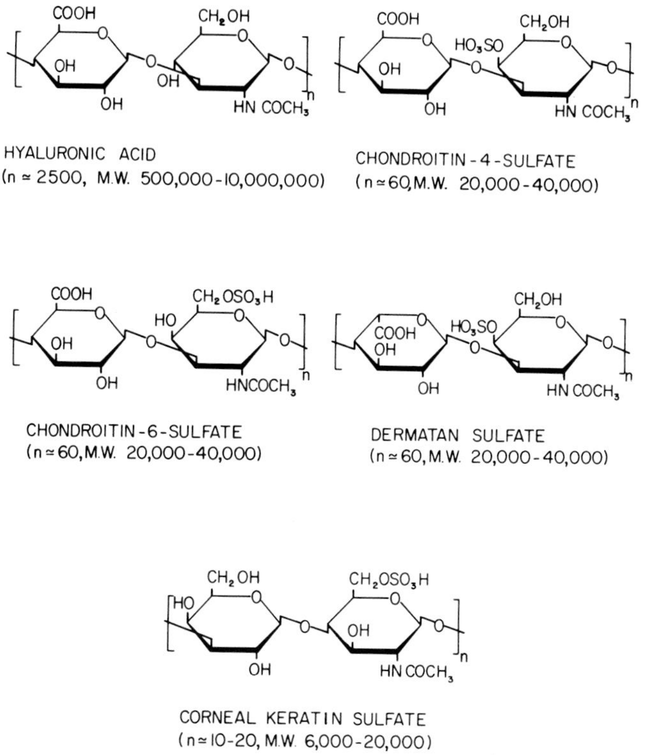

HYALURONIC ACID
(n ≈ 2500, M.W. 500,000-10,000,000)

CHONDROITIN-4-SULFATE
(n ≈ 60, M.W. 20,000-40,000)

CHONDROITIN-6-SULFATE
(n ≈ 60, M.W. 20,000-40,000)

DERMATAN SULFATE
(n ≈ 60, M.W. 20,000-40,000)

CORNEAL KERATIN SULFATE
(n ≈ 10-20, M.W. 6,000-20,000)

Fig. 9. Repeating units of hyaluronic acid, chondroitin-4-sulfate, chondroitin-6-sulfate, dermatan sulfate, and corneal keratin sulfate.

Like those in other connective tissues, dermal GAG's do not occur as free polysaccharides but are bound to protein as protein-polysaccharides or proteoglycans. The proteoglycans consist of a protein core to which the polysaccharide side chains are covalently attached, sticking out perpendicularly like the bristles on a bottle brush (Mathews, 1967). Roden and co-workers (1964) established the sequence of residues in the protein-polysaccharide linkage region as galactosyl–galactosyl–xylosyl–serine in chondroitin sulfate A and C and heparin, and more recently also in dermatan sulfate. Linkage in these cases is mediated through the hydroxyl group of serine and threonine. The protein core differs from collagen in its amino acid composition. Unlike the other GAG's, hyaluronate is linked in nature to little or no protein. It is the major, if not the exclusive, GAG in semifluid connective tissue such as vitreous humor, synovial fluid, or the swollen sex skin of some primates.

Except for glycogen, most of the substance in dermal connective tissue that stains with PAS is glycoprotein. The exact nature of this material and of the other components of the ground substance is unknown. Glycoproteins consist of a mixture of heterogenous molecules that contain protein and about 10 to 40% carbohydrate (sialic acid, mannose, galactose, fucose, and hexosamines). Connective tissue glycoproteins, like serum glycoproteins, are probably branched structures with radiating oligosaccharide chains attached to a protein core. Some glycoprotein of connective tissue probably functions as a linkage molecule that binds a number of proteoglycan subunits into proteoglycan complexes (Hascall and Sajdera, 1969).

The metabolism of the glycosaminoglycans has been studied in various ways. Under the electron microscope, histochemical localization of intracellular acidic glycosaminoglycans with colloidal thorium shows these substances only in the large vacuoles of the Golgi system (Revel and Hay, 1963). Electron microscopic autoradiographs of chondrocytes show that radioactive precursors of the glycosaminoglycans, such as acetate, sulfate, glucose and galactose, are incorporated directly (within 5 minutes *in vitro*) into the Golgi apparatus without passing through the rough-surfaced endoplasmic reticulum. From the Golgi area, the radioactive precursors pass directly to the extracellular ground substance. After radio acetate incorporation, the radioactivity is not extractable with chloroform-methanol and is not rendered dialyzable after digestion with papain. These findings suggest that the Golgi apparatus of chondrocytes is the site of either the accumulation or the synthesis of the polysaccharide moiety of the ground substance proteoglycans (Revel, 1970).

In contrast to the observations on electron autoradiographs, studies of fractionated chick chondrocytes (Horwitz and Dorfman, 1968) indicate that the synthesis of the protein core and the initiation of the polysaccharide chain occur in the rough endoplasmic reticulum, where the acceptor proteins and the enzymes necessary for chain initiation, chain elongation, and sulfation of the proteoglycans reside. Enzymes involved in both chain extension and sulfation occur in the smooth microsomes. The initiation and elongation of the polysaccharide chain, then, begin in the rough endoplasmic reticulum, and chain elongation may continue in the smooth endoplasmic reticulum and Golgi apparatus where sulfation takes place.

During the biosynthesis of the glycosaminoglycans, glucose is converted into the polysaccharide chains via uridine diphosphate nucleotides of N-acetyl glucosamine, galactose, glucuronate, and iduronate. For its carbohydrate chain initiation, dermatan sulfate apparently requires protein synthesis since chain initiation can be abolished in cultured fibroblasts by the addition of puromycin or cycloheximide, both specific inhibitors of protein synthesis. These also inhibit, to a lesser extent, the formation of hyaluronate by fibroblasts. The data suggest that synthesis of a proteinous core onto which the polysaccharide chain can be built is an essential prerequisite for the initiation of the polysaccharide chain. After the attachment of xylose to the protein core, the next monosaccharide residue, galactose, and subsequent residues are added sequentially to complete the "glycan" portion of the proteoglycan. The system of enzymes capable of adding specific sugar residues specifies the alternation of residues, i.e., in chondroitin sulfate, glucuronate is never added to the chain if the terminal moiety of the forming chain is glucuronate. Sulfation by specific sulfotransferases occurs after the formation of the polysaccharide chain.

Under appropriate tissue culture conditions, cloned connective tissue cell lines may produce one or more polysaccharides, collagen, and other extracellular stromal components. Their genetic constitution remains stable in vitro and their progeny inherit their state of differentiation.

The degradation of the dermal ground substance proteoglycans has not been as extensively studied as their synthesis. Preparations made from snake venom, mammalian testes, resorbing tadpole tail fin, and various bacteria depolymerize hyaluronate. The specificity and mode of action of these "hyaluronidases" vary with their source. Hyaluronate degradation in vivo presumably results from the secretion or the activation of a lysosomal hyaluronidase, an enzyme that has not yet been isolated and identified. The usefulness to both bacteria and venomous reptiles of a "spreading factor" that can lower the viscosity of the dermal ground substance is obvious. Hyaluronidase can degrade chondroitin sulfate A and C but not chondroitin sulfate B (DS).

The breakdown of proteodermatan sulfate (proteochondroitin sulfate B) can be inferred from studies on cartilage degradation, since proteochondroitin sulfate is the predominant proteoglycan of cartilage. Only a limited proteolytic digestion of cartilage by cathepsin D results in the degradation of the protein core of proteochondroitin sulfate and the release of ChS from the cartilage matrix. A specific antiserum to cathepsin D completely inhibits this breakdown. Cathepsin D easily hydrolyzes proteochondroitin sulfate and to a lesser extent, various blood proteins (i.e., hemoglobin, seralbumin, and fibrinogen) and histones.

But since cathepsin D cannot degrade the glycoproteins of intact inter-cellular matrices, glycoproteins accumulate in the granulation tissue associated with subacute inflammations, where cathepsin activity is extremely high. However, the significance of a catheptic enzyme with maximal activity at unphysiological pH in the degradation of tissue proteoglycans is uncertain. Even though many questions remain regard-ing the specific action of the proteoglycan catabolic enzymes, the time course of proteoglycan degradation is well worked out. Bentley *et al.* (1970) showed that a ^{14}C-glucose label incorporated into dermal hyaluronate, heparin sulfate, chondroitin sulfate, and dermatan sulfate fractions disappeared from these fractions over a period of about 20 days, a rate of turnover markedly faster than that of collagen from the same tissue.

III. Cellular Components

A. *Fibroblasts*

Fibroblasts, under normal conditions the most numerous cells in connective tissue, are responsible for the formation of collagenous and elastic fibers and the amorphous ground substance. Other cells, such as chondroblasts, osteoblasts, and odontoblasts, are essentially specialized fibroblasts.

The appearance of fibroblasts varies according to their state of activity (Fig. 10). When active, fibroblasts contain a large prominent nucleus with one or more equally prominent nucleoli. The extensive rough endoplasmic (ER) reticulum is responsible for cytoplasmic basophilia. The rough ER may take the form of narrow, long interconnected sacs but at times the cisternae appear dilated and filled with filamentous or flocculent material and reflects the synthetic activity of the cell. The Golgi complex is often multiple and consists of flattened lamellae and an assortment of vesicles. Autoradiographic studies suggest that the Golgi is the site of polysaccharide synthesis. Fibroblasts have numerous cytoplasmic filaments and an assortment of vesicles and vacuoles, many of them near the cell periphery, and are thought to be related to the process of pinocytosis. Many cytoplasmic filaments (50–80 Å) are ran-domly distributed throughout the cell, and closely packed filaments are often located at the cell periphery; these filaments may play a role in fibroblast motility.

When inactive, fibrocytes are long and slender and have a nucleus

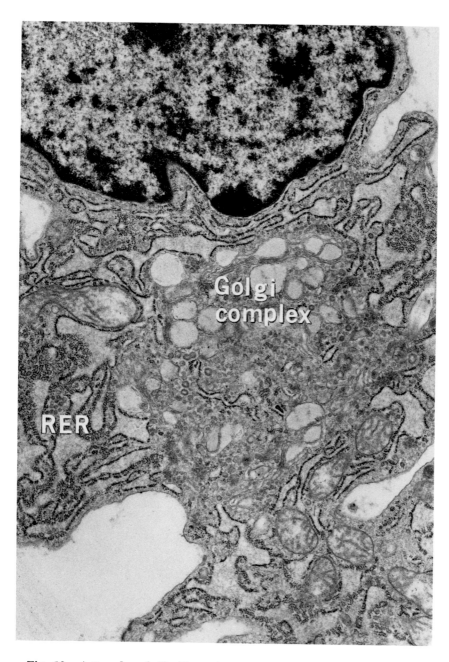

Fig. 10. Active dermal fibroblasts showing extensive arrays of rough-surfaced endoplasmic reticulum (RER) and Golgi complex. × 17,500. (Courtesy of Dr. M. Bell.)

that nearly fills the cell. The cytoplasm of the inactive cell is so scanty that the organelles described above are poorly developed.

Although they were once believed to have a hematogenous origin, several studies have shown that fibroblasts develop from tissue mesenchyme. Local irradiation shortly after wounding leads to a reduction of wound fibroblasts, an indication that they originate from mesenchymal cells in adjacent areas. When buffy coat prepared from blood collected free of extraneous connective tissue cells is cultured, the lack of fibroblasts and of connective tissue elements in the culture suggests that blood leucocytes are not the progenitor of fibroblasts (Ross and Lillywhite, 1965). Further evidence against a hematogenous origin was obtained in the following experiment. The femur of one member of parabiotic rats was shielded while both animals received 800 r of whole-body irradiation; incised skin wounds were made in each animal. The member with the shielded femur was then injected with ^{3}H-thymidine and the cross-circulation interrupted long enough to allow the label to clear the blood stream. Six days after labeling, radio-autography of the wound of the unshielded rat revealed that fibroblasts remained completely unlabeled despite the finding that various blood elements were labeled (Ross *et al.*, 1970).

The major function of fibroblasts is the production of fibrous and amorphous components in connective tissue. Few agree on the role of the Golgi complex in the pathway of collagen secretion. On the basis of radioautographic evidence, it was thought that precursors of collagen pass through the Golgi complex before being discharged extracellularly. It has also been proposed that collagen is secreted intermittently from vesicles which become confluent with the rough-surfaced endoplasmic reticulum and extracellular space (Ross, 1968). Once released into the extracellular spaces, fibrils are formed by aggregation.

In tissue culture, the cell organelles reflect the kinetics of collagen production by fibroblasts. Although the rate of collagen synthesis by an individual cell is fairly constant, in certain cell lines there is an inverse relationship between the growth rate in tissue culture and the rate of collagen biosynthesis by the culture. These cells do not begin to synthesize collagen until they have reached a state of confluent monolayer. Growing (log phase) cells are relatively uncommitted, but when growth stops (lag phase), differentiation and the consequent production of extracellular matrix materials ensue. However, the dissociation of cellular replication and specialized function, although frequently observed, is not obligate, and is not a fundamental biological principle (Priest, 1972).

In vitro collagen synthesis by many cell lines is as rapid as or more rapid in log phase growth than in lag phase, and hyaluronate synthesis in mouse embryo fibroblasts is maximal during the log phase of growth in tissue culture. With prolonged culture, many well-known, cloned fibroblast lines lose their ability to synthesize glycosaminoglycan even though they continue to make collagen.

Dermal fibroblasts grown *in vitro* produce both hyaluronate and various sulfated polysaccharides. The latter are retained in the cell-collagen layer, whereas the former are released into the medium. This may indicate a higher degree of interaction between sulfated polysaccharides and collagen than between hyaluronate and collagen. In evaluating the results of studies on cultured "fibroblasts," one must remember that, when cultured, connective tissue cells from various sources change into slender spindle-shaped cells that resemble fibroblasts.

B. Macrophages

Present throughout the animal kingdom are amoeboid cells, or macrophages, which remove effete autologous cells and "foreign bodies." Macrophages appear early in embryonic development and play a crucial role in tissue resorption during morphogenesis. Early in ontogeny they develop biosynthetic pathways that are later important to the mature animal. The phagocytic activity of these cells increases progressively throughout fetal and neonatal life.

Macrophages are ubiquitous in mammalian tissues. Dermal macrophages, which are typical of those in most tissue, are characterized by an eccentrically located indented nucleus (Fig. 11), which frequently has a concentration of chromatin at the periphery and one or two nucleoli. During the inactive phase, the most prominent cell organelle is the Golgi complex, with a few profiles of rough endoplasmic reticulum. During active phagocytosis, both the rough endoplasmic reticulum and the Golgi complex are large. Scattered throughout the cytoplasm are numerous lysosomes, phagosomes, and phagolysomes. Many residual bodies consisting of whorls of myelin are seen in the cytoplasm.

There is some controversy about the origin and fate of tissue macrophages. Monocytes derived from the blood transform to macrophages both *in vitro* (Fig. 12) and *in vivo*. However, it is still uncertain whether macrophages are also propagated by mitosis of resident tissue macrophages, migrate from other tissue, or are formed by the transformation of lymphocytes under the influence of polymorphonuclear leukocytes

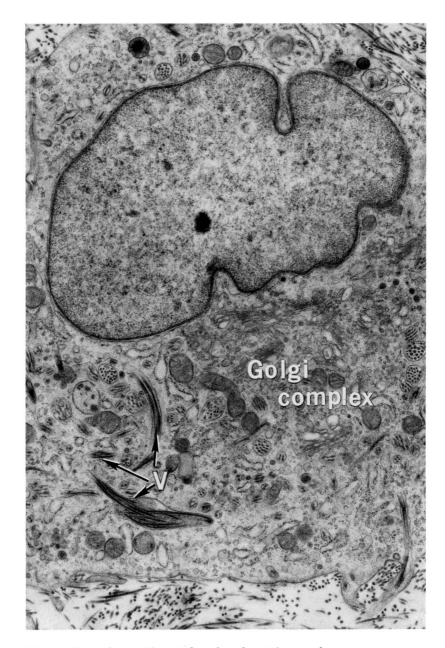

Fig. 11. Macrophage with an indented nucleus. The cytoplasm contains a prominent Golgi and many vacuoles (V) with phagocytosed collagen. × 14,000.

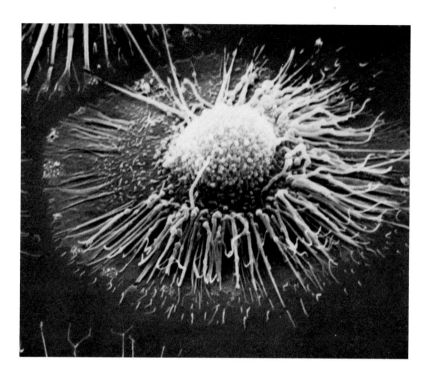

Fig. 12. Scanning micrograph of a cultured macrophage transformed from a human blood monocyte. The ruffled membranes and filopodia are arranged symmetrically around the cell. × 1000.

(Pearsall and Weiser, 1970). Epithelioid or giant cells are probably terminal forms of active macrophages.

The metabolism of macrophages reflects both their hematogenous origin and their phagocytic function. The energy required for phagocytosis is supplied primarily through glycolysis, as in peritoneal macrophages, or through oxidative phosphorylation and the hexose monophosphate shunt, as in alveolar macrophages. Besides energy-producing enzymes, macrophages must also produce the wide array of lysosomal hydrolytic enzymes needed to digest phagocytized materials. De novo synthesis of these enzymes can be induced experimentally in response to ingested substrates. The total lysozyme content of macrophages increases in response to infection.

Macrophages are active in both the synthesis and the degradation of lipids. Various lipidoses and problems with the degradation and synthetic utilization of cholesterol have been ascribed to aberrations in lipid metabolism by macrophages. Under normal conditions, phagocytosis by macrophages requires the formation of new phagosome phospholipid membranes. In turn, these cells take up and degrade cholesterol, fat granules, and the phospholipid membranes of other cells. Macrophages also produce interferon for protection against viral infection (Acton and Myrvilsk, 1966). The production of well-characterized interferon by rabbit peritoneal macrophages *in vitro* is blocked by the administration of actinomycin within 30 to 60 minutes after virus infection (Smith and Wagner, 1967). Finally, peritoneal macrophages produce certain serum globulins, including transferrin.

Macrophages have several important functions related primarily to their phagocytic capability. As scavengers, they dispose of foreign materials and effete autologous cells and degrade ingested materials. The products of such degradation can often be used by the body to benefit general body economy. Toxic substances ingested by macrophages are generally degraded to nontoxic forms for ultimate removal. Thus, macrophages are vital in wound healing, infection, tissue resorption, and the recycling of some tissue components. Substances released from dying macrophages stimulate fibroblast activity which results in fibrosis. Chronic infection or irritation from foreign material, as in tuberculosis, sarcoidosis, silicosis, and histoplasmosis, results in the accumulation of macrophages and in an often debilitating fibrosis.

A specific type of dermal macrophage, the melanophage, functions as a kind of "nurse," apparently supplying melanin to melanocytes. The specific role of the dermal macrophage in collagen resorption during hair cycles has been described (Fig. 11).

Macrophages recognize foreign materials and participate in the immune responses (Pearsall and Weiser, 1970). They are thought to influence the induction of delayed sensitivity reactions by trapping antigens and transporting them to the lymph nodes for processing. Nearly pure preparations of macrophages have been used successfully in the passive transfer of delayed hypersensitivity. Perhaps the transfer of information to inducible cells is mediated through contact with the antigen-carrying surfaces of the macrophage. In that case, cytophilic antibodies are produced which, by adsorbing to the macrophage cell membrane and binding to specific antigens, facilitate the transportation and phagocytosis of the antigens. Such antibodies may produce cellular immunity, and the lymphocytes themselves may act as effector cells in the immune response. Moreover, lymphocytes may interact with macrophages by producing substances chemotactic for macrophages or by immobilizing or perhaps directly activating macrophages at the antigenic site. Activated macrophages are essential in antimicrobial cellular immunity, either destroying or inhibiting the parasites intracellularly. During the rejection of allografts and the response to autochthonous tumors, they may also perform both effector and scavenger functions.

C. Mast Cells

Despite numerous scientific reviews and monographs, much uncertainty about the nature and biological properties of mast cells persists.

Mast cells were discovered and so named by Paul Ehrlich because their cytoplasm is filled with large metachromatic granules. They are ubiquitous in connective tissue and often arranged around the walls of small blood vessels. They are believed to have a hematogenous origin or to arise from perivascular connective tissue cells; they can multiply mitotically.

For the most part, mast cells are large and ovoid with a small nucleus and a cytoplasm filled with coarse membrane-bound granules which stain metachromatically with toluidine blue. The shape and structure of these granules differ according to the species of the animal. In man, they are electron dense and contain variously arranged lamellae which resemble "fingerprints" or "scrolls" according to the plane of sectioning (Fig. 13). Along the periphery the cells often have long microvilli.

Mast cells have been called unicellular glands and are known to degranulate under both physical (cold, heat, and ultraviolet and x-ray radiation) and chemical (protamine sulfate, peptides, compound 48/80, trypsin, and various bacterial or animal toxins) conditions. When degranulation occurs, the heparin, histamine, and serotonin (5 hydroxytryptamine) contained in the granules are released. After degranulation, the granules are regenerated.

The pharmacological actions of substances contained in mast cell granules have been studied extensively. Heparin is best known for the prevention of blood clotting except in cats where it is ineffectual. Very small concentrations of heparin greatly accelerate lipid transportation or clearance. Because of this property, mast cells are believed to participate in the regulation of lipid metabolism. Heparin is also believed to inhibit hyaluronidase. Mast cells are numerous in skin that contains large quantities of hyaluronate, such as the sex skin of primates, keloid scars, and myxedema lesions. Certain pathological accumulations of hyaluronate result from mast-cell dysfunction.

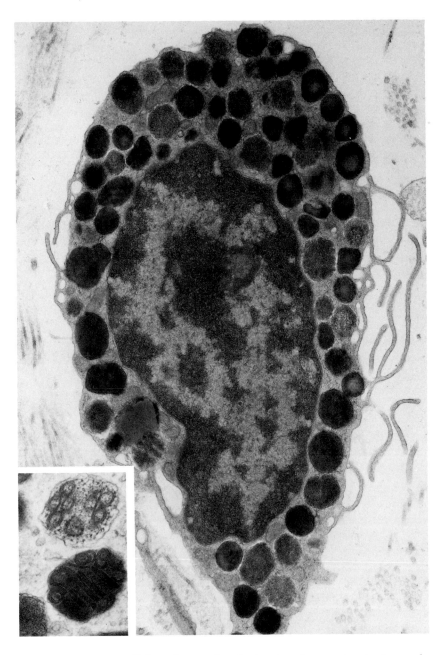

Fig. 13. A mast cell from human dermis showing the assortment of granules. × 12,800. The inset shows the fingerprints or scrolls of the granules. × 52,500. (Courtesy of Dr. M. Bell.)

Mast cell granules also contain histamine, which is released during degranulation. However, small amounts of histamine are sometimes liberated without degranulation. Histamine increases capillary permeability, causes smooth muscle fibers to contract, stimulates the phagocytic activity of blood and connective tissue cells, and promotes the secretion of various glands. Histamine may thus be responsible for the onset of inflammation, hyperemia, and the early increase in capillary permeability during the inflammatory process. It also affects the activity of the gastrointestinal tract, gastric hypersecretion resulting from the degranulation of cutaneous mast cells when skin comes into contact with cold water. The massive degranulation of mast cells in anaphylactic reactions leads to the release of sufficient quantities of histamine to produce broncho-constriction and death. Histamine also appears to elicit the sensation of itching.

In rats, which are quite resistant to the effects of histamine, serotonin takes over the pharmacological role of histamine. Degranulation of mast cells in rats is accompanied by a decrease in the tissue levels of serotonin, which, however, regain their former values when the granules are regenerated. The mast cell granules of mice, but apparently not those of man, may also contain serotonin.

Mast cells, then, perform many functions and participate in the initial response of connective tissue to various injuries.

IV. Age Changes in the Dermis

"Aging" may mean either growing old or maturation. Since the word usually connotes loss of function, so-called age changes often apply to degenerative alterations rather than to those that are an integral part of the normal development of tissues. In this discussion, age changes encompass all of these, from embryonic life through senescence.

Histologically and biochemically, the fibrous components of the dermal connective tissue undergo demonstrable changes. In fetal dermis, collagen fibers are fine and fibrillar, whereas in mature individuals they form coarse bundles. In man, the collagen fibers thicken progressively up to 20 years of age; the increments lessen after that. In fetal skin, elastic fibers are delicate filaments, found mostly around the pilosebaceous units; practically throughout the life of the individual they steadily increase in number and thickness.

Histologically, collagen is difficult to describe in biochemical terms because, barring extensive degradation, what occurs at any one moment depends on the relative rates of biosynthesis and cross-linking. Collagen synthesis, which can occur locally before total body growth, usually parallels growth (Gross, 1958; Wirtschafter and Bentley, 1962); it is rapid when growth is rapid, and stops when growth stops. Thus, the quantity of the most recently synthesized collagen, which is soluble, varies with the growth rate of the animal and decreases with aging only because of the decline in growth rate. Although total skin collagen increases until old age, it begins to decrease early in life on an absolute basis in relation to the mass of the other constituents (Zika and Klein, 1971). Thus, collagen is in a positive metabolic balance (i.e., it is synthesized continually), although its metabolic activity (both synthesis and degradation) sharply decreases with age. This is reflected in a reduced incorporation of radioactive precursors, in a longer half-life of the collagen, and in a decreased excretion of urinary hydroxyproline in the aged. An increased proportion of insoluble collagen also reflects the increased extent of collagen cross-linking with age. Furthermore, analysis of the denaturation components of skin collagen during aging shows a decrease in the ratio of uncross-linked to cross-linked chains and a progressive increase in the proportion of highly cross-linked chains.

In its early stages, elastin cross links by a mechanism similar to collagen, and the cross-links form relatively quickly. However, biochemical studies confirm the histological impression of increased numbers and diameters of elastic fibers during aging; the elastin content of human skin is said to increase fivefold from fetal to adult life (Sams and Smith, 1964).

Dermal ground substance undergoes little histological change with aging. Its apparent decrease could result from a reduction of extracellular, extrafibrillar space caused by an increase in collagen mass. However, measurements of the total acid glycosaminoglycan content (as uronic acid) of human skin from the third fetal month to the seventh decade show that the concentration in fetal skin at 3, 5.5, and 9 months of gestation is 20×, 5×, and 2× that of adult levels, respectively. For the most part, this drop is caused by the decrease in hyaluronate during fetal life (Breen *et al.*, 1970). Hyaluronate constitutes about 80% of the total cutaneous polysaccharides in the fetus, only about 30% in the adult. Dermatan sulfate, however, represents about 10% of the polysaccharide in embryonic skin and two-thirds of that in adult skin (Loewi and Meyer, 1958). Although skin from 5.5 months gestation to adult age contains hyaluronate, dermatan sulfate, and chondroitin

4 or 6 sulfate, skin from 3-month-old fetuses contains only hyaluronate and chondroitin 4- or 6-sulfate. Galactosamine is the specific hexosamine of dermatan sulfate and the chondroitin sulfates. Since the only other abundant glycosaminoglycan in the skin is hyaluronic acid, the hexosamine of which is glucosamine, the mole ratio of galactosamine to hexosamine has been used as an indication of the ratio of sulfated to total acidic glycosaminoglycans. In developing fetal skin there is a relative increase in the sulfated glycosaminoglycans from 3 months to term.

Qualitative and quantitative changes in skin glycosaminoglycans occur primarily during fetal life, infancy, or early childhood; very little change takes place after that. Hyaluronate falls from birth to infancy, and then levels out; its decrease throughout life is extremely low. Dermatan sulfate also drops precipitously after infancy but shows little or no change thereafter (Smith *et al.*, 1964). From birth to adult life, both the sulfated and the total acidic glycosaminoglycans in the skin decrease by about 50%, but the galactosamine:hexosamine ratio remains constant. From the fourth to the seventh decade, the total amount and distribution of the acidic glycosaminoglycans appear to be constant (Breen *et al.*, 1970), although the content of glucuronate sometimes rises after the sixth decade (Smith *et al.*, 1964). Analogous age changes have been found in chicken skin (Kondo *et al.*, 1971).

In old animals, the metabolic activity of the dermal polysaccharides is reduced. Radioactive glucose in both hyaluronate and dermatan sulfate pools turns over more rapidly in 4-week-old rats than in 12-month-old animals (Davidson and Small, 1963).

V. Biological Functions

The functions of this complex tissue are themselves complex, numerous, and diverse. Some of the natural functions of the dermis are obvious and can be measured. Most of them, however, can only be surmised from "test tube" behavior of the dermal constituents after chemical or physical purification. Such functions are, therefore, hypothetical but probably approximate those *in vivo*.

A. Protection

One of the most important functions of the dermis is to protect the body from external injury. Fresh skin, although remarkably strong, is supple and resilient. The dermis is uniquely constructed to resist tearing, shearing, and localized pressure but is also flexible enough to allow joint movement and local stretch.

Collagen provides resistance to mechanical stress. It is arranged in a loose fibrous network, in the meshes of which are the amorphous ground substance, which provides resistance to compression and bulk flow, elastic fibers, which are thought to restore the morphology of the collagen network after deformation, and reticulin, the function of which is unknown. For the dermis to possess the structural integrity essential for protection, there must be a high degree of physical interaction between its various components; however, for each component to fully express its individual contribution to the tissue a degree of latitude must also be built into these interactions.

We have seen that the physical properties of skin vary from one part of the body to another. Under chronic mild tension in most areas, it remains slack in others. In certain places it is bound tightly to the underlying subcutaneous tissues; in others it slips freely over them. Regional differences in mechanical characteristics adapt skin to local demands and may reflect, at least in part, the local architecture of the dermal collagen and elastin networks. The orientation of fibers in the dermis also varies from one area to the next. As we have mentioned in Chapter 1, the German physician Langer (1861) first showed that if the skin of a cadaver is punctured with a round awl, the holes are elliptical, the ellipses being oriented perpendicular to the lines of minimal extensibility of skin (Langer's lines). Along Langer's lines, collagen fibers do not have the typical meshwork arrangement but are oriented parallel to the lines. In human skin Langer's lines have been incompletely mapped, but they probably bear no relation to wrinkles.

When exposed to tension, skin "gives" until the slack is taken up; this is reflected in the stress-strain curve for skin (Fig. 14). This slack results from the loose random arrangement of the collagen network at rest. When a load is applied, the fibers within the network undergo parallel alignment in the direction of the pull. Once this slack is taken up, further extension of the skin requires much higher tension. Great force is required to tear skin. Skin that is kept taut for long periods gradually undergoes "slippage," as if fatigued, and stretching results. Contrary to the early reversible elastic phase of the stress-strain curve, slippage is irreversible; the *striae gravidarum* of pregnancy are a familiar example.

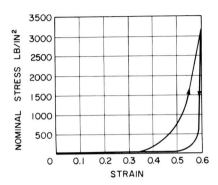

Fig. 14. Stress-strain curve (from Gibson and Kenedi, 1970).

The removal of fat and epidermis from excised skin has little effect on the elastic modulus or strength, an indication that strength rests in the dermis, particularly in the collagen (Tregear, 1966). If the elastic modulus of a piece of skin is divided by its collagen content, the values obtained approximate those of hydrated tendon, which is nearly pure collagen. When the skin is stretched and the slack taken up, the collagen fibers, normally running in various directions, become oriented to the force. Once the fibers have become so aligned skin exhibits the tensile properties of collagen alone. The tensile strength of skin, therefore, lies predominantly in its collagen. However, tensile strength is more than collagen *content*; in wound healing, for example, the tensile strength of an incised wound continues to increase long after the initial rise in collagen content has leveled off (Madden and Peacock, 1971). This indicates that the remodeling of the scar can contribute significantly to its tensile strength. Both the amount and the architecture of collagen, then, determine the tensile strength of the dermis. The orientation of the collagen fibers in the dermis and the manner in which deformation occurs under tension are not completely known; a number of models have been proposed (Tregear, 1966; Szirmai, 1970).

How the dermal collagen network regains its normal organization and returns to the resting state after mechanical distortion is also unknown; this may be the function of the branched, relatively thin elastic fibers.

When compressed, skin becomes thinner under the force and wells up around the force which serves to distribute the pressure. This phenomenon is familiar to anyone who has sat or fallen asleep on or leaned against a rough surface; when he arises, the pattern of that surface is imprinted on his skin. These impressions can be massaged away and in time disappear of their own accord.

Skin is constantly exposed to but unaffected by the high pressure of one atmosphere, 14 lb/in². If a normal man balanced his weight on a block one-inch square, he would generate a local pressure of 10 to 14 atmospheres. This would create an unpleasant sensation and when he stepped off the block there would be a blanching of the compressed area (ischemia), later replaced by a transient flush (reactive hyperemia); permanent tissue damage probably would not result. If, however, a paraplegic remained in bed for a week or two, a seemingly mild pressure by comparison, and were not turned frequently to redistribute the pressure, he might develop areas of decubitus ulcers, or bedsores from pressure necrosis. These examples illustrate the two most important considerations in pressure damage: (1) the *duration* of exposure to the force, and (2) the *distribution* of the force. More damage results from increasing the duration of exposure to pressure than from increasing the magnitude of the pressure (Husain, 1953; Exton-Smith and Sherwin, 1961). In a healthy individual, the duration of pressure is generally insufficient to produce damage. Pressure receptors in the skin and subcutaneous tissues produce awareness of external pressure, and duration is limited by discomfort. However, severe point pressure of even short duration can damage both cutaneous and underlying tissues. The distribution of pressure is a function of the dermal proteoglycans. These substances are restrained by neighboring molecules *in situ* and are hydrophilic. Thus they resist bulk flow or dehydration by compression and thereby resist the compression itself. Through this mechanism, a localized potentially damaging pressure applied to the surface of the skin is transformed into a well-distributed, nondamaging pressure. This protects the underlying fat and muscle, neither of which contains much proteoglycan, from focal pressure damage.

B. Epitheliomesenchymal Interaction

Another important function of the dermis is its role in inductive interaction with the epidermis, which, despite much ingenious research, remains poorly understood. The dermis provides the necessary substrate for the epidermis; Wessells (1964) has shown in tissue culture that the basal layer of epidermis spread on a Millipore filter became oriented and incorporated tritiated thymidine only where it was in contact with the filter and not over the pores. Van Scott and Flaxman (1968) have shown *in vitro* that outgrowths from explants of adult skin on a glass or plastic surface maintain "proliferation maturation gradients" even after the removal of the original explant with dermis. Both orientation and proliferation, then, require a firm substrate. However, no culture medium alone can maintain the epidermis without dermis for prolonged intervals. This shows that the dermis provides more than physical support to the epidermis. Chick and mouse epidermis separated enzymatically from the dermis (Moscona, 1956; Szabo, 1962) and cultured alone keratinizes and dies within a few days without undergoing mitosis. Kidney and salivary glands (Grobstein, 1953), thymus (Auerbach, 1960), limb bud (Saunders, 1958; Zwilling, 1961), tooth germ (Koch, 1967), and lens (Muthukkaruppan, 1966) also depend on their mesenchyme.

Although epithelia require mesenchyme, the specificity of this requirement varies. In studies of tissue recombination, some epithelia, e.g., salivary glands, underwent normal morphogenesis only in the presence of their own mesenchyme (Grobstein, 1953). Pancreatic epithelium, on the other hand, has no specificity and combines with any other embryonic mesenchyme. Skin and most other tissues have intermediate specificity; the epidermis grows and differentiates on various mesenchymes but responds in different ways to each (McLoughlin, 1968). For example, when grown on bird preventriculus, it fails to keratinize, secrete mucus, or become ciliated. On heart myoblasts, it spreads rapidly as an attenuated single layer of squamous cells that do not keratinize; but grown on limb mesenchyme, it keratinizes normally. Sengel (1964) has shown the regional specificity of the inductive capacity of the dermis by combining dorsal dermis with tarsometatarsal epidermis; the epidermis differentiated according to the site of origin of the dermis.

Inductive interaction is not unidirectional; the dermis and epidermis probably alternate as inductive and reactive tissues during cutaneous morphogenesis. Two observations illustrate epithelial participation. First, the apex of the developing chick limb bud is capped by a thickened ridge of ectoderm, called the apical ectodermal ridge (AER). If this ridge is excised during limb elongation, defects occur in the terminal limb. Conversely, an additional AER grafted onto the limb apex gives rise to a double limb (Saunders, 1948). Second, during morphogenesis of the chick corneal stroma, the epithelium produces a highly structured collagenous matrix beneath its basal layer. This matrix then serves as a template for the subsequent orientation and deposition of the secondary corneal stroma by mesenchymal cells (Trelstad and Coulombre, 1971).

In epitheliomesenchymal interaction, the dermis must be considered an inductive tissue whose specific physical or diffusible components are the active agents that influence epidermal behavior.

Collagen, the dominant dermal protein, may be involved in epithelial morphogenesis. The epithelium from the rudiments of the salivary gland of mice, separated from its mesenchyme by a Millipore filter, can be grown *in vitro*. Such cultured epithelium soon develops an increased organization of cell organelles which coincide with the appearance of collagen fibrils along the basal lamina (Kallman and Grobstein, 1965). Mesenchyme, prelabeled with radioactive proline or glycine before culturing, shows an accumulation of label at the basal surface of the epithelium. If the epithelium is prelabeled with the same compounds, however, there is no accumulation. When cultured with salivary gland mesenchyme, the epithelium branches dichotomously to form secretory ducts and acini. If the epithelium is treated with collagenase, which removes the labeled collagen, morphogenesis is destroyed but growth remains unaffected. These observations may have a bearing on such epithelial invasions as are seen in old burn scars, chronically sun-damaged skin, carcinogen-painted skin, or wound healing, all of which show degenerating collagen.

Collagen, however, can also exert an altogether different influence on epithelial growth. In necrotic wounds, collagenolysis often precedes infiltration of the wound by polymorphonuclear leukocytes. In a Boyden Chamber *in vivo*, 1000 to 30,000 MW peptides, derived from the digestion of soluble collagen with a highly purified cutaneous collagenase, were highly "leucotactic" (Houck and Chang, 1971).

There is abundant evidence for diffusible inductors; a number of endogenous compounds exert specific growth-promoting and morphogenetic influences on various embryonic and adult target tissues. For example, epidermal growth factor (EGF) (Cohen, 1959), a protein extracted from the salivary glands of mice, produces a precocious opening of the eyelids and the premature eruption of teeth in young mice and rats. This appears to result from the stimulation of epidermal development and keratinization. Another cell-free inductor, which duplicates the activity of mesenchyme in stimulating the growth and differentiation of pancreatic epithelium, promotes the normal orientation and proliferation of epidermal basal cells on a suitable substrate (Wessels, 1964). The growth-promoting effects of embryo extract on cell cultures are now well established. Although the dermal glycosaminoglycans and proteoglycans are not diffusible and not capable of physical interactions, they may play an inductive role in the skin; however, reports of their influence on cell growth have been equivocal.

At an early stage, epithelial cells are transformed from nondifferentiated to differentiated, and what has been called induction may be a nonspecific unmasking of covert differentiation. For example, although the mechanism is unknown, Na^+, K^+, Mg^{2+}, or HCO_3^- ions may elicit the differentiation of presumptive epidermis into nerve, pigment, or ciliated cells (Hauschka and Konigsberg, 1966). The function of these substances may be to release or cue preprogramed integrated sets of synthetic processes and allow the cells to express potentials already determined or programed in them (Holtzer, 1968). There may exist chains of inductors with serially connected activities so that within a given system subsequent inductions depend on previous ones. Such inductive interaction in skin must be as necessary for the maintenance of tissue equilibrium as it is for its initial establishment.

C. Barrier to Infection

Undoubtedly dermal connective tissue participates in other activities that are less defined. The ground substance, or more specifically its hyaluronic acid, acts as a barrier to bacterial penetration. A number of bacteria produce hyaluronidase, which digests hyaluronic acid and chondroitin sulfate A and C but not chondroitin sulfate B (dermatan sulfate). Hyaluronidase was formerly called "spreading factor" (Duran-Reynals, 1942) because, when injected into a connective tissue, it facilitated the penetration of dye. When it was discovered that dermal connective tissue contains hyaluronic acid, a specific substrate for the enzyme, it was concluded that bacteria spread by digesting their way through the connective tissue. Early studies related the increasing virulence of pneumococcus type I upon serial passage through mice to an increased ability of these bacteria to produce hyaluronidase (McClean, 1936). However, a direct relationship between virulence and the production of hyaluronidase by bacteria has not been substantiated and the role of dermal ground substance as a barrier is still conjectural.

D. Water Binding

Hyaluronic acid also binds water and maintains dermal turgor in tissues. Unlike other areas of skin, the sex skin of female baboons and of some macaques in estrus (Fig. 1) doubles its concentrations of hyaluronate in response to high blood titers of estrogen though the sulfated glycosaminoglycans undergo little change. During anestrus the content of hyaluronate does not differ from that of other skin areas. A mere doubling of hyaluronate, however, does not seem sufficient to account for the dramatic swelling of sex skin until the relation between hyaluronate concentration and the osmotic pressure of solutions of this glycosaminoglycan is examined. This relationship (Ogston, 1966) indicates that merely by doubling the hyaluronate concentration, a fivefold increase in water-binding capacity results. The increase in water binding, which characterizes certain pathological states, may also relate to increased concentrations of hyaluronic acid. In localized myxedema, the involved skin contains 6 to 16 times as much glycosaminoglycan, mostly hyaluronic acid, as normal skin (Sisson, 1968).

Because hyaluronic acid binds water, the dermis may function as a water-storage organ. More water enters skin than other organs and conversely, during water deprivation, skin loses it more quickly than other tissues. Since skin represents a significant proportion of the total body weight, its function in water storage may help to maintain proper hydration of other tissues under mild water deprivation; under severe water deprivation, the dermal supply would be rapidly depleted.

E. Ion Exchange

To a minor degree the dermis also functions as a cation "sink." Although the cation concentration of such connective tissue as cartilage can be many times that of the blood, dermal cation concentrations are generally only slightly above those of the blood. Such a reservoir may provide a homeostatic function in maintaining proper levels of blood cation and in supplying mineral metabolism in general. The ability of a given connective tissue to bind cations depends on its charge density, which, at least in part, depends on the predominant glycosaminoglycan species. Cartilage, which has a high content of chondroitin sulfate, with two anionic groups per repeating disaccharide unit, "binds" more cations than the dermis, the predominant glycosaminoglycan of which, hyaluronic acid, has only one anionic function per repeating unit.

F. Fibrillogenesis

The proteoglycan component of the dermis may also participate in the control of collagen fibrogenesis. *In vitro*, collagen fibers are formed in two distinct phases, initiation or nucleation and growth. For a given solution of collagen, the greater the numbers of nuclei that form, the less collagen per nucleus and the thinner the fibrils; the more nuclei formed, the greater the rate of fibril formation. Chondroitin sulfates and keratin sulfate accelerate the rate of collagen fibril formation *in vitro* by producing fine regular fibrils, whereas heparin retards collagen deposition and produces thick irregular fibrils. Dermatan sulfate and hyaluronate are without effect (Wood, 1964). The glycosaminoglycans may alter collagen nucleation by virtue of both their charge density and their ability to alter the shape of the forming nuclei. In the cornea, where the predominant species of glycosaminoglycan is keratin sulfate, and in the cartilage, with a predominance of chondroitin sulfate, collagen fibers are characteristically thin. In the papillary layer of the dermis, which contains mostly hyaluronate and dermatan sulfate, the collagen fibers are considerably thicker. In the reticular layer of the dermis, with a relatively much higher concentration of heparin, the collagen fibers are thicker still. Finally, in the granulation tissue of wounds where the glycosaminoglycan population differs from that of the initial normal tissue (Bentley, 1968), the scar subsequently formed by collagen shows aberrations in both fibril thickness and orientation (Forrester, 1968).

References

Acton, J. D., and Q. N. Myrvils. 1966. Production of interferon by alveolar macrophages. *J. Bacteriol.* **91**: 2300–2304.

Auerbach, R. 1960. Morphogenetic interactions in the development of the mouse thymus gland. *Develop. Biol.* **2**: 271–284.

Bellamy, G., and P. Bornstein. 1971. Evidence for procollagen, a biosynthetic precursor of collagen. *Proc. Nat. Acad. Sci. U.S.* **68**: 1138–1142.

Bentley, J. P. 1968. Mucopolysaccharides and wound healing. *In* "Repair and Regeneration: The Scientific Basis of Surgical Practice" (J. E. Dunphy and W. Van Winkle, Jr., eds.), pp. 151–160. McGraw-Hill, New York.

Bentley, J. P., R. L. Wuthrich, and A. M. Van Bueren. 1970. Lathyrism and mucopolysaccharide metabolism in aorta, skin and cartilage. *Atherosclerosis* **12**: 159–172.

Bornstein, P., H. P. Ehrlich, and A. W. Wylie. 1972. Procollagen: conversion of the precursor to collagen by a neutral protease. *Science* **175**: 544–546.

Bouteille, M., and D. C. Pease. 1971. Tridimensional structure of native collagenous fibrils, their proteinaceous filaments. *J. Ultrastruct. Res.* **35**: 314–338.

Breen, M., H. G. Weinstein, R. L. Johnson, A. Veis, and R. T. Marshall. 1970. Acidic glycosaminoglycans in human skin during fetal development and adult life. *Biochim. Biophys. Acta* **201**: 54–60.

Cohen, S. 1959. Purification and metabolic effects of nerve growth promoting protein from snake venom. *J. Biol. Chem.* **234**: 1129–1137.

Davidson, E. A., and W. Small. 1963. Metabolism *in vivo* of connective tissue mucopolysaccharides. *Biochim. Biophys. Acta* **69**: 445–452.

Dehm, P., S. A. Jiminez, B. R. Olsen, and D. J. Prockop. 1972. A transport form of collagen from embryonic tendon: Electron microscopic demonstration of an NH_2-terminal extension and evidence suggesting the presence of cystine in the molecule. *Proc. Nat. Acad. Sci. U.S.* **69**: 60–64.

Dorfman, A. 1953. The effects of adrenal hormones on connective tissues. *Ann. N. Y. Acad. Sci.* **56**: 698–703.

Duran-Reynals, F. 1942. Tissue permeability and the spreading factors in infection. A contribution to the host: parasite problem. *Bacteriol. Rev.* **6**: 197–252.

Eisen, A. Z., E. A. Bauer, and J. J. Jeffrey. 1971. Human skin collagenase. The role of serum alpha-globulins in the control of activity *in vivo* and *in vitro*. *Proc. Nat. Acad. Sci. U.S.* **68**: 248–251.

Exton-Smith, A. N., and R. W. Sherwin. 1961. The prevention of pressure sores. Significance of spontaneous bodily movements. *Lancet* **ii**: 1124–1126.

Forrester, J. 1968. Correlations of mechanical and biochemical factors in repair. *In* "Repair and Regeneration: The Scientific Basis of Surgical Practice" (J. E. Dunphy and W. Van Winkle, Jr., eds.), pp. 71–86. McGraw-Hill, New York.

Gibson, T., and R. M. Kenedi. 1970. The structural components of the dermis and their mechanical characteristics. *In* "Advances in Biology of Skin. The Dermis" (W. Montagna, J. P. Bentley, and R. L. Dobson, eds.), Vol. 10, pp. 19–38. Appleton, New York.

Grant, M. E., and D. J. Prockop. 1972. The biosynthesis of collagen. *N. Engl. J. Med.* **286**(4): 194–199, 242–249, 291–300.

Grobstein, C. 1953. Epithelio-mesenchymal specificity in the morphogenesis of mouse submandibular rudiments *in vitro*. *J. Exp. Zool.* **124**: 383–414.

Gross, J. 1958. Studies on formation of collagen. II. The influence of growth rate on neutral salt extracts of guinea pig dermis. *J. Exp. Med.* **107**: 265–277.

Gross, J., and C. M. Lapiere. 1962. Collagenolytic activity in amphibious tissues: A tissue culture assay. *Proc. Nat. Acad. Sci. U.S.* **48**: 1014–1022.

Harper, E., K. J. Block, and J. Gross. 1971. The zymogen of tadpole collagenase. *Biochemistry* **10**: 3035–3041.

Hascall, V. C., and S. W. Sajdera. 1969. Protein polysaccharide complex from bovine nasal cartilage. The function of glycoprotein in the formation of aggregates. *J. Biol. Chem.* **244**: 2384–2396.

Hauschka, J. D., I. R. Konigsberg. 1966. The influence of collagen on the development of muscle clones. *Proc. Nat. Acad. Sci. U.S.* **55**: 119–126.

Hodge, A. J. 1967. Structure at the electron microscopic level. *In* "Treatise on Collagen. Chemistry of Collagen" (G. N. Ramachandran, ed.), Vol. 1, pp. 185–205. Academic Press, New York.

Holtzer, H. 1968. Induction of chondrogenesis: A concept in quest of mechanisms. *In* "Epithelial-Mesenchymal Interactions" (R. Fleischmeyer and R. E. Billingham, eds.), pp. 152–164. Williams & Wilkins Co., Baltimore, Maryland.

Horwitz, A. L., and A. Dorfman. 1968. Subcellular sites for synthesis of chondromucoprotein of cartilage. *J. Cell Biol.* **38**: 358–368.

Houck, J., and C. Chang. 1971. Chemotactic properties of products of collagenolysis. *Proc. Soc. Exp. Biol. Med.* **138**: 69–75.

Husain, T. 1953. An experimental study of some pressure effects on tissues, with reference to the bed-sore problem. *J. Pathol. Bacterial* **66**: 347–358.

Jackson, D. S., and J. P. Bentley. 1960. On the significance of the extractable collagens. *J. Biophys. Biochem. Cytol.* **7**: 37–42.

Jackson, D. S., and J. P. Bentley. 1968. Collagen-glycosaminoglycan interactions. *In* "Treatise on Collagen. Biology of Collagen" (B. S. Gould, ed.), Vol. 2, part A, pp. 189–214. Academic Press, New York.

Kallman, F., and C. Grobstein. 1965. Source of collagen at epitheliomesenchymal interfaces during inductive interaction. *Develop. Biol.* **11**: 169–183.

Koch, W. E. cited in Grobstein, C. 1967. Mechanisms of organogenetic tissue interaction. *Nat. Cancer Inst. Monogr.* **26**: 279–299.

Kohn, R. R., and E. Rollerson. 1960. Aging of human collagen in relation to susceptibility to the action of collagenase. *J. Gerontol.* **15**: 10–14.

Kondo, K., N. Seno, and K. Anno. 1971. Mucopolysaccharides from chicken skin of three age groups. *Biochim. Biophys. Acta* **244**: 513–522.

Kühn, K. 1969. The structure of collagen. *In* "Essays in Biochemistry." (P. N. Campbell and G. D. Greville, eds.), Vol. 5, pp. 59–87. Academic Press, New York.

Langer, K. 1861. Zur Anatomie und Physiologie der haut. I. uber die Spaltbarkeit der Cutis. *S. B. Acad. Wiss. Wien.* **44**: 19.

Loewi, G., and K. Meyer. 1958. The acid mucopolysaccharides of embryonic skin. *Biochim. Biophys. Acta* **27**: 453–456.

Madden, J. W., and E. E. Peacock. 1971. Studies on the biology of collagen during wound healing. 3. Dynamic metabolism of scar collagen and remodeling of dermal wounds. *Ann. Surg.* **174**: 511–520.

Mathews, M. B. 1967. Biophysical aspects of acid mucopolysaccharides relevant to connective tissue structure and function. *In* "The Connective Tissue" (B. M. Wagner, and D. E. Smith, eds.), pp. 304–329. Williams & Wilkins Co., Baltimore, Maryland.

McClean, D. 1936. A factor in culture filtrates of certain pathogenic bacteria which increased permeability of the tissue. *J. Pathol. Bacterial* **42**: 477–512.

McLoughlin, C. B. 1968. Interaction of epidermis with various types of foreign mesenchyme. *In* "Epithelial-Mesenchymal Interactions (R. Fleischmeyer and R. E. Billingham, eds.), pp. 244–251. Williams & Wilkins Co., Baltimore, Maryland.

Miller, E. J., G. R. Martin, C. E. Mecca, and K. A. Piez. 1965. The biosynthesis of elastin cross-links. The effect of copper deficiency and a lathyrogen. *J. Biol. Chem.* **240**: 3623–3627.

Moscona, A. A. 1956. Development of heterotypic combination of dissociated embryonic chick cells. *Proc. Soc. Exp. Biol. Med.* **92**: 410–416.

Muthukkaruppan, V. 1966. Inductive tissue interaction in the development of the mouse lens *in vitro*. *J. Exp. Zool.* **159**: 269–287.

Ogston, A. G. 1966. On water binding. *Fed. Proc. Fed. Amer. Soc. Exp. Biol.* **25**: 986–989.

Parakkal, P. F. 1969. Role of macrophages in collagen resorption during hair growth cycle. *J. Ultrastruct. Res.* **29**: 210–217.

Partridge, S. M. 1970. Isolation and characterization of elastin. *In* "Chemistry and Molecular Biology of the Intercellular Matrix. (E. A. Balazs, ed.), Vol. I., pp. 593–616. Academic Press, New York.

Pearsall, N. N., and R. S. Weiser. 1970. "The Macrophage." Lee & Febiger, Philadelphia, Pennsylvania.

Priest, R. E. 1972. Cellular replication and specialized function of fibroblasts. *J. Invest. Dermatol.* **59**: 35–39.

Ramachandran, G. N., and V. Sasisekharan. 1961. Structure of collagen. *Nature (London)* **190**: 1004–1005.

Revel, J. P. 1970. Role of the golgi apparatus of cartilage cells in the elaboration of matrix glycosaminoglycans. *In* "Chemistry and Molecular Biology of the Intercellular Matrix. Vol. 3. (E. A. Balazs, ed.), pp. 1485–1502. Academic Press, New York.

Revel, J. P., and E. L. Hay. 1963. An autoradiographic and electron microscopic study of collagen synthesis in differentiating cartilage. *Z. Zellforsch. Mikrosk. Anat.* **61**:110–144.

Rich, A., and F. H. Crick. 1961. The molecular structure of collagen. *J. Mol. Biol.* **3**: 483–506.

Roden, L. 1964. The protein carbohydrate linkage in protein complexes of acid mucopolysaccharides. Absts 148th Meeting of Amer. Chem. Soc., Chicago. Aug.–Sept., p. 12c.

Rosenberg, L., W. Hellmann, and A. K. Kleinschmidt. 1970. Macromolecular models of protein-polysaccharides from bovine nasal cartilage based on electron microscopic studies. *J. Biol. Chem.* **245**: 4123–4130.

Ross, R. 1968. The connective tissue fiber forming cell. *In* "Treatise on Collagen. Biology of Collagen (B. S. Gould, ed.), Vol. 2, pp. 1–75. Academic Press, New York.

Ross, R., and E. P. Benditt. 1965. Wound healing and collagen formation. V. Quantitative electron microscope radio-autoradiographic observations of proline-^{3}H utilization by fibroblasts. *J. Cell Biol.* **27**: 83–106.

Ross, R., and P. Bornstein. 1969. The elastic fiber. I. The separation and partial characterization of its macromolecular components. *J. Cell Biol.* **40**: 366–381.

Ross, R., and J. W. Lillywhite. 1965. The fate of buffy coat cells grown in subcutaneously implanted diffusion chambers. A light and electron microscopic Study. *Lab. Invest.* **14**: 1568–1585.

Ross, R., N. B. Everett, and R. Tyler. 1970. Wound healing and collagen formation. VI. The origin of the wound fibroblast studied in parabiosis. *J. Cell Biol.* **44**: 645–654.

Sams, W. M., and J. G. Smith. 1964. Alterations in human dermal fibrous connective tissue with age and chronic sun damage. *In* "Advances in Biology of Skin. Aging (W. Montagna, ed.), Vol. 6, pp. 199–210. Pergamon, Oxford.

Sandberg, L. B., N. Weissman, and W. R. Gray. 1971. Structural features of tropoelastin related to the sites of cross-links in aortic elastin. *Biochemistry* **10**: 52–56.

Saunders, J. W. 1948. The proximo-distal sequence of origin of the parts of the chick wing and the role of the ectoderm. *J. Exp. Zool.* **108**: 363–403.

Saunders, J. W. 1958. Inductive specificity in the origin of integumentary derivatives in the fowl. *In* "The Chemical Basis of Development" (W. D. McElroy and B. Glass, eds.), pp. 239–253. Johns Hopkins, Baltimore, Maryland.

Sengel, P. 1964. The determinism of the differentiation of the skin and the cutaneous appendages of the chick embryo. *In* "The Epidermis (W. Montagna, and W. C. Lobitz, Jr., eds.), pp. 15–34. Academic Press, New York.

Sisson, J. C. 1968. Hyaluronic acid in localized myxedema. *J. Clin. Endocrinol. Metab.* **28**: 433–436.

Smith, J. G., E. A. Davidson, and R. W. Taylor. 1964. Human cutaneous acid mucopolysaccharides: The effects of age and chronic sun damage. *In* "Advances in Biology of Skin. Aging (W. Montagna, ed.), Vol. 6, pp. 211–218. Pergamon, Oxford.

Smith, T. J., and R. R. Wagner. 1967. Rabbit macrophage interferons. I. Conditions for biosynthesis by virus-infected and uninfected cells. *J. Exp. Med.* **125**: 559–577.

Speakman, P. T. 1971. Proposed mechanism for the biological assembly of collagen triple helix. *Nature (London)* **229**: 241–243.

Szabo, G. 1962. Cultivation of skin, pure epidermal sheets, and tooth germs *in vitro*. *In* "Fundamentals of Keratinization" (E. O. Butcher and R. F. Sognnaes, eds.), pp. 45–60. Publ. No. 70 of Am. Assoc. Adv. Sci., Washington, D. C.

Szirmai, J. A. 1970. The organization of the dermis. *In* "Advances in Biology of Skin. The Dermis" (W. Montagna, J. P. Bentley, and R. L. Dobson, eds.), Vol. 10, pp. 1–17. Appleton, New York.

Tregear, R. T. 1966. "Physical Functions of Skin," pp. 157–180. Academic Press, New York.

Trelstad, R. L., and A. J. Coulombre. 1971. Morphogenesis of the collagenous stroma in the chick cornea. *J. Cell Biol.* **50**: 840–858.

Van Scott, E. J., and B. A. Flaxman. 1968. Environmental control of epithelial cells *in vivo* and *in vitro*. *In* "Epithelial-Mesenchymal Interactions" (R. Fleischmeyer and R. E. Billingham, eds.), pp. 280–293. Williams & Wilkins Co., Baltimore, Maryland.

Wessells, N. K. 1964. Substrate and nutrient effects upon epidermal basal cell orientation and proliferation. *Proc. Nat. Acad. Sci. U.S.* **52**: 252–259.

Windrum, G. M., P. W. Kent, and J. E. Eastoe. 1955. Constitution of human renal reticulin. *Brit. J. Exp. Pathol.* **36**: 49–59.

Wirtschafter, Z. T., and J. P. Bentley. 1962. The influence of age and growth rate on the extractable collagen of skin of normal rats. *Lab. Invest.* **11**: 316–320.

Wood, G. C. 1964. The precipitations of collagen fibers from solution. *In* "International Review of Connective Tissue Research." (D. A. Hall, ed.), Vol. 2, pp. 1–3. Academic Press, New York.

Zika, J. M., and L. Klein. 1971. Relative and absolute changes in skin collagen mass in rat. *Biochim. Biophys. Acta* **229**: 509–515.

Zwilling, E. 1961. "Limb Morphogenesis. Adv. Morphogen," Vol. 1, pp. 301–330. Academic Press, New York.

5

Blood Supply

I. Introduction

Some years ago, Winkelmann *et al.* (1961) lamented that textbooks of anatomy represent cutaneous vascular patterns in schematic diagrams based largely on the delineations of Spalteholz. Actually, Spalteholz (1927) constructed his diagrams to generalize and summarize his findings and did not intend that they be interpreted literally. Statements to the contrary notwithstanding, cutaneous blood vessels are not regular and do not form geometric patterns. In fact, regularities are the exception. Even a cursory glance at the different influences that affect vascularization will tell why. The kinds of cutaneous vascular beds are determined by (1) the kinds of skin they perfuse whether in the same or different individuals, (2) the thickness of the various dermal and hypodermal layers, (3) the types and numbers of appendages present, and (4) the specific relation of the skin to the bones and muscle fascias under it. Attempts to reconstruct patterns of cutaneous blood vessels from thinly cut histological preparations of dubious quality and unknown orientation only add to the confusion.

Only rarely is it appreciated that there is little relationship between the quantity of blood vessels in skin and the functioning cutaneous tissue. Thus these vessels are nearly always greatly in excess of the biological needs of the tissue but they do have a function of their own (Winkelmann *et al.*, 1961; Saunders, 1961; Burton, 1959, 1961; Moretti, 1968). As we shall see, the masses of interlacing small vessels in the dermis function primarily for thermal regulation and secondarily for nutrition of the tissues.

II. The Vascular Patterns

For descriptive purposes, it is convenient to follow the path of cutaneous vessels from their origin, before they enter the skin, to their termination in the capillary networks. To avoid burdening the reader with unnecessary nomenclature, we divide the major vessels into three categories—segmental, perforator, and cutaneous—and remind the reader that all have venous counterparts.

The *segmental* vessels follow the metameric pattern of the embryo and are primarily related to the underlying peripheral nerves; as they develop, many of the bones and muscles shift from their origin, and their segmental pattern is lost. From the embryonic dorsal aorta emerge 30 rows of bilaterally paired segmental arteries, whose ventral rami later become the intercostal and lumbar arteries. When these paired arteries anastomose on the ventral midline, they form the mammary and epigastric arterial system. The vasculature of the limbs, which is not segmental, accompanies distally the major nerves of the limbs to form the axial vessels. This, then, is the segmental-anastomotic-axial system, referred to as *segmental vessels*. Most segmental arteries emerge from the aorta and together with peripheral nerves are located deep to the muscles. They form the intercostal, lumbar, and internal mammary arterial system (Fig. 1).

Branches from the segmental arteries pass through muscles as *perforator arteries*, give off muscular vessels, and continue peripherally to the skin (Fig. 2); these are *musculocutaneous vessels*. Skin is also supplied by a few *direct cutaneous* vessels (Morujo, 1961) that perforate the muscles and go straight to the skin. These vessels are found mostly in the extremities and thorax and supply relatively small areas. The major trunks of direct cutaneous vessels travel parallel to the surface (Webster, 1937).

Blood is said to be distributed from the major vessels to the surrounding dermis by way of superficial and deep plexuses. Winkelmann *et al.*

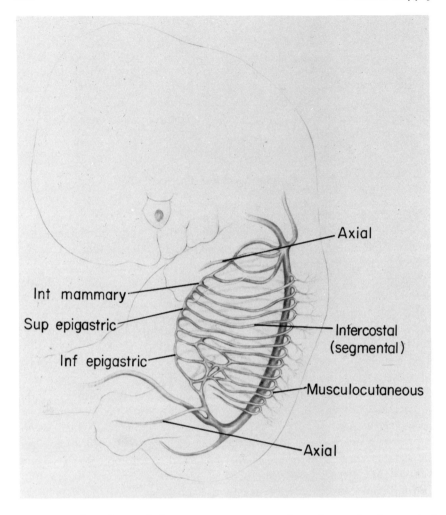

Fig. 1. The anlagen of the major cutaneous vessels as they spring from paired intercostal (segmental) arteries. Segmental arteries meet on the ventral midline forming the mammary, above, and the epigastric arterial system, below.

(1961) and Saunders (1961), however, showed that the apparently separate plexuses are really interconnecting vessels of different sizes at all levels of the dermis. Even though discrete networks can be seen around hair follicles and sweat glands, the general distribution is really one unit.

Except for pigs, whose cutaneous vasculature resembles that of man, loose-skinned mammals are vascularized mostly by direct cutaneous

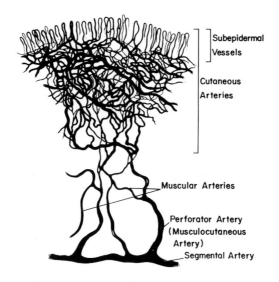

Subepidermal
Vessels

Cutaneous
Arteries

Muscular Arteries

Perforator Artery
(Musculocutaneous
Artery)
Segmental Artery

Fig. 2. Diagram showing the patternless cutaneous vascular system, the branching of the segmental artery into the perforator (musculocutaneous) artery, and the latter, in turn, branching into muscular and cutaneous arteries.

arteries. For example, Saunders (1961) showed abundant vasculature in the ears of rabbits, lambs, and other mammals. The main trunk of these arteries runs superficial to the muscles, parallel to the skin's surface. From this conduit arise perpendicular cutaneous vessels that go directly to the skin where, depending on location, they form more or less complex meshworks.

The dermis and hypodermis, then, are perforated by a continuous arteriovenous network with meshes of different sizes and with various spatial relationships. Every level of the tissue, every cutaneous appendage, is vascularized by vessels coming and going in various directions. Exact "vascular trees" with progressive arterioles or degressive venules that split dichotomously do not exist. There are no specific "plexuses" since all levels of the dermis are rich in vessels (Fig. 3). Observations of skin injected with radio-opaque substances and viewed with an x-ray microscope show an extensive interconnection of macro- and micro-networks (Saunders *et al.*, 1957; Saunders, 1961). In such preparations, the cutaneous vascular trees appear to be continuous meshworks composed of vessels of different diameters. Despite this reticular effect, two general interconnected vascular areas can be recognized: one includes

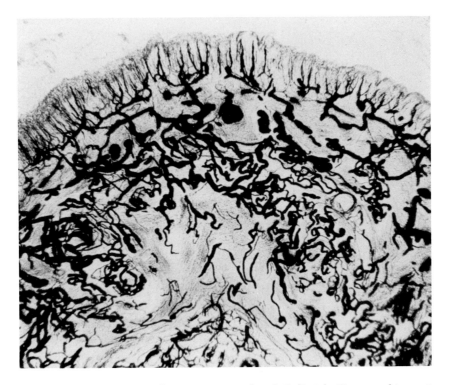

Fig. 3. 1 100-μm section from a toe injected with India ink. Compare this specimen with the diagram in Fig. 2. (Courtesy of Dr. R. K. Winkelmann.)

all of the vascular beds of the dermis (Moretti, 1968), the other the perifollicular network. The former vascularizes every entity of the skin, including the periadnexal networks; detailed descriptions of the latter are given in the chapter on the pilary system. Briefly, vascular baskets surround discrete hair follicles or groups of them; the diameter of the vessels and the extent of the vascularity depend on the size and depth of the follicles.

The complex, sometimes seemingly chaotic, distribution of vessels in the dermis is well adapted to the various changes and stresses to which the skin is exposed. For example, being more or less serpentine, the major vessels allow the skin to distend without appreciably interfering with the circulation of blood. This adaptability is particularly evident during the various phases of the hair growth cycles (Montagna and Ellis, 1957; Ellis and Moretti, 1959); in catagen, for example, the hair follicle withdraws upward, leaving the lower part of the perifollicular vascular basket collapsed below it. When the follicles become active again, they must grow through these collapsed sleeves of vessels.

Morphologically, the vascular beds differ widely in different areas (Fig. 4). For example, in the dermis the long, narrow papillary ridges lying side by side in the olecranon and patellar regions and, to an even greater extent, those on the palms and soles contain long and relatively straight capillary loops. Those on the trunk are more shallow and more broad and have more meandering, less continuously identifiable loops. A schematic diagram of the pattern of terminations of cutaneous vessels (Zweifach, 1949) is shown in Fig. 5. Metarterioles or "preferential" thoroughfare channels appear to emerge from terminal arterioles and to be surrounded by one layer of smooth muscle. The most direct channels from arterial to venous circulation give off side branches called precapillary sphincters, which control the flow of blood into the capillaries proper. Sometimes the capillaries, consisting only of an endothelial tube, anastomose and then join the collecting venules. Thus, capillary blood flow can be regulated by the contraction and dilatation of the venules and the precapillary sphincters. If only the precapillary sphincters contracted, blood would be shunted through a preferential channel and would bypass the various lateral capillary networks. Contraction of the metarteriole would shunt the blood through arteriovenous anastomoses directly to muscular venules. Unlike the venules in other tissues, those in the skin have remarkable ability to contract and dilate (Zweifach, 1959).

The major role of the cutaneous vascular meshwork is to regulate heat and blood pressure. Burton (1959) stated that ". . . in its blood flow the skin is peculiarly the servant of the whole organism and is less endowed with autonomous control than other tissues. Possibly this view is important in understanding the susceptibility of the skin to ischemia in abnormal conditions, and worth remembering in considering management of disturbed skin functions."

III. Arteriovenous Anastomoses (AVA) or Shunts

As a rule, blood flows from arteries and arterioles, through capillaries, to venules and veins. Skin, however, has certain mechanisms that enable blood to pass directly from arteries to veins through structures called *arteriovenous anastomoses* or *shunts*. In man, such anastomoses are found predominantly in the upper parts of the reticularis dermis of the fingertips and nail bed; descriptions of them elsewhere in the skin are

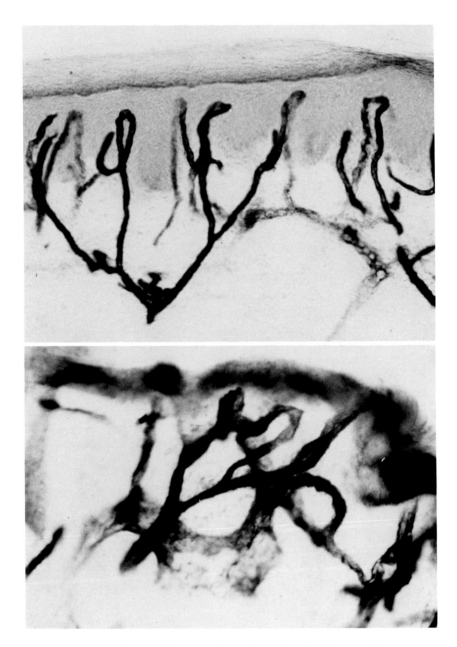

Fig. 4. Superficial arterial loops and arcades in the skin from the labia minora. The figure above shows simple loops, that below more intricate (anastomosing) ones. Thick frozen sections treated with the alkaline phosphatase technique.

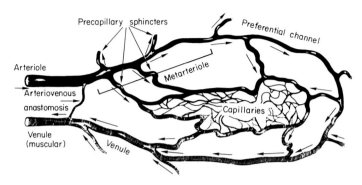

Fig. 5. Zweifach's (1949) concept of the basic structural pattern of the terminal cutaneous vessels. (Reproduced by permission of the author and the Josiah Macy, Jr. Foundation, New York.)

widely divergent (Moretti, 1968). Reports on the numbers of AVA in man are greatly discrepant, perhaps because there are individual differences. In the skin of other mammals, they are particularly numerous in the ears, where blood vessels abound (Clara, 1959).

The shape of digital AVA varies from tortuous (usually in infants) to straight. Generally plump, roundish, and discrete, they have been called *glomi* or organs of Hoyer-Grosser. Regardless of size or shape, AVA's originate as branches of arteries or arterioles and connect directly with the accompanying vein or venule. At the point of their origin, AVA's have a typical arterial wall, and at the efferent end they resemble veins. Thickening of the muscular and adventitial layers of the intervening segment gives them their characteristic structure. Here the endothelium rests on thin reticulum fibers and a greatly thickened media that consists largely of tightly packed smooth muscle fibers aligned longitudinally. There is no internal elastic membrane. The lumen is usually so small as to be inconspicuous. The layer of circular muscle fibers in the media is also thick as is the adventitial tissue.

The histochemical properties have been studied by Mescon *et al.* (1956) and Beckett *et al.* (1956). The most significant findings are that the muscle fibers are different from those in the other vessels and that the AVA's are heavily innervated by acetylcholinesterase-containing nerves, like the vessels in the external genitalia (Erickson and Montagna, 1973). The glomus, then, is well endowed with contractile tissue and abounds in nerves from the autonomic system, probably both sympathetic and parasympathetic.

Burton (1961) believes that glomus bodies act as "open–shut" stopcocks. Because of their vasomotor tone, they have a tendency to close completely (critical closure) if the blood pressure (transmural pressure) in the small arteries does not exceed a critical value. Their opening during the phenomenon of cold-induced vasodilation can be prevented by an increased tissue pressure, which in turn reduces transmural pressure. The opening of countless shunts in the skin of the limbs could be responsible for the great increase of blood flow in the skin in a warm environment.

The major role of arteriovenous anastomoses is obviously in temperature regulation. However, arteriovenous shunts are numerous in the stomach and the mesenteries where they can play no role in temperature regulation. Here all shunts open and close automatically; sudden closure of the capillary bed, in response to a circulating vasoconstrictor agent or local stimulus, causes them to open. Thus Burton (1959) suggested a second function that services the whole organism rather than just the skin. He compared the shunts to the "safety valves on a boiler." A sudden rise of blood pressure brought about by vasoconstriction anywhere raises the pressure in their lumina which automatically opens them and thus prevents to some degree a rise in pressure. Pressure regulation, then, is one of the functions of the skin that is not usually mentioned.

Unlike capillary flow, increases in blood flow through shunts do not supply oxygen and nutrition to the skin. Thus, agents applied to skin to increase its circulation do so without a comparable improvement in the oxygenation of the tissues unless the metarterioles that supply the capillaries are opened as well as the shunts.

IV. Cutaneous Blood Flow and Its Significance

From the point of view of metabolic need, then, skin is vastly overperfused. The minimum value, measured in fingers during strong vasoconstriction caused by exposure to cold, is about 0.5 to 1.0 ml/minute/100 ml tissue (Burton, 1961). Even when such vasoconstriction is maintained, the tissue does not necessarily become necrotic although it can become damaged when rewarmed. This *minimum* flow, then, is adequate for the physiological demands of the skin itself. The necrosis that accompanies peripheral vascular diseases may represent a complete cessation of blood flow or a complete spasm of the controlling arterioles (Burton, 1961).

The cutaneous *mean flow* and *maximum flow* are considerably higher than the minimum physiological flow. The mean flow expressed for

100 ml of tissue is 20 to 30 times greater than the minimum, and the maximum flow during vasodilatation can be more than 100 times (Burton, 1961). These differences reemphasize that cutaneous blood flow serves the entire organism and is not governed solely by the metabolic requirements of the skin which probably vary only with changes in temperature and require at most only a 2- or 3-fold increase over the minimum flow.

The cutaneous vascular system is a prime regulator of total body temperature. Steady fluctuations in the middle range of the mean blood flow are a result of periodic, rhythmic responses to discharges of nervous impulses from the sympathetic vasoconstrictor nerves. This mechanism maintains a proper value to transport an adequate amount of heat for regulating body temperature. Without such sympathetic peripheral vaso-constrictor activity, thermal fluctuations in the environment would result in marked variability of body temperature (Burton, 1961).

Cutaneous blood vessels are also affected by such vasoactive substances as circulating adrenaline and acetylcholine, but they can be completely dominated by neural control. If one inhales the powerful vasodilator amyl nitrite, an intense vasoconstriction occurs in the digits (Burton, 1961). This paradoxical effect is probably brought about by "buffer reflexes" (e.g., the carotid sinus) that fall below the effective point because many blood vessels elsewhere in the skin are chemically dilated.

Not all areas of the skin have the same patterns of circulatory response. The blood vessels in the blush area (upper thorax, neck, and head) are controlled mostly if not entirely by circulating pressor agents. After the inhalation of amyl nitrite, these areas have a deep flush at the same time that the extremities show strong reflex vasoconstriction. Furthermore, unlike the vessels in other parts of the body, those in these areas do not respond to vasoconstriction (Froese and Burton, 1957).

The phenomenon of the dilatation of the finger vessels when they are exposed to cold is called the "hunting reaction." When a finger is immersed in extremely cold water, there is immediate vasoconstriction, the skin temperature falls to near freezing, and there is much pain. (The vasoconstriction that occurs in the other hand is probably brought about by spinal reflexes.) After 4 to 5 minutes, the blood flow suddenly increases in the cooled finger, the skin feels warm, and the pain subsides. This is followed by a second vasocontriction and repeated cycles of constriction and dilatation. What causes this phenomenon is not known.

V. Changes in Skin Circulation

For all their apparent fragility, cutaneous vessels have a remarkable capacity for adaptation. They usually repair well and with dispatch after injury and adapt accordingly in response to climatic and other circumstances (Burton, 1961). In the ears of rats, the number of capillary endothelial cells in mitosis increases when the animals are being acclimatized to cold (Héroux, 1959a,b, 1960), an example of the physiological adaptation of cutaneous circulation. When the maximal blood flow of the fingers of subjects chronically exposed to climatic changes was measured, it showed a steady increase when exposed to heat for several days (Burton *et al.*, 1939). Again, under certain conditions, many new channels open up in the superficial veins of the arms and chest. How these and other changes are effected, however, is not easily understood.

Although much is yet to be learned about this subject, the easy regeneration and increased vascularity of cutaneous vessels in response to abnormal conditions suggest an important local autonomy.

Not much is known either about the behavior of cutaneous vessels during aging. Some authors (e.g., Bellocq, 1925) have claimed that a progressive rarefaction of the vessels occurs in the dermal meshwork, but Spalteholz (1927) denied this. However, a diminution of superficial as well as of deep blood vessels does occur in the aging balding scalp (Chiale, 1927; Ellis, 1958), where the flattening out of the underside of the epidermis is accompanied by a pronounced decline in the number of capillary loops and other vessels in the dermal vascular meshworks (Fig. 6) (Ellis, 1958). Very few recent studies have explored this relatively unknown terrain. Even at this late date, more systematic studies are needed for a better understanding.

VI. The Lymphatic System

Cutaneous lymphatics are probably as extensive as blood vessels, but they have not been shown satisfactorily. After the injection of radioopaque substances, radiological techniques reveal a series of capillaries

Fig. 6. Aging changes in the superficial capillary loops in the human scalp, demonstrated with the alkaline phosphatase technique. Above is the scalp of a 3-year-old boy, in the center is that of a man 22 years old, and below, the bald scalp of a man 69 years old.

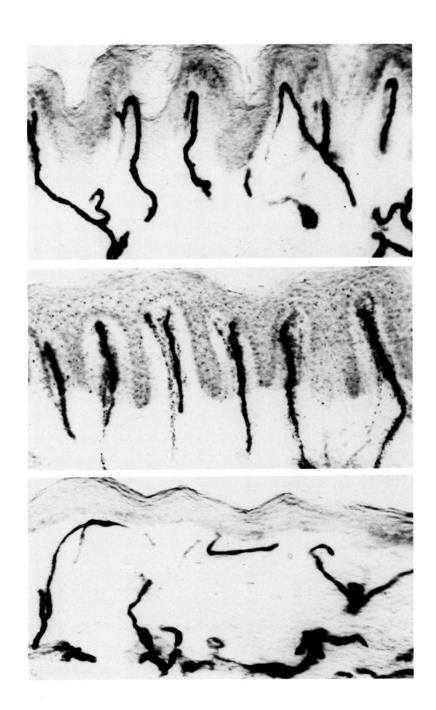

dead-ending in the papillary dermis and draining into a "subpapillary lymphatic plexus" that finally empties into deeper plexuses with valves. Lymph vessels progress centripetally via progressively larger ones, and filtering through lymph nodes, empty into the thoracic duct. We suggest rather that these blind-ending lymphatic capillaries empty into a vast, unstructured network throughout the dermis.

Identifying lymphatic vessels is difficult because despite their similarity to capillaries and veins, they are extremely frail and usually collapse in histological tissues. Though thin, the larger vessels in the deeper portions of the dermis and in the hypodermis can often be recognized by their numerous valves.

The lymphatic system is indispensable to blood circulation, which is at a relatively high pressure. Fluid passes out of arterial capillaries into the surrounding tissue because the pressure within the vessels is greater than that in the tissue. Lymph fluids pass back into venous capillaries where pressure is low. Furthermore, because the osmotic pressure of plasma proteins is greater in the arterial vessels, they leak out into the tissues and cannot pass back into the circulation except through the lymph vessels. When plasma proteins accumulate in tissues, the consequent rise in osmotic pressure disturbs the balance of capillary filtration and causes the fluid to remain in the tissues; the result is edema. The principal role of the lymphatic system, then, is to remove plasma proteins from extracellular spaces. Secondarily, it removes particulate and antigenic materials from tissues.

The flow of lymph is slow and varies according to muscular activity. The one-way flow from the tissues into the lymphatic vessels is maintained by the direction of the valves.

All in all, we are particularly interested in the lymphatic system of the skin as it functions in the skin. Its structure is flimsy and easily observed.

References

Beckett, E. B., G. H. Bourne, and W. Montagna. 1956. Histology and cytochemistry of human skin. The distribution of cholinesterase in the finger of the embryo and the adult. *J. Physiol. (London)* **134**: 202–206.

Bellocq, P. 1925. "Etude Anatomique des Arteres de la Peau Chez L'homme." Masson, Paris.

Burton, A. C. 1959. Physiology of cutaneous circulation, thermoregulatory functions. *In* "The Human Integument" (S. Rothman, ed.), Publ. No. 54, pp. 77–88. Amer. Assoc. Advance. Sci., Washington, D. C.

Burton, A. C. 1961. Special features of the circulation of the skin. *In* "Advances in Biology of Skin. Blood Vessels and Circulation (W. Montagna and R. A. Ellis, eds.), Vol. 2, pp. 117–122. Pergamon, Oxford.

Burton, A. C., J. C. Scott, B. McGlone, and H. C. Bazett. 1939. Slow adaptations in the heat exchange of man to changed climatic conditions. *Amer. J. Physiol.* **129**: 84–101.

Chiale, C. 1927. Delle modificazioni dei vasi cutanei inerenti all'eta. *G. Ital. Dermatol. Sifilol.* **68**: 1625–1645.

Clara, M. 1959. "Le Anastomosi Arteriovenosi." Casa Editrice Dr. Francesco Vallardi, Milano.

Ellis, R. A. 1958. Ageing of the human male scalp. *In* "The Biology of Hair Growth" (W. Montagna and R. A. Ellis, eds.), pp. 469–485. Academic Press, New York.

Ellis, R. A., and G. Moretti. 1959. Vascular patterns associated with catagen hair follicles in the human scalp. *Ann. N. Y. Acad. Sci.* **83**: 448–457.

Erickson, K. L., and W. Montagna. 1973. New observations on the anatomical features of the female genitalia. *J. Amer. Med. Women's Ass.* **27**: 573–581.

Froese, G., and A. C. Burton. 1957. The heat losses of the human head. *J. Appl. Physiol.* **10**: 235–241.

Héroux, O. 1959a. Comparison between seasonal and thermal acclimatization in white rats. Surface temperature, vascularization, and *in vitro* respiration of the skin. *Cancer J. Biochem. Physiol.* **37**: 1247–1253.

Héroux, O. 1959b. Histological evidence for cellular adaptation to nonfreezing cold injury. *Cancer J. Biochem. Physiol.* **37**: 811–820.

Héroux, O. 1960. Mitotic rate in the epidermis of warm- and cold-acclimatized rats. *Cancer J. Biochem. Physiol.* **38**: 135–142.

Mescon, H., H. J. Hurley, Jr., and G. Moretti. 1956. The anatomy and histochemistry of the arteriovenous anastomosis in human digital skin. *J. Invest. Dermatol.* **27**: 133–145.

Montagna, W., and R. A. Ellis. 1957. Histology and cytochemistry of human skin. XIII. The blood supply of hair follicles. *J. Nat. Cancer Inst.* **19**: 451–463.

Moretti, G. 1968. The blood vessels of the skin. *In* "Handbuch der Haut und Geschlechtskrankheiten (O. Gans and G. K. Steigleder, eds.), Vol. 1/1, pp. 491–623. Spinger-Verlag (Bergmann), Berlin and New York.

Morujo, A. A. 1961. Terminal arteries of the skin. *Acta Anat.* **58**: 289–295.

Saunders, R. L. de C. H. 1961. X-ray projection microscopy of the skin. *In* "Advances in Biology of Skin. Blood vessels and Circulation" (W. Montagna and R. A. Ellis, eds.), Vol. 2, pp. 38–56. Pergamon, Oxford.

Saunders, R. L. de C. H., J. Lawrence, and D. A. Maciver. 1957. Microradiographic studies of the vascular patterns in muscles and skin. *In* "X-Ray Microscopy and Microradiography" (E. V. Coslett, A. Engström, and H. H. Pattee, Jr., eds.), pp. 539–550. Academic Press, New York.

Spalteholz, W. 1927. Blutgefässe in der Haut. *In* "Handbuch der Haut und Geschlechtskrankheiten" (J. Jadassohn, ed.), pp. 379–433. Springer-Verlag, Berlin and New York.

Webster, J. P. 1937. Thoraco-epigastric tubed pedicles. *Surg. Clin. N. Amer.* **17**: 145–184.

Winkelmann, R. K., S. R. Scheen, Jr., R. A. Pyka, and M. B. Coventry. 1961.
Cutaneous vascular patterns in studies with injection preparation and alkaline
phosphatase reaction. *In* "Advances in Biology of Skin. Blood Vessels and
Circulation" (W. Montagna and R. A. Ellis, eds.), Vol. 2, pp. 1–19. Pergamon
Press, Oxford.
Zweifach, B. W. 1949. Basic mechanisms in peripheral vascular homeostasis. Trans.
3rd Josiah Macy, Jr. Conf. on Factors Regulating Blood Pressure.
Zweifach, B. W. 1959. Structural aspects and hemodynamics of microcirculation in
the skin. *In* "The Human Integument" (S. Rothman, ed.). Amer. Assoc. Advance.
Sci., Washington, D. C.

6

Cutaneous Innervation

The largest sense organ of the body, interposed between the organism and its environment, skin must maintain that organism in a constant state of awareness of all environmental changes. Because of its enormous importance for survival and its ready availability for study, skin has been the object of numerous pharmacological, electrophysiological, and anatomical studies. Yet precious few of the problems that involve cutaneous innervation and specific cutaneous sensibilities have been solved, let alone the clinical problems that are in some way related to the peripheral nervous system.

Hence any book that deals with the skin must at least touch on these aspects of the general subject. By way of introduction, it should be noted that all cutaneous nerves are sensory except those that subserve the glands, muscle fibers, and blood vessels, which belong to the autonomic nervous system.

I. Effector Cutaneous Nerves

The effector nerves to the skin are postganglionic fibers of the paravertebral chain ganglia. Anatomically, they belong to the sympathetic division of the autonomic nervous system; yet at least some of their

fibers are cholinergic and must be classified physiologically as parasympathetic, since they have acetylcholine as their neurohumoral trans-
mitter. Histochemical techniques for acetylcholinesterase admirably
demonstrate these nerves around eccrine sweat glands, blood vessels,
and arrectores pilorum muscles. However, cholinesterase in a nerve does
not prove that it is cholinergic since these enzymes are also present in
sensory nerves. In reality, it is morphologically and histochemically
impossible to separate effector from sensory nerve fibers in the skin
since their fiber size, like their histological and histochemical properties,
overlaps. Another conundrum is the fact that although many nerves can
be demonstrated with both the cholinesterase and silver techniques,
others can be shown only with one or the other.

Information on the physiology and pharmacology of cutaneous effector
nerves is sketchy, controversial, and paradoxical. Eccrine sweat glands,
as we shall see, respond not only to cholinomimetic but also to
adrenomimetic drugs. Hence some authors have argued that they have
a double innervation or that the same nerves can respond to either
stimulation. On the other hand, the nerve fibers that supply cutaneous
blood vessels and arrectores pilorum muscles, which are also postganglionic fibers of the paravertebral ganglia, respond primarily, if not
entirely, to adrenomimetic substances. These discrepancies are not easily
explained; those interested in a thorough account of the physiology of
the sympathetic nervous system are referred to Koelle's (1970) review
article.

A puzzling action of the effector nerves, what Langley and Anderson
(1894) called the axon reflex, is elicited by a local impulse (e.g., injury
or drugs) which travels centripetally along a number of the terminal
efferent axon branches up to the highest point of ramification. Here,
instead of continuing centripetally to the cyton, it reverses its course
and passes centrifugally to all the other axon branches to stimulate the
entire area supplied by them, which is sometimes some distance from
the point of stimulation. This phenomenon can be generated by the
stroke stimulus, which in normal individuals produces a clearly defined
red line. However, in some cases of cutaneous abnormalities, within a
few seconds the original red line suddenly spreads to the surrounding
skin (Rothman, 1954). Another example of axon reflex is drug action
on eccrine sweat glands. When pressure, such as a beaker, is applied
to the skin and a sudorific drug is injected on one side of a pressure
barrier, sweating also occurs for a short distance on the other side of
the barrier. If the axon reflex does not exist (and there are some who

question it), this phenomenon remains unexplained. At any rate, in such a reflex, effector fibers must be able to carry the sensory impulse away from the stimulated area toward their cytons in the paravertebral ganglia. Furthermore, perhaps the cyton is not necessary for the relay of response impulses, and the highest points of ramification in such axons have an autonomous motor relay function comparable to that of ganglion cells. All of this assumes, of course, that no sensory fibers intimately accompany the effector fibers, whose terminal branches sometimes cover a large area of skin.

II. The Sensory or Afferent Nerves

The nerve receptors of cutaneous sensation are the objects of much confusion and disagreement. Investigators of cutaneous sensory mechanisms generally belong in one of two groups: (1) those who subscribe to Max von Frey's theory of specific nerve energies, or single fiber specificity, which assumes that there are different kinds of end organs and that their nerves carry different sensory impulses; and (2) those who believe in pattern responses. Despite wide acceptance, von Frey's theory has never been satisfactorily demonstrated. The many neuro-anatomists concerned with "different" kinds of sensory receptors have compiled long lists of apparently anatomically different entities which they believe to be specific receptors of the different modalities of cutaneous sensibilities.

To pinpoint the nerve endings that respond to the sensation of warmth and cold, Nafe and Wagoner (1937) excised small areas of skin that were especially sensitive to heat and cold and examined them for special end organs or terminal structures. They found nothing characteristic in any of the tissues. Moreover, it is a fact that one feels cold during fear and anxiety, chills during fever, and warmth in moments of passion. Thus, if warmth and cold are experienced without changes in temperature, they do not depend on a structure that specifically records temperature changes but must be due to action induced by temperature or bodily changes. According to Nafe (1969), specificity lies in the nonneural tissues that contain the nerve terminals and not in the nerves themselves. When an object comes into contact with the skin, the tissues are moved or disturbed, and so are the nerve fibers they contain. The result is that a volley of impulses is discharged over many fibers and is translated as a feeling of pressure that includes intensity, shape, and temporal course. Each volley has an order determined by the conditions of stimulation. Moreover, the frequency of discharge in each fiber at any given point in any given moment varies with the intensity of the stimulation, and the number of fibers activated may vary from moment to moment. The relative density of discharging fibers within the activated area is still another variable; whereas the activity in a nerve fiber is not known to excite another fiber, it may well influence it. Furthermore, from the moment the volley of nerve impulses is aroused, it is subject to inhibitory and reinforcing forces and to aroused efferent impulses which may alter the conditions of stimulation.

Therefore, regardless of "types" of endings, afferent nerve fibers appear to be essentially alike; they conduct impulses and do not transmit messages. Whether or not a nerve fiber responds to a specific type of stimulation depends on the relation of the fiber to the tissue in which it is found, and it is the tissue, not the nerve, that adapts to stimuli.

There are innumerable modalities of cutaneous sensations that go beyond the basic ones of touch, hot, cold, pain, and itch. The results of studies on some theses indicate that the various parts of the body differ in cutaneous sensibility. Sensation is not a unitary function of sex, laterality, and body part. Although women are more sensitive than men to pressure, everyone seems to have his own gradient in sensitivity, particularly in the extremities where the most distal parts are more sensitive than the proximal ones (Weinstein, 1969).

Nearly every new histological technique has demonstrated different "kinds" of sensory nerve endings never before described, and each of these has been assigned some specific receptor function. Many of these "different" end organs are probably artifacts produced by shrinking or compression, by the thinness of the sections, and by the plane of cut. Depending on the latter, for example, a dense, discretely condensed nerve net in thin sections can look like a variety of mucocutaneous corpuscles. Most of the histological techniques in use have been singularly unreliable and often unrepeatable. Even the leucomethylene blue methods, used with great success by Tamponi (1940), Arthur and Shelley (1959), and Miller *et al.* (1960), do not always work in the same way and often stain other tissues so densely that nerves are difficult to identify with certainty.

Taking all these conditions into consideration, we can describe the afferent nerve endings in the skin only in a general way. We recognize a superficial *dermal nerve network*, probably the chief sensory receptor; a *hair follicle network* or end organs, highly specialized and organized structures to be described later; *mucocutaneous end organs, Meissner corpuscles,* and *Vater-Pacini corpuscles.* All others are probably artifacts.

III. Dermal Nerve Network

Where the papillary dermis joins with the reticularis dermis, countless fine, always wavy nerve fibers, for the most part oriented horizontally, form a feltwork of different densities depending on its location (Fig. 1). This mass of nerves, found everywhere in cutaneous and mucous surfaces, was first described in detail by Tamponi (1940), then by Miller *et al.* (1960) and Winkelmann (1960a), who called it the dermal nerve network.

Tamponi (1940) described nerve nets that extend throughout the dermis far beyond the simple patterns described by others. Beginning at the base of the reticular dermis, interconnected, superimposed nerve nets become progressively denser at higher levels. At about the base

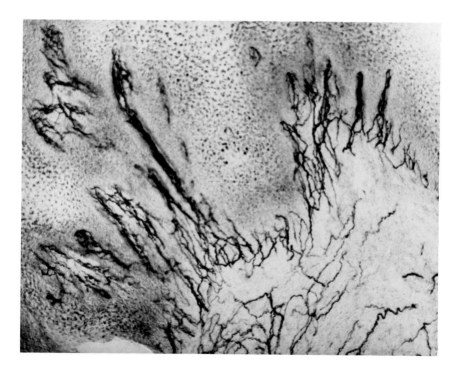

Fig. 1. Superficial nerve network in the labia minora prepared to show the presence of acetylcholinesterase.

of the papillary dermis, these nerves form networks of different densities. Since these networks are present in all skin, hairy or glabrous, they are probably the principal sensory receptors (Tamponi, 1940; Winkelmann, 1960a,b). In mucous membranes and in the cornea, networks of nerve fibers, but no end organs, are found. Winkelmann (1960a,b) believes that the arrangement of these networks around the hair follicles in hairy skin is similar to that in glabrous skin, except that the fibers around the outer root sheath of the follicle form an orderly structured pattern. But this is an oversimplification (cf. Chapter 7). Special end organs, such as those at the borders of mucocutaneous tissues, the friction surfaces, and erogenous areas, are also believed to have a similar arrangement except that the nerve filaments are rolled into balls or coils rather than loosely associated with an epithelial structure.

Among the more paradoxical patterns of innervation are those in the nipples of females and around the Meibomian glands. In the former, Montagna (1970) and Montagna and Yun (1972) found mucocutaneous end organs only around the dilated terminal parts of the lactiferous ducts (Fig. 2). Throughout their length, the ducts proper were com-

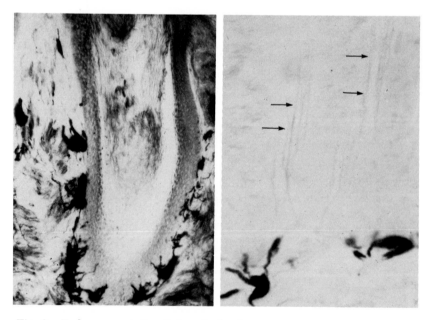

Fig. 2. End organs at tip of the nipple of a young woman on the right, and similar ones on the snout of pig. Note the intraepidermal nerves in the pig snout (left, arrows). Both preparations treated for acetylcholinesterase.

pletely entwined with nerves (Fig. 3), which must be sensory since apparently the duct is physiologically inert. The numerous nerves around the lobules of Meibomian glands could be but probably are not effector fibers.

During fetal development, the nerve net is the first ordered structure to appear in the dermis and may therefore influence the development of cutaneous appendages (Tello, 1923–1924; Winkelmann, 1959a,b; 1960a). Its distribution varies with the density of the cutaneous appendages: where hair follicles are dense, most of the cutaneous sensory nerves are distributed around them; but where follicles are sparse or absent, the networks are prominent.

IV. Intraepidermal Neurites

Assertions and denials that nerve fibers transverse the epidermis are rampant. These fibers can be demonstrated in the nose of moles (Giacometti and Machida, 1965), opossums (Munger, 1965) (Fig. 4),

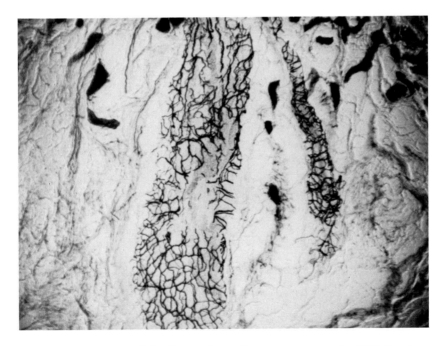

Fig. 3. Nerves around lactiferous ducts of a young woman. Acetylcholinesterase technique.

and other mammals; but not even their strongest proponents have produced convincing evidence of their presence in man. Yet despite the difficulty of demonstrating them, nerve fibers are present in human epidermis (Fig. 5). In fetal skin intraepidermal nerves are numerous (Fitzgerald, 1969), but in adult tissue they are found predictably only in the eyelids (Montagna and Ford, 1969) and around the genitalia, notably the clitoris. Nerve fibers are said to be more numerous in the epidermis of psoriatic skin (Weddell *et al.,* 1965) and of healing wounds than in normal skin, although neither is more sensitive than normal skin. Intraepidermal nerve fibers appear to have a turnover rate, particularly where fibers are numerous and long. Munger (1965) believes that at least in the nose of the opossum these intraepidermal neurites are associated with Merkel cells which act as transducers.

V. Mucocutaneous End Organs

Structurally organized, sometimes encapsulated, end organs are found underneath glabrous cutaneous epidermis and mucocutaneous surfaces with a ridged underside; only occasionally are they found in sparsely haired areas. From hairy skin and across the mucocutaneous junctions to the mucous membranes, there is a gradual waning of hair follicles and other appendages, with sudden appearances and disappearances of mucocutaneous end organs.

These structures consist mostly of coils or rolls of fine nerve fibers from which issue myelinated A fibers (Fig. 6). When prepared for cholinesterases, they appear sausage-shaped or like intensely arachnoid bodies (Fig. 7). Thus, except for the follicle, they resemble the basket-shaped end organs around hair follicles. Winkelmann contends that without the follicle, the nerve net could form a ball and like a muco-cutaneous end organ rise up higher in the dermis. This happens, no doubt, in the labia minora of girl children where the numerous well-innervated hair follicles of the fetus and infant disappear and are later replaced by the genital corpuscles, which are, in fact, mucocutaneous end organs.

VI. Meissner Corpuscles

Meissner corpuscles are found only in the volar surfaces of the pes and manus of primates; in other mammals, similar structures are called

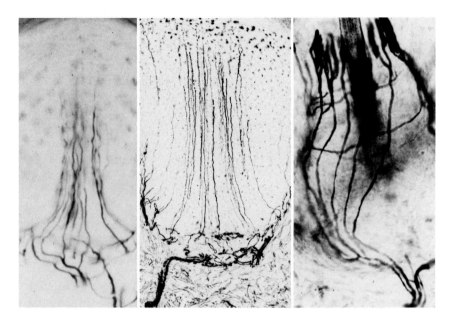

Fig. 4. From left to right, an Eimer corpuscle from the nose of a mole, an opossum, and a human hair follicle end organ. All preparations treated with Winkelmann's silver impregnation technique. Note the striking similarity in structural patterns and the long intraepidermal nerves from the corpuscles of Eimer.

mammalian end organs (Winkelmann, 1960a). The corpuscles are not appreciably different from mucocutaneous end organs except that they are better encapsulated and the nerve terminals appear to be flattened (Fig. 8). Both types are associated with myelinated A fibers and both contain nonspecific cholinesterase. In all mammals that walk on thick epidermal pads, including man, unencapsulated, so-called hederiform endings are found at the base of the epidermal ridges (Miller *et al.*, 1960); similar structures, also associated with heavy A fibers, are found deeper in the dermis (Miller *et al.*, 1958). Especially common in all primates, these structures are not different end organs but probably modifications of the dermal nerve network.

VII. Vater-Pacini Corpuscles

These are the most highly structured and largest end organs in skin or, for that matter, in any other part of the body. They are located

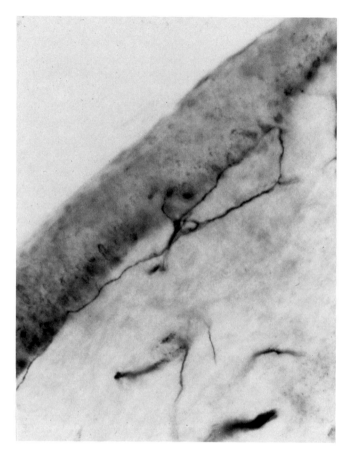

Fig. 5. Intraepidermal neurites in human epidermis from the eyebrow. Acetyl-cholinesterase technique.

deep in the dermis, along fascial and connective tissue planes, near nerve trunks, and in joints and mesenteries. Depending on their location, they vary from 0.2 to 1 mm. In skin they are most numerous in the digits and clitoris. Each corpuscle has a stalk in which the receptor, a heavily myelinated nerve, makes 4 to 16 right-angled turns before it enters an oval capsule shaped somewhat like an onion. This capsule is composed of concentric fibrous laminae with fluid in the spaces between the layers (Winkelmann, 1960a). The nerve constricts abruptly as it enters the capsule.

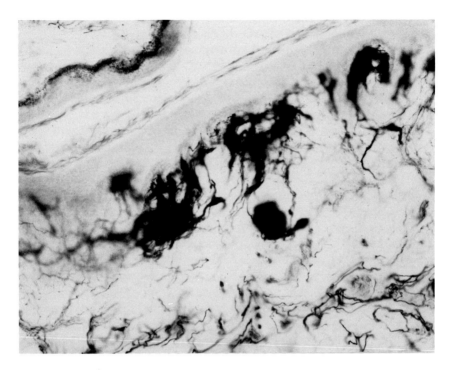

Fig. 6. Mucocutaneous end organs in the clitoris of a young woman. Acetylcholinesterase technique.

The function of the corpuscles is believed to be the reception of deep pressure stimuli. Slowly adapting organs, they discharge with an all-or-none response when a definite threshold of receptor potential is reached; they are poor transmitters of pressure but are extremely sensitive to vibrations. However, the response to pressure is rapidly adapting. Since they are closely associated with blood vessels, they probably signal local changes in blood supply and the control of arteriovenous anastomoses (Sinclair, 1967).

Pacinian corpuscles are usually found deep in the dermis of glabrous skin and have a distinctive structural organization and characteristic physiological properties (Iggo, 1969). In other sensitive cutaneous areas or "touch spots" that contain the so-called Merkel's discs (nerve network), 30 to 40 endings converge toward one myelinated axon. However, only one or very few nerves form the myelinated axon of a Pacinian receptor. The nonmyelinated central bulbs of Pacinian corpuscles are

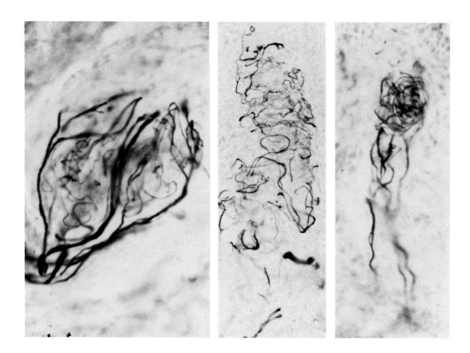

Fig. 7. Mucocutaneous end organs from the labia minora of a young woman. Winkelmann's silver impregnation technique.

long, slender rods, somewhat uniform in shape but with different-sized diameters. The other endings are less uniform. The lamellated structure of the corpuscle functions as a mechanical filter that restricts the range of response, and the nerve terminal also imposes its own restraints on the frequency of discharge. The structure of this receptor, then, determines its response. Perhaps the accessory structures of all nerve terminals determine their differential sensitivity (Loewenstein and Skalak, 1966).

VIII. Comments

All recognizable sensory receptors are basically similar. Even the inner bulb of the Vater-Pacini corpuscle, stripped of its surrounding laminar capsule and wound into a ball, has some resemblance to mucocutaneous end organs.

In man all cutaneous sensory endings contain nonspecific cholinesterase (Montagna, 1960; Winkelmann, 1960b; Cauna, 1968). When abundant,

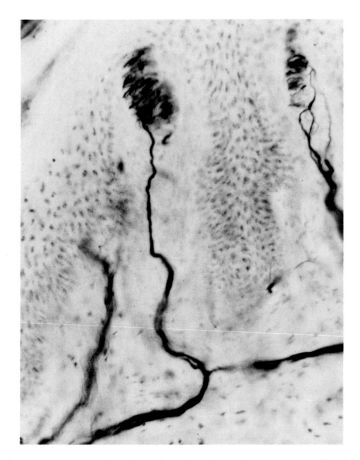

Fig. 8. Meissner corpuscles from the index finger of a young man. Winkelmann's silver impregnation technique.

as in the clitoris and labia minora, the nerves of the dermal neural network are also rich in specific cholinesterase, and most of the nerves around hair follicles contain specific cholinesterase. Around the upper third of the follicle, between the bulge and the entrance of the sebaceous ducts, the enzymes usually diffuse away from the nerve terminals, giving the impression of being surrounded by a collar or stockade. The central bulbs of Vater-Pacini corpuscles, Meissner corpuscles, and mucocutaneous end organs contain nonspecific cholinesterase, but little is known about the function of this enzyme.

Since skin possesses unlimited modalities of sensation, it is unlikely that each modality is exclusively subserved by an anatomically distinct end organ. Perhaps the encapsulated Meissner corpuscles and muco-cutaneous end organs subserve acute touch in glabrous skin; in hairy skin the same service may be performed by the highly complex arrangement of nerves around the hair follicle. Like touch, other sensations are probably not transmitted by anatomically specific end organs. Furthermore, if special nerve endings were tailored to perceive only certain sensations, they would have to be present in more than the few locations mentioned. Thus, the "different" end organs are probably modified according to their specific region and to the degree of local sensitivity, not according to the function they subserve. Perhaps the best examples of special adaptation are found in the receptors on the lips of the platypus, the snout of the opossum and mole, and the "lips" of the elephant's proboscis.

The only relationship to specific sensations for which there is any evidence is found in the Vater-Pacini corpuscle. Stimulation of this end organ gives the sensation of pressure but does not discharge a receptor potential until a definite threshold is reached (Gray and Sato, 1955). It is also the nerve end organ responsible for energy summation in cutaneous tissues (Verrillo, 1969).

References

Arthur, R. P., and W. B. Shelley. 1959. The innervation of human epidermis. *J. Invest. Dermatol.* **32**: 397–411.

Cauna, N. 1968. Light and electron microscopical structure of sensory end-organs in human skin. *In* "The Skin Senses" (D. R. Kenshalo, ed.), pp. 15–37. Thomas, Springfield, Illinois.

Fitzgerald, M. J. T. 1969. The innervation of the epidermis. *In* "The Skin Senses" (D. R. Kenshalo, ed.), pp. 61–83. Thomas, Springfield, Illinois.

Giacometti, L., and H. Machida. 1965. The skin of the mole *(Scapanus townsendii).* *Anat. Rec.* **153**: 31–40.

Gray, J. A. B., and M. Sato. 1955. Movement of sodium and other ions in Pacinian corpuscles. *J. Physiol. (London)* **129**: 594–607.

Iggo, A. 1969. Electrophysiological and histological studies of the cutaneous mechano-receptors. *In* "The Skin Senses" (D. R. Kenshalo, ed.), pp. 84–111. Thomas, Springfield, Illinois.

Koelle, G. B. 1970. Neurohumoral transmission and the autonomic nervous system. *In* "The Pharmacological Basis of Therapeutics" (L. S. Goodman and A. Gilman, eds.), pp. 402–465. Macmillan, New York.

Langley, J. N., and H. K. Anderson. 1894. On reflex action from sympathetic ganglia. *J. Physiol. (London)* **16**: 410–440.

Loewenstein, W. R., and R. Skalak. 1966. Mechanical transmission in a Pacinian corpuscle. An analysis and a theory. *J. Physiol. (London)* **182**: 346–378.

Miller, M. R., H. J. Ralston III, and M. Kasahara. 1958. The pattern of cutaneous innervation of the human hand. *Amer. J. Anat.* **102**:183–217.

Miller, M. R., H. J. Ralston III, and M. Kasahara. 1960. The patterns of cutaneous innervation of the human hand, foot and breast. *In* "Advances in Biology of Skin. Cutaneous Innervation" (W. Montagna, ed.), Vol. 1, pp. 1–47. Pergamon, New York.

Montagna, W. 1960. Cholinesterase in the cutaneous nerves of man. *In* "Advances in Biology of Skin. Cutaneous Innervation" (W. Montagna, ed.), Vol. 1, pp. 74–87. Pergamon Press, Oxford.

Montagna, W. 1970. Histology and cytochemistry of human skin. XXXV. The nipple and areola. *Brit. J. Dermatol.* **83**: 2–13.

Montagna, W., and D. M. Ford. 1969. Histology and cytochemistry of human skin. XXXIII. The eyelid. *Arch. Dermatol.* **100**: 328–335.

Montagna, W., and J. S. Yun. 1972. The glands of Montgomery. *Brit. J. Dermatol.* **86**: 126–133.

Munger, B. L. 1965. The intraepidermal innervation of the snout skin of the opossum. *J. Cell Biol.* **26**: 79–97.

Nafe, J. P. 1969. Neural correlates of sensation. *In* "The Skin Senses" (D. R. Kenshalo, ed.), pp. 5–14. Thomas, Springfield, Illinois.

Nafe, J. P., and K. S. Wagoner. 1937. The insensitivity of the cornea to heat and pain derived from high temperature. *Amer. J. Psychol.* **49**: 631–635.

Rothman, S. A. 1954. "Physiology and Biochemistry of the Skin." Univ. of Chicago Press, Chicago, Illinois.

Sinclair, D. 1967. "Cutaneous Sensations." Oxford Univ. Press, London.

Tamponi, M. 1940. "Strutture Nervose Della Cute Umana." Editore L. Cappelli, Bologna, Italy.

Tello, J. F. 1923–1924. Génése des terminaisons motrices et sensitives. II. Terminaisons dans les poils de la souris blanche. *Trav. lab. Rech. Biol. Univ. Madrid* **21**: 257–384.

Verrillo, R. T. 1969. A duplex mechanism of mechanoreception. *In* "The Skin Senses" (D. R. Kenshalo, ed.), pp. 139–159. Thomas, Springfield, Illinois.

Weddell, G., M. A. Cowan, E. Palmar, and S. Ramaswamy. 1965. Psoriatic skin. *A.M.A. Arch. Dermatol.* **91**: 252–266.

Weinstein, S. 1969. Intensive and extensive aspects of tactile sensitivity as a function of body part, sex, and laterality. *In* "The Skin Senses" (D. R. Kenshalo, ed.), pp. 195–222. Thomas, Springfield, Illinois.

Winkelmann, R. K. 1959a. The erogenous zones: Their nerve supply and its significance. *Proc. Staff Meet. Mayo Clin.* **34**: 39–47.

Winkelmann, R. K. 1959b. The innervation of a hair follicle. *Ann. N.Y. Acad. Sci.* **83**: 400–407.

Winkelmann, R. K. 1960a. "Nerve Endings in Normal and Pathologic Skin." Thomas, Springfield, Illinois.

Winkelmann, R. K. 1960b. Similarities in cutaneous nerve end-organs. *In* "Advances in Biology of Skin. Cutaneous Innervation" (W. Montagna, ed.), Vol. I, pp. 48–62. Pergamon, New York.

7

*The Pilary Apparatus**

I. Introduction

Bonnet (1892) and others called mammals "Trichozoa" (hair animals) and "Pilfera" (hair bearers) since only they have hairs. Little is known for certain about the phylogenetic origin of hair, nor has anyone established a satisfactory relation between hairs and homologous epidermal appendages in other vertebrates. Nevertheless we have advanced considerably in our knowledge of the types and distribution of hair, of its structure and cyclical patterns of growth, of its metabolism and biochemistry, and of how they are affected by genetic, hormonal, and environmental factors. Such a plethora of data on all these facets of hair is available that we shall content ourselves with summarizing much of it here, reserving the rest of the chapter for a more expanded treatment of less well-known subjects such as growth and differentiation.

To begin at the beginning, then, hairs consist of compactly cemented keratinized cells produced by follicles, i.e., epidermal appendages sunk

---

* With contributions by Giuseppe Moretti, Università di Genova, Genova; Kenji Adachi, Veterans Administration Hospital, Miami; Vicente Pecoraro, Catedra de Dermatologia, Rosario, Argentina; Enrico Rampini, Università di Genova, Genova; and Franco Crovato, Università di Genova, Genova.

into the dermis. These follicles, together with the sebaceous glands that grow from their sides, form what are known as pilosebaceous units.

Hairs have been divided into three main types according to size: overhairs, underhairs, and vellus hairs. Overhairs, which are usually long and coarse, variably rigid, and pigmented, include the spines and quills of some insectivores, rodents, and spiny anteaters; the manes of horses and lions, and the bristles of pigs. The soft, thin underhairs have a somewhat uniform length, are easily bent, and often wavy; they include the terminal hair of man, wool, and the fur of many other mammals (Danforth, 1925). Soft, silky, usually unpigmented vellus hairs—short, unmedullated, and very fine—are found on the seemingly hairless areas of the human body such as the forehead, eyelids, bald scalp, and most of the areas erroneously referred to as glabrous.

The wide range of color in human hair, from "coal black" to the lightest towhead, is due to the different amounts, distribution, and types of pigment in them, as well as to variations in surface structure, which cause light to be reflected in different ways.

In man, most body hairs grow to predictable lengths that are peculiar to the body area as well as to the individual. When they undergo periodic shedding, they are subsequently replaced by other hairs of about the same length. Hairs do not grow continuously since the follicles that produce them have precisely controlled periods of growth and rest, collectively known as hair-growth cycles. In mice and rats, hair grows synchronously in waves, dorsally and posteriorly from the throat region so that at any given time all the follicles in any given area are in about the same growth phase. In man and many other mammals, however, follicles grow and rest independently and as a result hair grows in asynchronous cycles. In some animals, all hairs are replaced seasonally: twice a year, almost simultaneously old hairs are shed and replaced by new ones. In elephant seal, hair is shed in patches held together by sheets of horny layer from the epidermis (Ling, 1965).

Depending on the species of animal and specific body areas, hairs usually emerge from the skin in predictable patterns such as rows or clusters; sometimes, however, it is difficult to distinguish specific patterns. As a rule, they are slanted away from the head, toward the tips of the appendages, but in some areas the direction changes. In the vertex, sacral regions, and umbilicus, for example, they grow in whorls.

The distribution of hairs per surface area of skin varies so much that accurate counts are impossible; hence reports vary widely. How, for example, can we be sure that we have accounted for all follicles when the hairs they produce vary from 4 to 100 μm in diameter and from less than a millimeter to a meter or more in length? Most of the reports of hair counts are so discrepant that they should be discounted. Furthermore, club hairs sometimes fall out of hair follicles; hence a trichogram could easily give a distorted picture of the follicle population.

About 90% of the hair population of the chest, trunk, shoulders, legs, and arms of men consists of terminal hairs compared with less than 35% in women (Danforth, 1925; Pinkus, 1927). About 18% of terminal hairs are reported to grow singly, the rest in groups or in clusters, but since they often grow in mixed patterns, these figures are irrelevant. The number of hairs and their distribution and patterns of growth are the same in men and women; the difference lies in the quantity of hair produced. In young adults regardless of sex, 90% of the 100,000 to 150,000 scalp hairs are said to be growing and 10% resting. If this were so, 10,000 to 15,000 follicles would be quiescent for about 100 days, and the club hairs they contain would be shed on an average of 100 per day (Orentreich, 1969). However, even these figures are not significant and the reported percentages of quiescent follicles are probably low (Barman *et al.*, 1965; Kligman, 1959, 1961). Actually, the ratio of growing to resting follicles varies considerably according to age and the regions of the body.

Scalp hair grows faster in women than in men (Myers and Hamilton, 1951; Saitoh *et al.*, 1969): discounting diurnal variations, almost 0.44 mm per day on the vertex and 0.39 mm on the temple (Saitoh *et al.*, 1969). The robust eyebrow hairs, which are flattened and curved, are about 1 cm long. Except in very blond and fair-skinned persons, the cilia (or eyelashes) are the darkest hairs on the body. Like the eyebrows, they are flattened and curved and their shafts vary between 20 and 120 μm in diameter. After a growth period of about 30 days, cilia take 15 days to become quiescent, and then rest for 105 days (Pinkus, 1927). Beard hairs, the coarsest of the body, grow 0.27 mm per day (Saitoh *et al.*, 1969) and if left uncut can grow as long as 30 cm. With aging, many of the generally nonpigmented vellus hairs in the ears develop into long, coarse terminal hairs, usually on the tragus and antitragus (Montagna and Giacometti, 1969). Axillary hair is characteristically curled or twisted around it axis and varies between 1 and 60 mm in length. Pubic hair in women grows in an inverted triangular pattern; in men it grows in a rhomboidal shape, the apex of the long axis pointing toward the umbilicus. However, these patterns vary widely and are interchangeable. Like axillary hairs, pubic hairs are almost always curled around their axis and are reportedly more luxuriant in men than in women.

This brief resumé has served to point up some of the discrepancies in the published reports on hair growth and population. The problem is that these studies have been limited and have often been guilty of errors of omission. When all the reliable data have been sifted, remarkably few *facts* can be harvested. A brief summary of these facts will conclude this introductory section on hair.

Even in extremely hirsute persons, human body hair has comparatively little protective value. Admittedly, of course, eyelashes and eyebrows, the hairs inside the external ears and nostrils, and those around the anogenital orifices have obviously useful functions, and scalp hair is often thick and long enough to afford some protection from the elements. Like the beard and mustache, axillary and pubic hairs provide ornament as well as erogenous stimulation. If frequent bathing is neglected or omitted, these hairs can become encrusted with the odoriferous substances produced by the glands in these areas.

Perhaps the most important function of human hair is its role as a sensory mechanism. All hair follicles are surrounded by sensory nerves which react to any pressure on the hair shaft. Highly specialized sensitive hairs (vibrissae) surround the eyes, lips, and muzzle of all mammals except man. Particularly large in nocturnal mammals, these "tactile" hairs emerge from follicles that abound in nerves and are encircled by an erectilelike tissue filled with blood. Human hair follicles, especially those on the face and anogenital areas, are generously supplied with nerves. This fact may explain why man's skin is probably more sensitive than that of any other mammal.

Unlike the hair follicles in the rest of the body, those in the human scalp sometimes grow uninterruptedly for years and thus produce hair of considerable length. Surprisingly, however, it is the scalp, with its dense and vigorous population of follicles, that becomes bald to variable degrees in all individuals. Contrary to common belief, baldness is not a disease but a systematic involution of hair follicles, which without any real diminution in numbers, eventually revert to their primitive embryonic state. This topic will be treated in detail at the end of this chapter.

The structural characteristics of hairs vary from one species to another, from one region to another, and even within any given bodily area of the same animal (Fig. 1). Hairs range the whole gamut from spiny and

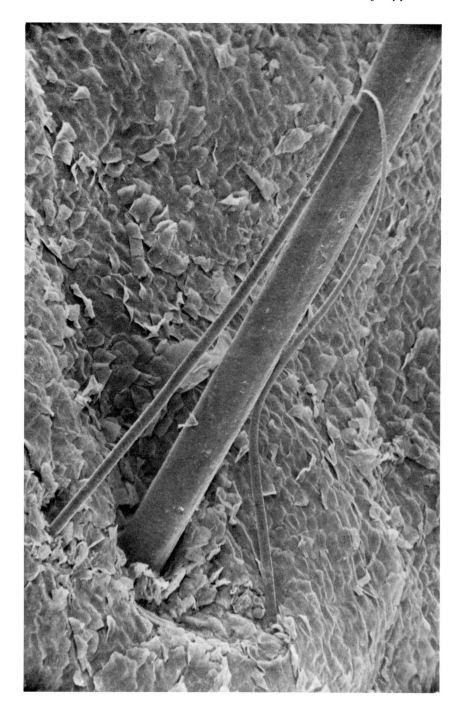

stiff to soft and silky; from extremely long to very short; from dense and woolly to sparse; and from variably colored to white. And even within the same region, hairs can differ in length, texture, and color. In man, the longer and coarser hairs are called terminal; the very short, soft, and often colorless ones are called vellus. In addition to their surface coat, many mammals have a thick, woolly underfur of variable length and color. Despite many attempts to do so, hairs cannot be precisely classified into types on the basis of length alone. The hairs of the scalp, for example, are characteristically longer than those of the trunk, but sometimes the terminal hairs on the trunk are almost as long as some hairs on the scalp, and both terminal and vellus hairs are found together in both regions.

When first formed in utero, all follicles produce lanugo hairs that resemble the vellus hairs of postnatal life. During their life cycle, some of these follicles later form terminal hairs and later still vellus hairs. In man, many of the follicles that earlier produced vellus-like hairs enlarge at puberty and produce coarse hairs. The axillae of infants and children have relatively large, arrested follicles that produce vellus hairs until adolescence. Some of these form a loose core of keratinous debris that sheds at the surface. Conversely, in baldness, the large scalp follicles revert to produce vellus hairs; this phenomenon has also been observed in nonhuman primates (Montagna *et al.*, 1966). When female hamsters are castrated at 2 weeks of age, the costovertebral melanotic spot has at best very fine, short black, vellus-type hairs. When as adults the same animals are treated with androgens, these follicles produce extremely large coarse hairs (Hamilton and Montagna, 1950). The follicles that produced bristles in the merino lamb produce wool in the adult sheep.

Just as hairs in the same body area differ in length, texture, and color, so do they vary widely in diameter and shape. Generally oval or round, hairs can be so flattened as to resemble ribbons. When such ribbonlike hairs are twisted along the longitudinal axis, they give the impression of varying greatly in diameter, but except in certain patho- logical conditions, the diameter remains the same. The diameter of human scalp hairs increases rapidly and uniformly during the first 3 or 4 years after birth, less rapidly during the next 6 years, and scarcely at all from 12 years on (Trotter, 1930; Trotter and Duggins,

Fig. 1. Group of three hairs from the scalp of a 35-year-old-man. The one to the right is a terminal, the middle is a stout terminal, and the one to the left is a full-grown vellus hair. Scanning electron micrograph under low power. (Courtesy of Dr. W. H. Fahrenbach.)

1950). Beard and other body hairs do not attain their full growth until much later in life.

What determines whether hairs will be straight, wavy, or crimped is not known, but the observation of Mercer (1953) and others about the crimping of wool fibers may shed some light. Apparently this crimping is caused by the deflection of the hair bulb and the eccentric disposition of the fiber in the follicle. The consequent asymmetric keratinization begins on the thin side of the inner sheath, proceeds across the fiber, and is completed on the thick side (Auber, 1952). These differences in the time sequence of keratinization of the two halves of crimped fibers are reflected in the differences in their staining properties. The disposition of these two elements (the ortho- and paracortex) in each crimp is always the same: one is on the inside of it, the other on the outside (Fraser and Rogers, 1953).

The hair shaft consists of an outer *cuticle*, a central *medulla*, which is lacking in many hairs, and a *cortex* between the two. The cuticle is a single layer of imbricated scales, with the free margins directed toward the tip of the hair. The number of cuticular scales per unit area usually drops slightly in the first year of life but later does not follow any specific trend. Although the cuticle is single layered, the individual cells are so elongated that many overlap at certain places and form stacks.

Cuticle scales are translucent and nonpigmented. Inside the follicle, the cuticle cells are interlocked with those of the inner root sheath, an arrangement that firmly anchors the hair in the follicle. In addition, the cuticle binds the cortex, which without this protection becomes frayed and split or falls apart, as can be seen at the ends of long scalp hairs (Fig. 2). A thin layer of lipid and carbohydrate is said to envelop the cuticle and to protect the hair from physical and chemical agents. Cuticle cells that completely surround the hair are called *coronal*; those that do not are called *imbricate*. They can be further divided according to size and shape into elongate, acuminate, ovate, and flattened (Fig. 3). The free edges of cuticle cells are either simple, dentate, or serrated. In coarse hairs, the free margins are not raised very high and the hairs cannot interlock and be woven into textiles. Such hairs have a high luster because their relatively unbroken surface reflects the light. In

Fig. 2.   Three views of the same scalp hair about 12 inches long seen under the scanning electron microscope: (A) the base of the hair has clean, undamaged cuticles; (B) in the middle portion, the cuticular scales are worn and flaking off; and (C) at the tip there are no cuticles and the hair cortex is evidently frayed. (Courtesy of Dr. W. H. Fahrenbach.)

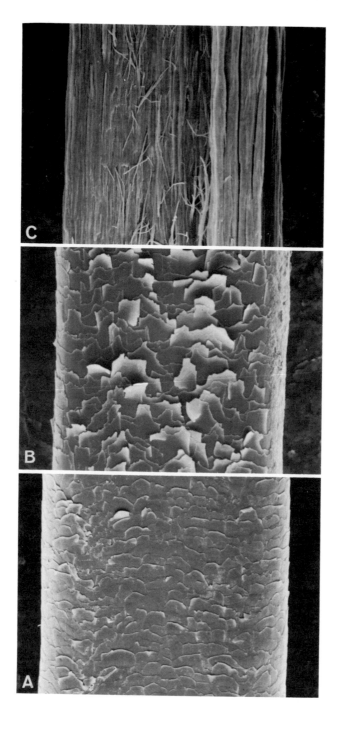

long fine hair such as wool, on the other hand, the free margins are raised so that the hairs interlock and are easily woven into textiles. These hairs are dull since the reflected light is broken by the raised edges.

In the center of all hair shafts, except the extremely fine ones, is a continuous or discontinuous (fragmented) medulla composed of large, loosely connected keratinized cells. Sometimes the medulla is only one or two cells in diameter and escapes detection by the ordinary light microscope. In coarse hairs the medulla is generally continuous, but may be fragmented. In man and many other mammals, the medulla forms only a small part of the hair; but in rodents, some carnivores, prosimians, and others, hairs consist mostly of large, vacuolated medulla cells held together by a thin tube of cortex. Large intra- and intercellular air spaces in the medulla greatly affect the color tones of the hair.

The number of medullated hairs at birth is slight but increases rapidly during the first 7 months. Then from 7 months to the second year, the percentage of medullated hairs decreases and a period of great irregularity follows. The percentage of medullated hairs is slightly higher in female infants than in male infants. From 2 to 6 years, the percentages are similar, and from 6 to 14 are higher in boys. Boys have more scalp hairs with broken or discontinuous medullas than girls, and Negro children have more medullated hairs than white children.

Most hairs are composed chiefly of cortex. In the cells of the cortex, melanin granules are aligned longitudinally in pigmented hairs; thus in the absence of pigment, hair appears dull white or translucent. Variable numbers of delicate air spaces called fusi are interspersed among the cemented, keratinized cells of the cortex. Before the hair is fully keratinized, these fusi are filled with fluid (Hausman, 1932, 1944); later, as the hair grows and dries out, the fluid is replaced by air.

Before closing this summary of the known facts on hair, we should reemphasize the enormous structural variation that characterizes not only the hairs of all individuals but also the hairs in any given body region of the same person. We do so primarily to discredit the common forensic practice of using hairs as means of personal identification. The time is long overdue for such unscientific practices to cease. If the diameter of hairs in the same group can vary so much (Fig. 1); if because of wear and damage the cuticle of the hairs from the same

Fig. 3.   Hairs of different animals photographed with the scanning electron microscope. These examples show different patterns of cuticle structure: (a) cape hunting dog; (b) red fox; (c) tiger; (d) opossum; (e) sloth bear; (f) kincajou. (Courtesy of Dr. W. H. Fahrenbach.)

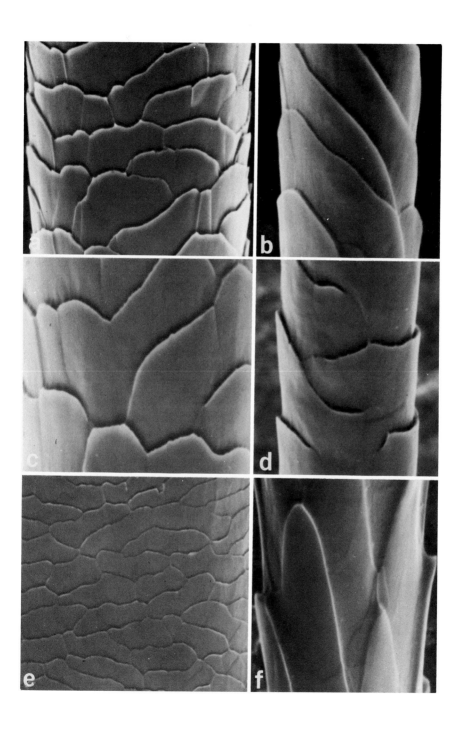

individual, or, indeed, from the same hair, can vary so much; and if the presence or absence of a medulla and the distribution of pigment can be so unpredictable, how can any one be certain that a hair or even any number of hairs belongs to this or that individual? Furthermore, human beings shed some hairs nearly all the time. Hence it would be nearly impossible to vacuum an office, home, or automobile carpet without recovering some pubic or other body hairs whose structures give no clue to the identity of the owner, except where unique congenital anomalies or unusual structural and color properties exist.

II. Development

The development of human hair follicles has been described in detail by Pinkus (1958) and others. More recently, studies with the electron microscope (Breathnach and Smith, 1968; Hashimoto, 1970) have confirmed these earlier histological descriptions. Hence only a summary treatment will be given here.

The first cutaneous appendages appear as primordial hair follicles on the eyebrows, upper lip, and chin at the end of the second and the early part of the third fetal months. Later on, follicles appear simultaneously on the forehead and scalp and in the fourth and fifth month spread throughout the body in cephalocaudal progression. The development, therefore, is asynchronous: by the time the scalp follicles have become fully formed and are producing hairs, those on the back, abdomen, and limbs are undergoing various stages of differentiation. Even within a specific body region the follicles are not equally differentiated. As the skin expands, new follicles are formed as primary or secondary follicles between already existing ones. Secondary follicles develop on each of the established follicles and form groups, usually of three.

The first indication that a hair follicle is about to form is a crowding of cells at spaced intervals in the basal layer of the still relatively undifferentiated bilayered epidermis. This cell crowding causes a slight bulge on the underside of the epidermis, the primitive hair germ (Fig. 4). Whether these first visible epidermal changes are preceded, accompanied, or followed by inductive influences from the underlying mesenchymal cells is not known. The consensus is that they precede the mesenchymal changes; however, in such specialized hairs as vibrissae, mesenchymal changes seem to occur first (Pinkus, 1958).

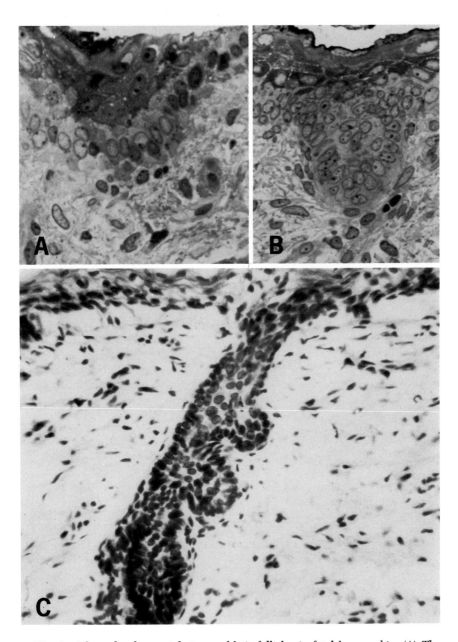

Fig. 4. Three developmental stages of hair follicles in fetal human skin. (A) The basal epidermal cells have become columnar and bulge slightly into the dermis. (B) A more advanced stage than in (A), showing the aggregation of mesenchymal cells at the base of the primary hair germ. These are the presumptive cells of the dermal papilla. (C) An elongated hair germ with the humps on one side. The upper one will develop into a sebaceous gland, the lower into the bulge proper.

The crowded basal cells and their nuclei, which are smaller and stain darker than the surrounding ones, elongate perpendicularly to the epidermis, and the hair germ bulges conspicuously into the dermis. A group of mesenchymal cells, the presumptive dermal papillae, gather at the bottom of the hair germ (Fig. 4). The anterior face of the hair germ cells is at a right angle to the basal layer of the epidermis, whereas the posterior one slopes gradually. As the hair germ continues to grow into the dermis, it forms a column or peg of cells slanted anteroposteriorly. The large columnar cells at the periphery of this peg are arranged radially to the center, whereas those in the center are aligned longitudinally. The free end of the peg becomes progressively clavate, and the middle of its bulbous end indents and gradually grows around the dermal papilla. Glycogen is abundant first in the central, later in the peripheral cells of the follicle.

At this stage, two solid epithelial swellings begin to grow at the posterior side of the follicle. At first, the lower one, the "Wulst" (Fig. 4), is the larger of the two, but later it becomes relatively smaller. The upper one, the anlage of the sebaceous gland, contains central cells with a foamy cytoplasm that indicates the synthesis and accumulation of lipid. Subsequent differentiation of sebaceous cells proceeds centrifugally.

Just below the sebaceous gland, mesenchymal cells arrange themselves linearly in a slender band parallel to the posterior border of the follicle. These cells gradually extend downward toward the bulge and give rise to the arrector pili muscle. The follicle is still a solid epithelial structure surrounded by a mesenchymal sheath.

Once all the component parts of the pilosebaceous units have been established, differentiation ensues. The bulb, i.e., the expanded distal part of the follicle, encloses the dermal papilla, which remains attached by a narrow stalk to a *basal plate* of dermal cells. In earlier development, visible pigment cells are found throughout the bulb, but later they are limited almost entirely to about the upper two-thirds of the bulb, above the dermal papilla. The lower bulb contains mostly undifferentiated glycogen-free matrix cells and an occasional pigment cell.

The cells that give rise to the inner root sheath differentiate first and can be traced to the bulb cells around the dermal papilla, where they

Fig. 5. The bulb of four hair follicles from the scalp separated from the dermis with the potassium bromide method. Note particularly the pore through which a stalk of connective tissue connects the dermal papilla inside the bulb and the basal plate.

align themselves longitudinally and acquire trichohyalin granules. Later these cells form the hair cone which is composed of several layers of inner root sheath cells that contain trichohyalin granules of various sizes (Robins and Breathnach, 1969, 1970; Hashimoto, 1970). Just below the advancing hair cone, cortical cells begin to differentiate. By the time the hair cone has reached the isthmus and infundibulum, the intraepidermal hair canal has already been formed by the keratinization of the canal wall. The tip of the emerging hair consists of pigment-free cortical cells without a medulla.

Hair follicles are surrounded by a hyalin or glassy membrane (the basal lamina) and a two- or three-layered sheath of connective tissue made up of collagen fibers and fibrocytes roughly arranged orthogonally.

The connective tissue cells at an angle to the bulb become elongated, increase in size, and differentiate into the smooth muscle fibers of the arrectores pilorum muscles. Earlier in differentiation these cells have many of the characteristics of fibroblasts, with well-ordered arrays of rough endoplasmic reticulum and a prominent Golgi complex. Later they form myofilaments that resemble smooth muscle cells. At each end, the muscle band interdigitates with elastic fibers which anchor the muscle fibers to the bulge of the hair follicle, the insertion, and to its origin, the superficial dermis (Bell, 1969).

After a period of growth, the hair above the level of the bulb becomes clubbed, and most of the follicle below the bulge is resorbed; this is called the period of quiescence.

All of the sequences involved in the formation of the first follicle—the production of the first hair and the conversion from active to quiescent follicles—begin as early as 4½ to 5 months of gestation and are the templates of follicular behavior in postnatal life.

III. Hair Cycle

As we have just seen, hair grows cyclically, with alternating periods of growth and quiescence. During the growth phase, follicles are said to be in anagen (Fig. 6); during the subsequent resting phase, they contain club hairs and are in telogen (Fig. 8). The transition period, when follicles are reorganizing into an inactive quasi embryonal state, is called catagen (Fig. 7). When a quiescent follicle grows again, it forms a new hair which either dislodges the club hair (which is then shed) or grows alongside it.

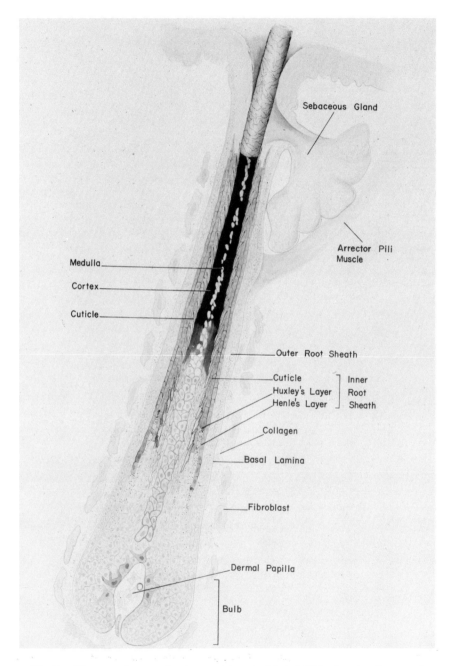

Fig. 6. Diagram of a growing (anagen) terminal hair follicle showing the complex, relative position of the various layers.

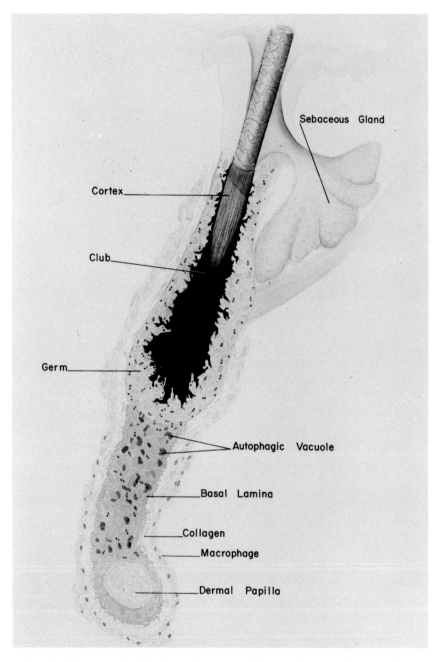

Fig. 7. Diagram of a follicle in the transitional stage (catagen) showing the corrugation of the basal lamina and the resorption of the epithelial and connective tissue elements.

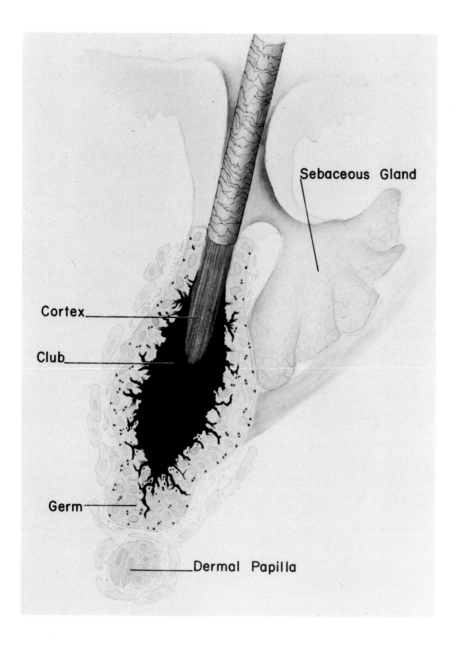

Fig. 8. Diagram of a quiescent (telogen) terminal hair follicle showing the relationship between the club and the germ. The dermal papilla is now outside the follicle.

This pattern of growth and rest and the rate of growth vary from species to species and sometimes in the same animal from one part of the body to another. In mammals that molt annually, the club hairs are shed more or less at once, at a certain time of the year when they are replaced by new hairs. In rats and mice, frequent growth periods occur in waves that move anteroposteriorly and ventrodorsally over the animal's body. In any specific area of the growth wave, all of the follicles are more or less in the same growth phase (Butcher, 1934, 1951; Chase and Eaton, 1952; Chase, 1954). The total growing period of each follicle is 17 to 20 days in mice and 21 to 26 days in rats; in both, hairs grow about 1 mm a day. In man, each follicle has its own growth cycle that is largely independent of others nearby. Surprisingly, not much is known about hair growth in man mostly because the investigation of this phenomenon is tedious and unexciting and has therefore failed to capture the interest of investigators. Human hair grows about 0.3 to 0.4 mm per day (Myers and Hamilton, 1951; Saitoh *et al.*, 1969), but this rate is affected by many variables. In man quiescent periods range from extremely short in the scalp to long in the general body surface.

If the changes that occur during hair growth cycles are to be appreciated, the structure of the follicle during each of the growth stages must be understood.

A. *Anagen*

During anagen follicles have a bulbous base of mitotically active, pluripotential matrix cells that produce the cells of the medulla, cortex, cuticle, and inner root sheath (Fig. 6). Once the cells of the medulla, cortex, and cuticle are completely keratinized, they form the hair proper whereas the cells of the inner root sheath are shed into the pilary canal at the level of the sebaceous gland. The outer root sheath on the outside of the follicle is multilayered in the upper part of the follicle and continuous with the wall of the pilary canal; it is much thinner around the lower end of the bulb. The matrix, medulla, cortex, and inner and outer root sheaths of the follicle are derived from the ectoderm, the dermal papilla, and the connective tissue sheath from the dermis. The dermal papilla is enclosed by the bulb and attached by a narrow stalk to a connective tissue basal plate (Figs. 5, 6, 9).

1. THE BULB

The thickest part of a follicle is its onion-shaped bulb (Figs. 5, 6, 9). An imaginary line drawn across the widest diameter of the dermal papilla and dividing the bulb into two distinct regions has been called "the critical level" by Auber (1952) (Fig. 10). The part below this

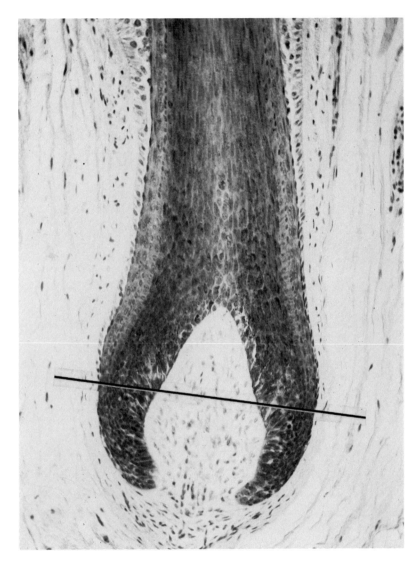

Fig. 9. Longitudinal section through the bulb of a scalp terminal hair follicle. The black line indicates the critical level which separates the upper and lower bulb.

level consists mostly of rapidly dividing undifferentiated cells (Fig. 11). Most mitotically active cells are located here, but studies with colchicine (Kligman, 1959) and with tritiated thymidine (Epstein and Maibach, 1969) showed some mitotic activity in the upper part of the bulb as well. The cells in the undifferentiated matrix have a high turnover; for

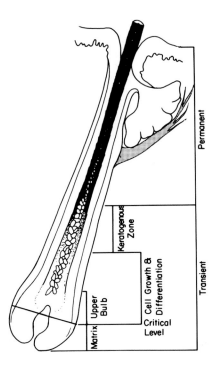

Fig. 10. Diagram of a growing hair follicle indicating the various regions discussed in the text.

example, in mice injected with Colcemid, every matrix cell divided about every 13 hours (Bullough and Laurence, 1958).

As we mentioned at the beginning of this section, the growth and differentiation of the matrix cells give rise to all the cells of the hair shaft (medulla, cortex, and cuticle) and of the inner root sheath (Figs. 6, 10). From the matrix, cells move up in rows to the upper bulb, the *preelongation region,* and elongate vertically (Auber, 1952). In the *elongation region* proper, where the bulb constricts, the cells are long and thin with very sharp boundaries, the result of fibrils piling up against them. Farther up, in the cortical *prekeratinization region,* the fibrils are coarse and numerous and stain with basic dyes. Immediately above, in the *keratogenous zone,* the cells are hyalinized, and distinct fibrils can be seen only with the electron microscope. These four regions have been given other, less well-known names (Mercer, 1949).

Matrix cells have a characteristic ultrastructure (Birbeck and Mercer, 1957a; Parakkal, 1969a; Roth, 1967). A spherical nucleus occupies almost the entire cell; the scant cytoplasm has numerous ribosomes and some

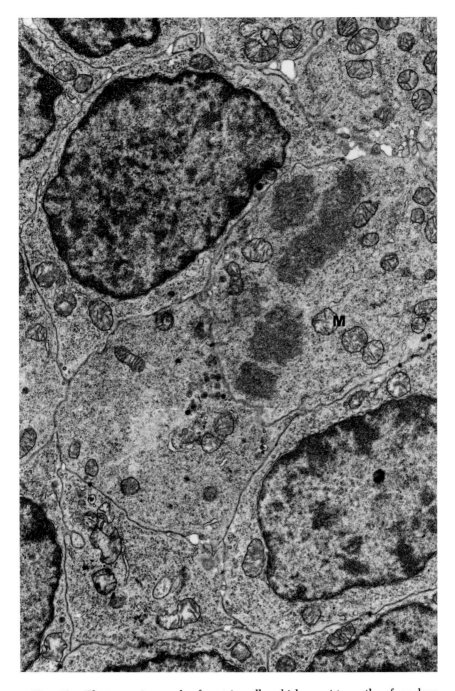

Fig. 11. Electron micrograph of matrix cells which consist mostly of nucleus. The available cytoplasm contains ribosomes and mitochondria. The plasma membranes have only a few desmosomes. The cell in the middle of the field is undergoing mitosis (M). × 12,000.

mitochondria (Fig. 11). A few profiles of rough-surfaced endoplasmic reticulum, a compact Golgi zone usually close to the nucleus, and some arrays of cisternae and vesicles make up the rest of the cell. The membranes of adjacent matrix cells lie close to one another, but occasionally small, empty, intercellular spaces lie between. At intervals between the cells, these membranes are modified to desmosomes, and filaments from the inner aspects of these desmosomes extend into the cytoplasm of the respective cells.

2. THE MEDULLA

The medulla originates from the matrix cells at the apex of the dermal papilla. As these cells move upward, they begin to differentiate; the profiles of rough-surfaced endoplasmic reticulum increase slightly and the Golgi elements become more prominent; and the cells begin to develop dense spherical, medullary granules about 300 to 500 Å in diameter (Fig. 12). Farther up in the follicle, granules, now several micrometers in diameter, fill most of the cytoplasm; these granules are never restricted by a membrane (Parakkal, 1969a; Roth and Helwig, 1964; Puccinelli et al., 1967). Medullary cells also produce some filaments which are aggregated into bundles distributed randomly in the cytoplasm where numerous spherical, membrane-limited, apparently empty vesicles 1 to 2 μm in diameter also appear. At this stage of differentiation, large glycogen granules are aggregated near the nucleus, and multiple rows of them in concentric whorls form glycogen-membrane complexes (Fig. 12). Higher up in the follicle and at advanced stages of differentiation, the nucleus and other cytoplasmic organelles begin to disintegrate: the mitochondria become characteristically swollen, the cristae lose their orientation, and the density of the matrix decreases. Later the mitochondria become vacuolated like empty vesicles.

By the time the medullary cells are fully differentiated, the small vesicles of the earlier stages have coalesced into larger vacuoles, and the once discrete medullary granules have fused into an amorphous mass concentrated at the cell periphery with random bundles of filaments scattered throughout. When almost fully formed, medullary cells are wedged between projections of cortical cells, and mature cells are

---

Fig. 12. A row of medullary cells aligned one above the other. The most characteristic features of these cells are medullary granules of different sizes and empty vacuoles. Glycogen is often arranged in concentric patterns as in the cell in the center of the field. × 14,000.

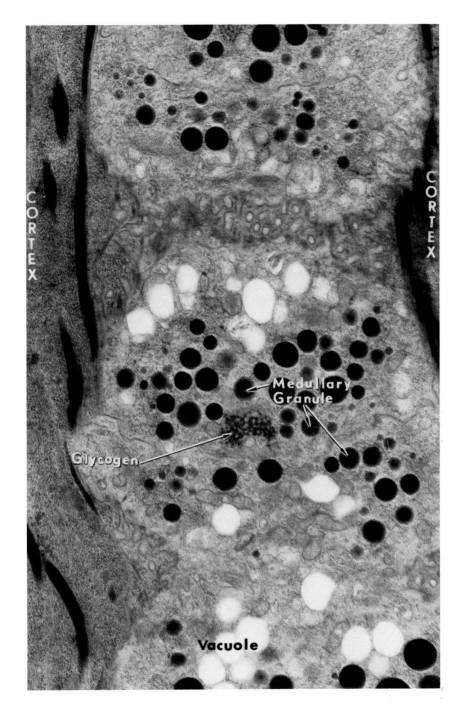

arranged along the core of the hair in a ringlike fashion with spaces between them.

Auber (1952) believed that the amorphous medullary granules are trichohyalin, and Mercer (1961) believed that they undergo a fibrous transformation similar to that of the trichohyalin granules of the inner root sheath. Although the amorphous protein that is the major component of medullary cells is not like the fibrous protein of the inner root sheath, the cells of both the medulla and of the inner root sheath contain citrulline. Thus there is some chemical similarity between the two (Rogers, 1964a,b, 1969); Harding and Rogers, 1971).

3. THE CORTEX

Human hair is composed mostly of cortex, a compact aggregate of cemented fusiform, keratinized cells. As the presumptive cortical cells move up from the bulb, they become progressively fusiform, elongated, and keratinized. The first detectable sign of differentiation is the appearance of wispy filaments 70 to 80 Å in diameter which are clumped into loose bundles (Birbeck and Mercer, 1957a) and later arranged symmetrically at the periphery of adjacent cortical cells. Later still, the filament bundles become scattered throughout the cytoplasm. Cross sections of the filaments show them to be unstained circular profiles surrounded by a stained matrix, like islands surrounded by ribosomal particles and some mitochondria. As they mature, the filament bundles gradually fill the whole cell, and the ribosomes and mitochondria disintegrate and are eliminated. Fully keratinized cortical cells are so packed with filaments that they appear to be scarcely stained. These filaments are surrounded by a dense matrix which gives the cells their typical "keratin pattern" (Orfanos and Ruska, 1970) (Fig. 13).

Because thiol groups have an affinity for osmium, Birbeck and Mercer (1957a) concluded that the matrix is sulfur-rich and consists of high sulfur-proteins. However, various other stains such as uranyl acetate, potassium permanganate, and silver nitrate seem to be localized in the matrix as well (Fraser *et al.*, 1972). Even though these stains are not specific for thiol groups, the matrix is generally believed to be rich in cystine (Lundgren and Ward, 1963).

The cortex of wool fibers has a continuous bilaterality running from the root to the tip (Horio and Kondo, 1953; Mercer, 1953; Chapman

Fig. 13. Fully keratinized cortical cells showing the characteristic keratin pattern with the arrangement of electron light filament surrounded by dense matrix. (Courtesy of Dr. C. Orfanos.) × 120,000.

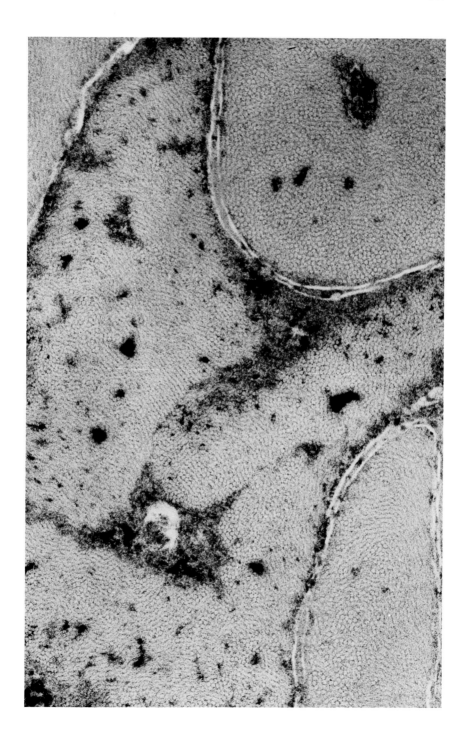

and Gemmell, 1971). Cross sections of whole fibers stained with Janus green or other basic dyes show about half of each fiber stained and the other half unstained. The stained (basophilic) segment is the orthocortex, the nonstained part the paracortex. Both the paracortex and the orthocortex maintain a constant spatial relationship to the natural crimp or wave of the fiber, the paracortex being located on the inner side of the curvature of the crimp and the orthocortex on the outside. In wool, which has a slight crimp, the distribution of the two cortices is variable and sometimes the cells of the paracortex surround the orthocortex completely (Fraser and MacRae, 1956; Fraser and Rogers, 1965).

During differentiation, cell membranes also undergo change (Birbeck and Mercer, 1956). In differentiating cortical cells, the plasma membranes are about 80 to 90 Å thick and about 100 Å from adjacent ones, which they contact by means of scattered desmosomes. Later in differentiation, the dilated spaces between adjacent cell membranes are filled with an amorphous substance.

In the heavily medullated hair of some animals, the cortical cells next to the medulla project into it. Seen in longitudinal section, these projections give the medulla a ladderlike appearance, the "rungs" of opposite cortical cells sometimes meeting and thus causing the medulla to become discontinuous.

Morphologically, cortical and epidermal cells appear to mature in a similar way, but there are differences. For example, when first formed, the filaments of the cortex are already arranged into bundles and have the "keratin pattern" that characterizes fully cornified epidermal cells. Moreover, the nuclei of keratinized cortical cells are not visibly lost like those of the epidermis even though there appears to be no DNA. Unlike epidermal cells, which form and discharge the membrane-coating granules that compose their intercellular matrix, cortical cells contain no membrane-coating granules but do possess an intercellular amorphous substance. Finally, there are no keratohyalin granules in the cortical cells.

4. THE HAIR CUTICLE

The cells of the hair cuticle move upward in a single row from the matrix and can be recognized in the upper part of the bulb. Midway up the bulb, these cells are cuboidal with strongly basophilic cytoplasm stippled with darker basophilic granules; unlike the cells of the cortex, however, these contain no melanin. In the upper part of the bulb and for a short distance above, the cells are tall and columnar and the long axis is oriented radially; higher up, their outer edges begin to tip upward. This tipping automatically causes the tall, greatly attenuated cells to become imbricated (Birbeck and Mercer, 1957b), and the change from horizontal to vertical is accompanied by a flattening of cuticle cells, which in vertical sections resemble pointed scales (Figs. 14, 15). The reorientation is complete below the midway mark of the follicle. As cuticle cells become hyalinized, their nuclei disappear and the cells adhere to the cortex. Such an imbricated pattern results from a series of morphological movements. As cuticle cells arise from the matrix, they move up in the bulb, increase in volume, and expand horizontally. They are more firmly attached to the cortex on the axial side than laterally to the cuticle cells of the inner sheath. Since the latter appears to grow at the same rate as the hair, its cells probably increase in volume at a faster rate than those of the hair cuticle and in the reorientation process pull the lateral borders of the cuticle cells of the cortex upward. This explanation, however, is conjectural since morphogenetic potentials inherent in the cells themselves could bring about these movements without the aid of extraneous physical factors.

The staining properties of the cuticle cells differ from those in the surrounding layers. When thin and reoriented, the cuticle cells of the cortex from just above the bulb stain a brilliant green with red nuclei with Altmann's acid fuchsin-methyl green. At the level of the sebaceous ducts, cuticle cells stain red. Parenthetically, medullary cells stain green with this dye until they are cornified in the upper half of the follicle. Although totally nonspecific, this staining reaction indicates that the cuticle cells are not fully keratinized until they reach the upper third of the follicle. The cells of the inner sheath can be recognized by their brilliant fuchsinophilia once trichohyalin granules appear in them.

The cuticle cells are the last of the cell lines in the hair follicle to differentiate. A cross section of the bulb that cuts through the upper part of the dermal papilla shows filament bundles in the cortical cells and filaments and trichohyalin granules in the inner root sheath cells. At this level, the cuticle cells between them contain no differentiation product.

Farther up in the follicle and sometimes in association with the vesicles of the Golgi zone, small granules about 300 to 400 Å in diameter (Fig. 15) are located above the flattened nuclei of cuticle cells. These granules increase in number and grow to about 5000 Å in diameter and aggregate next to the cell membrane that faces the inner root sheath (Fig. 15). Still farther up, they coalesce into a compact mass that fills the flattened overlapping cuticle cells (Birbeck and Mercer, 1957b; Happey and Johnson, 1962).

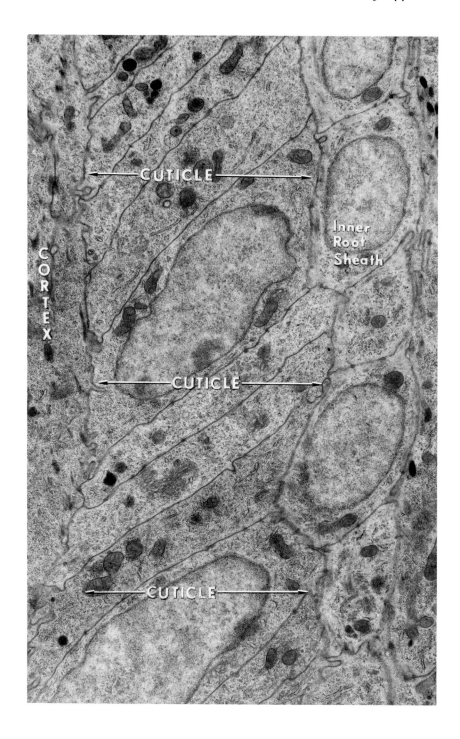

The changes in the membranes of the cuticle cells during maturation are similar to those in the cortical cells. In early differentiation, the 90-Å thick membranes are separated by an intercellular space about 80 Å wide. By the time the cuticular granules become aligned, the intercellular space is about 300 Å wide and is filled with amorphous material.

Completely keratinized cuticular cells have a distinct outer (*exocuticle*) and inner (*endocuticle*) layer. The outside of the exocuticle, the α-layer (Lagermalm, 1954) has a high affinity for osmium tetroxide and silver nitrate and can be easily demonstrated after reduction (Rogers, 1959; Kassenbeck, 1961; Dobb and Sikorski, 1961); thus the α-layer must be rich in cystine.

5. THE INNER ROOT SHEATH

Between the hair shaft and the outer sheath is the inner root sheath, composed of cuticle next to the hair shaft, and Huxley's and Henle's layers in the center and outside, respectively (Fig. 16). All three layers arise from the peripheral and lateral mass of matrix cells. Although their rate of differentiation differs, they undergo identical sequences of changes (Birbeck and Mercer, 1957c; Roth and Helwig, 1964) which begin in Henle's layer, then advance to Huxley's layer and the cuticle. Keratinization, on the other hand, begins in Henle's layer, proceeds to the cuticle, and ends in Huxley's layer.

During early differentiation, filaments 70 to 80 Å thick develop in the cells of the inner root sheath with small trichohyalin granules that appear as condensations here and there on the filaments. As the cells move upward, the filaments increase in number and eventually fill the cytoplasm; simultaneously, the trichohyalin granules become more numerous and larger and occupy large areas of the cell (Fig. 17). Seen under the electron microscope at high magnification, the granules appear as complex structures of fibrous and amorphous substances not limited by a membrane, and the filaments appear to pass through them. These granules are aggregates of spicular particles closely associated with the 70-Å thick filaments (Steinert et al., (1971). When differentiation is nearly complete, the nuclei, mitochondria, and ribosomes disintegrate and are partially eliminated from the cell so that fully mature cells have no discernible trichohyalin granules and are filled instead with a filamentous matrix complex.

Early in differentiation, the membranes of inner root sheath cells are

Fig. 14.   Presumptive cuticle cells slanting toward the cortex as they differentiate. At this stage, differentiation products are not evident. × 8700.

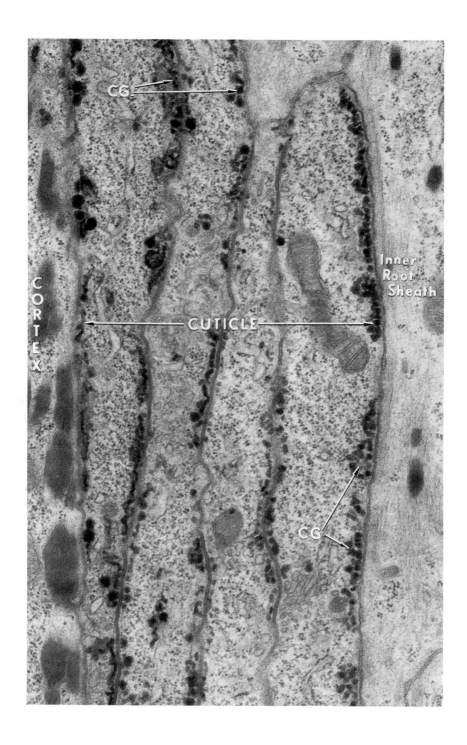

separated by narrow intercellular spaces with only a few scattered desmosomes which later increase in number. By the time the cells are nearly mature, the membranes are about 150 Å thick, and the widened intercellular spaces between are filled with an amorphous material.

In the follicles of mice, rats, and sheep, and a few other mammals except man, the inner root sheath becomes corrugated into horizontal folds at the bottom of the pilary canal near the entrance of the sebaceous glands (Gemmell and Chapman, 1971).

The inner root sheath is ultimately discharged inside the pilosebaceous canal. Detailed explanations for this disappearance are lacking, but it could be due to chemical changes that culminate in reabsorption, dissipation, or exfoliation inside the pilary canal. The main function of the inner root sheath may be to shape or contour the hair shaft since as the first structure to keratinize in the follicle, it forms the outer boundaries through which the developing hair shaft is funneled during its formation.

The keratohyalin of the epidermis and the trichohyalin of the inner root sheath appear to be morphologically but not biochemically identical; only trichohyalin contains an arginine-rich protein (Rogers, 1963).

6. THE OUTER ROOT SHEATH

In large follicles, the thickness of the outer root sheath generally varies and the hair is eccentrically located. In human hair, the outer sheath of most follicles has some degree of swelling on the side of the bulge, but in sheep, both inner and outer sheaths have pronounced lateral swellings (Auber, 1952). When the bulb is bent or curved, as in the follicles of Negroes, the outer sheath is thicker on the convex than on the concave side.

The outer sheath can be divided into three distinct segments: the part above the opening of the duct of sebaceous glands, the pilary canal, is continuous with the superficial epidermis; the middle part extends from the orifice of the sebaceous gland down to the neck of the bulb; and the lower part surrounds the bulb. Around the bulb, the outer root sheath is only 1 or 2 cells thick (Fig. 9); the outer ones are slightly elongated, those on the inner surface are greatly flattened. Just above the bulb, the sheath has three layers and a third of the way up becomes

---

Fig. 15. Partially differentiated cuticular cells so imbricated as to give a false impression of being multilayered. The majority of the granules (CG) are aligned near the plasma membrane facing the inner root sheath. × 3000.

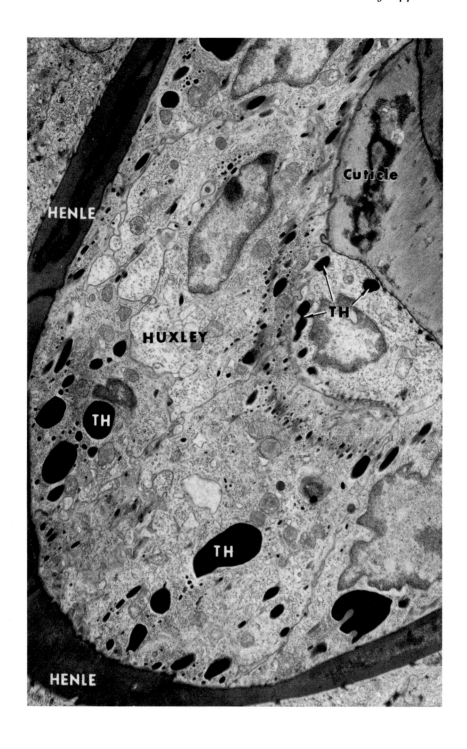

gradually pluristratified and thickest. The tall columnar cells at the outer periphery are oriented perpendicular to the axis of the follicle. As they grow centripetally, the cells are increasingly flattened along the long axis of the follicle. Below the level of the sebaceous gland, they store large amounts of glycogen (Figs. 18, 19) and in histological preparations appear vacuolated and distorted. Lesser amounts of glycogen are found in the cells around and immediately above the bulb.

Whether the cells of the outer sheath actively move upward along with those of the inner sheath is not known. In mice and merino sheep, the outer root sheath does move upward. In mice, mitotic figures are numerous in the lower part of the outer sheath, and the keratinized cells in the upper part are probably sloughed off into the pilary canal below the orifice of the sebaceous gland duct (Straile, 1962; Gemmell and Chapman, 1971).

Even though the outer root sheath and the epidermis are continuous and basically similar in proliferation, keratinization, and exfoliation, they differ nonetheless. For example, in the epidermis keratohyalin granules are irregular and large but rounded and small in the pilary canal. In the outer sheath, an increasing number of membrane-coating granules (MCG) are apparently discharged into the intercellular spaces at the junction of the granular and horny layers. Moreover, the horny cells do not seem to be tightly attached to each other and are easily exfoliated. As the pilary canal reaches the sebaceous orifice, keratohyalin granules and MCG decrease in size and number (Knutson, 1973). (The changes that take place in the pilary canal during acne will be dealt with in the chapter on sebaceous glands.)

7. THE DERMAL PAPILLA

The major connective tissue inside the bulb is the dermal papilla (Fig. 9). When various types of hair follicles are compared, the size and shape of their papillae and those of the bulbs can be correlated (Auber, 1952; Pinkus, 1958; Durward and Rudall, 1958). Moreover, the follicles that produce large hairs have large bulbs and dermal papillae, and vice versa (Schinckel, 1961). In a meticulous study of the matrices of the bulb and the dermal papilla in scalp follicles, Van Scott and Ekel (1958) found the volume of the matrix to be about ten

---

Fig. 16. Henle's and Huxley's layers and the cuticle of the inner root sheath. The cells in Henle's layer and the cuticle are completely differentiated and filled with a filamentous matrix complex. The cells of Huxley's layer contain trichohyalin granules (TH). × 7500.

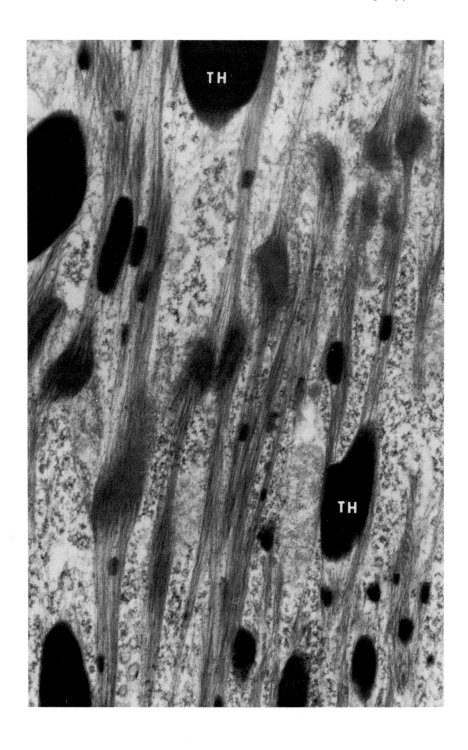

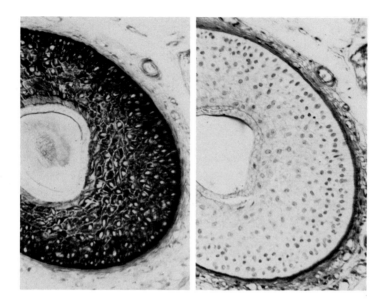

Fig. 18. Consecutive transverse section through a follicle treated with PAS technique. In the specimen to the left, the cells are full of glycogen; in that to the right, which has been previously digested with diastase, glycogen has been eliminated.

times that of the dermal papilla. They also found that the ratio between the number of cells in the papilla and those dividing in the matrix was about nine to one. Hair follicles with more than one medulla have a papilla split distally with as many apices as there are medullae. In the pig, the onion-shaped papilla terminates in a long central pinnacle surrounded by rows of lesser spires. When hairs in a single follicle occur in duplicate, there is a corresponding duplication of the dermal papilla. Such close correspondence between subdivisions of the papilla and the number of hairs formed in a single follicle focuses attention on the dermal papilla, each lobe of which must induce hair formation from a common matrix.

In growing follicles, the dermal papilla is attached to a basal plate of connective tissue by a narrow stalk. The vascularity of the papilla depends on its size; the small ones of small follicles sometimes have no visible vascular supply, whereas larger ones have variable numbers

Fig. 17. Enlarged view of different-sized trichohyalin granules (TH) and the filaments with which they are associated. × 43,000.

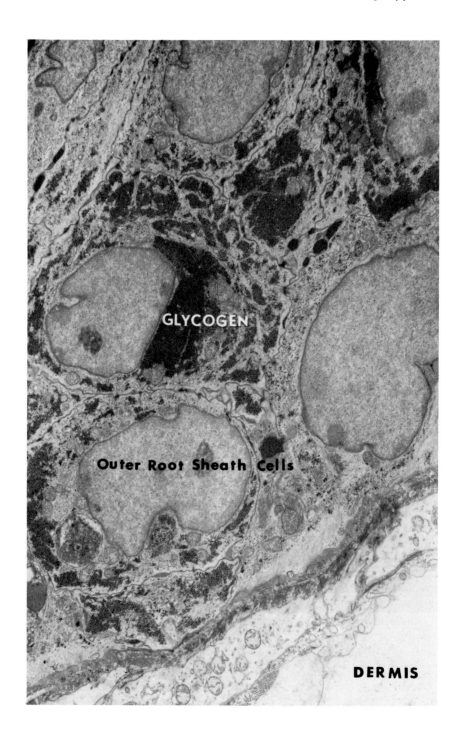

of blood vessels (Durward and Rudall, 1949; Montagna, 1962). Fibro-blasts are the most numerous cells in papillae (Fig. 20). The noncellular part consists mostly of ground substance through which float wisps of collagen fibers. A basement membrane (basal lamina) about 500 Å thick separates the papilla from the cells of the bulb. The fibroblasts have well-developed rough-surfaced endoplasmic reticulum which consists of numerous profiles of membranes studded with ribosomes. A large and prominent Golgi complex is near the nucleus (Fig. 20).

Most biologists are convinced that the dermal papilla induces the formation of hair follicles. However, experiments by Oliver (1967a) on rat vibrissae have shown that if the dermal papilla is removed, a new one may be formed. In addition, transplanted into the bases of the superficial or distal half of vibrissa follicles, the papilla induces the growth of whiskers (Oliver, 1967b). What light these experiments shed on the induction of other follicle growth is not known.

8. THE CONNECTIVE TISSUE COMPONENTS

The two connective tissue elements that surround hair follicles are a noncellular hyalin membrane (vitreous, glassy, or basement membrane) of varied thickness, and a connective tissue sheath of variable numbers of layers in the different regions of growing follicles. The hyalin mem-brane, which stains pale pink with the PAS and Masson's triacid methods, is thick and conspicuous around the lower third of the follicle, but thin in the upper part. It is actually the same 500-Å-thick basal lamina that separates the follicular cells from the connective tissue sheath (Rogers, 1957; Parakkal, 1969b). In the lower third of the follicle, two layers of collagenous fibers surround the basal lamina; the fibers of the inner layer are oriented parallel to the long axis of the follicle, those of the outer layer are perpendicular to it. Outside these orthog-onally arranged collagen fiber layers, fibroblasts and other connective tissue cells can be seen where the connective tissue blends into that of the dermis.

B. Catagen

1. EPITHELIAL CHANGES

Some animals have follicles that grow almost continually. For example, some follicles in the human scalp produce hair for 8 years or more and

Fig. 19. Accumulations of glycogen granules in the cells of the outer root sheath. × 6400.

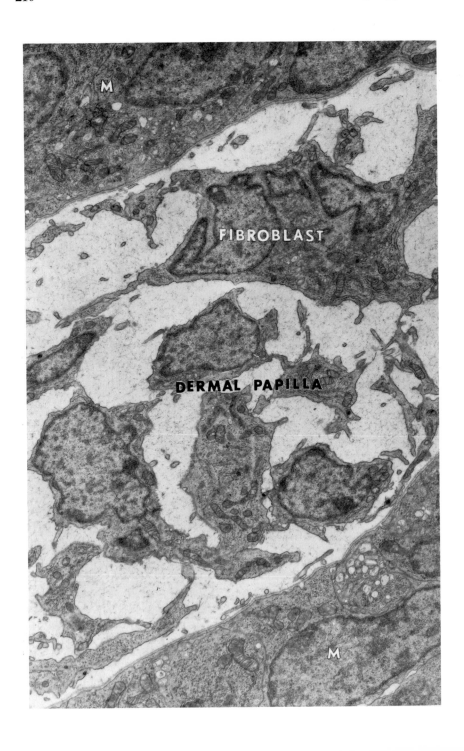

then rest for variable periods. In sheep, short resting periods are followed by long growth periods.

Follicular growth comes to a halt gradually and in orderly fashion through a phase called catagen, during which the follicles undergo striking morphological and functional changes (Kligman, 1959; Montagna, 1962) (Fig. 21). First, they regress to about ⅓ their growing length and form a brushlike club hair. When this process is complete, they rest or go into telogen. The club hair formed during catagen normally remains attached to what remains of the follicle, but it may fall out if the period of telogen is long.

The onset of catagen is heralded by the cessation of melanogenesis in the melanocytes of the bulb. They contract the dendrites and no more pigment granules are transferred to the still-growing cortex; the last part of each hair, then, is white. Cell proliferation in the matrix decreases and finally ceases. The cells in the upper part of the bulb, however, continue to move up and to differentiate into a hair shaft, consisting only of cortex and inner root sheath, until all that remains of the bulb is a flimsy, disorganized column of cells. The hair club and the nonpigmented shaft above it consist only of cortical cells. The last step in follicular differentiation is the formation of the club and the keratinous rootlets that attach it to what remains of the follicle (Fig. 22). When the club first begins to be formed, differentiating cells have bundles of filaments 80 Å wide arranged at random in their cytoplasm. With continued differentiation, they increase numerically and eventually fill the cytoplasm. On one side, the club cells are attached tightly to the cortical cells of the hair shaft and on the other are anchored firmly to the surrounding germ cells by interdigitating, modified desmosomal attachments (Fig. 22).

At about the middle of growing follicles, the germ cells around the club derive from the cells of the outer root sheath which at this level is multilayered. The outer, basal cells are cuboidal but progressively flatten centripetally along the long axis of the follicle. Unlike the inner root sheath, which at this level has scarcely any glycogen, all the cells of the outer root sheath are replete with glycogen, a characteristic feature. In addition, they have ribosomes, a few profiles of rough-surfaced endoplasmic reticulum throughout the cytoplasm, and a compact Golgi zone, but the often convoluted cell membranes have only a few desmosomes.

Fig. 20. Dermal papilla cells in a growing (anagen) follicle surrounded by matrix (M). These cells are similar to very active fibroblasts with well-developed endoplasmic reticulum and a Golgi complex. × 9600.

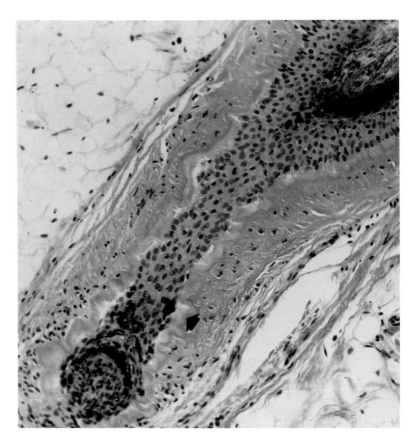

Fig. 21.   A scalp follicle in catagen. The glassy membrane (between the arrows) and connective tissue sheath are greatly thickened. The epithelial portion of the bulb is undergoing resorption. In the upper right corner is the well-formed club hair. (Courtesy of Dr. A. M. Kligman.)

In early catagen, when the cells of the outer root sheath begin to transform into germ cells, they acquire autophagic vacuoles of various sizes and shapes which contain mitochondria, ribosomes, endoplasmic reticulum, and glycogen (Fig. 23). The progressive degeneration of these organelles is mediated by acid hydrolysis, specifically acid phosphatase and esterases, whose reaction products have been histochemically demonstrated inside the vacuoles. Concurrently, cytoplasmic

Fig. 22.   A follicle in telogen containing a club hair and its keratinized rootlets anchored into the surrounding germ cells. × 8000.

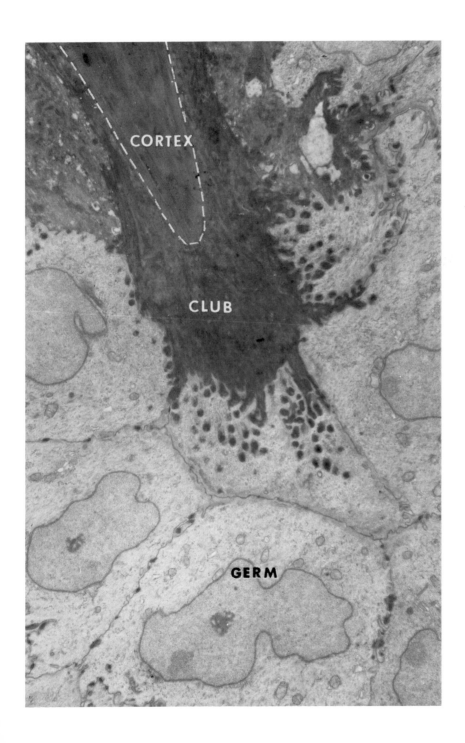

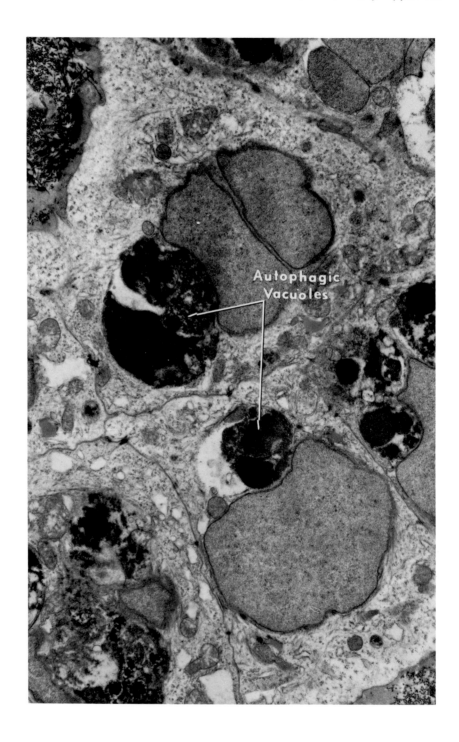

filaments gradually increase in number and eventually fill most of the cells. Plasma membranes and those facing the dermis then develop numerous desmosomal attachments and hemidesmosomes respectively whose most characteristic feature is the 80-Å filaments that occur singly and in bundles in the cytoplasm. Ultimately 2 or 3 layers of germ cells, which resemble basal epidermal cells, surround the hair club, forming a capsule or germ sac. These seeds for the next generation of hair are the most important part of the resting follicles.

When the club and its capsule of germ cells are completely structured, most of the cells below them are resorbed, and all of them have auto-phagic vacuoles (Figs. 7, 23) that contain acid phosphatase and esterases. Later these cells disintegrate, and much of the area vacated by the bulb is then occupied by cell debris, myelin figures, and dense bodies and vesicles containing amorphous materials. Eventually, the connective tissue sheaths also undergo degeneration.

2. CONNECTIVE TISSUE CHANGES

a. The Dermal Papilla. The papilla of the growing follicle is large and its cells are separated by connective tissue. Its fibroblasts are obviously active, with well-developed endoplasmic reticulum and Golgi complexes (Fig. 20). At the onset of catagen, the basal lamina around the blood vessels of the dermal papilla invaginates into multiple layers (Puccinelli *et al.*, 1967; Parakkal, 1969b). Simultaneously the basal lamina that separates the papilla from the cells of the bulb becomes crinkled, perhaps as a result of the shrinking that follows upon the death and resorption of the endothelial and matrix cells, and is slowly resorbed. The rounded fibroblasts of the dermal papilla continue to lose cytoplasm until their nucleus occupies most of the cells (Roth, 1965; Parakkal, 1970) (Fig. 24). Finally these fibroblasts form a ball of cells with a negligible amount of connective tissue between them and no blood vessels (Fig. 24). Whether papilla cells decrease during catagen and then increase during the next anagen is not known (Wolbach, 1951). We do know that they do not degenerate appreciably and that many of the mitotic figures seen in early anagen are in the endothelial cells. The number of papilla cells, then, remains fairly constant; hence the changes in the size of the papilla are mostly due to the increases and decreases in the size of the capillary plexus and intercellular substances.

Fig. 23.   Degenerating cells in the lower portion of a catagen follicle. Every cell has large numbers of autophagic vacuoles. × 8000.

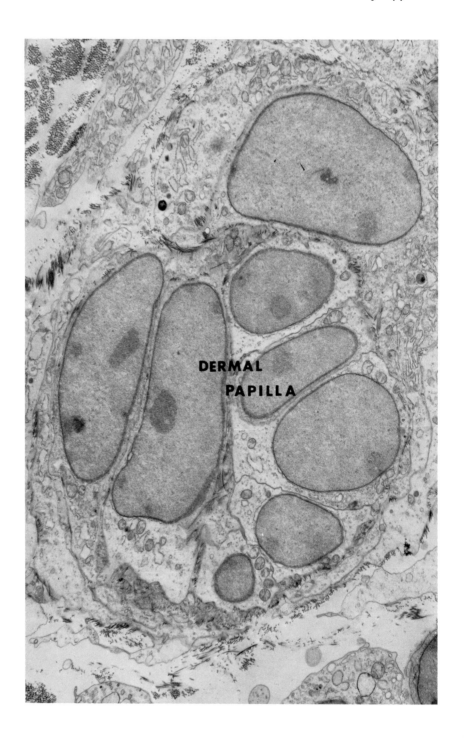

b. The Hyaline Membrane and Connective Tissue Sheath. During catagen, the lower part of the follicle degenerates through acid hydrolysis in the lysosomes. The basal lamina remains relatively intact (Figs. 21, 25, 26), but having lost contact with the shrinking outer root sheath, it becomes extensively pleated; this plication is seen as a hypertrophied, PAS-positive wrinkled sac around the waning lower part of the follicle. This sac is subsequently fragmented and resorbed.

During anagen, the connective tissue sheath which invests the hyalin membrane contains a loosely arranged layer of easily distinguished fibroblasts and macrophages. In both cells, the nucleus is in the central bulging portion and the cytoplasm is tapered on either side. Macrophages have clusters of ribosomes or polysomes in their cytoplasm, but fibroblasts have a well-developed, rough-surfaced endoplasmic reticulum. The plasma membrane of the macrophages is more or less smooth with only a few invaginations. During catagen, the orthogonally arranged collagen fibers around the basal lamina are broken down by the tightly packed layer of macrophages; these can be seen engulfing and degrading collagen fibers. The macrophages that have begun to phagocytose collagen have highly convoluted plasma membranes on the side facing the hair follicle and a comparatively smooth one on the opposite side (Fig. 25). Single groups of collagen fibers are sequestered by the invaginated plasma membrane and engulfed (Parakkal, 1969c). Longitudinal, oblique, or cross sections of collagen fibers, easily identified by their characteristic banding, can be seen undergoing degradation inside the vacuoles; the amorphous material inside some of these vacuoles may represent degraded collagen.

C. Telogen

Once they go into telogen, follicles have achieved a mature, stable stage of quiescence during which the hairs are anchored in the follicle by means of a club which sends keratinous rootlets between the germ cells of the epithelial sac. Structurally, these follicles differ considerably from growing follicles, being about ½ to ⅓ the length of the latter. The dermal papilla is reduced to a ball of cells located immediately below the capsule of germ cells. Most of the structures and cell layers that characterize growing follicles—the matrix, the inner and outer root sheaths, and the cuticle of the hair—are lacking in resting follicles.

Fig. 24. In a telogen follicle the cells of the dermal papilla are arranged as a ball at the base of the hair germ. The nucleus occupies most of the cell. A few delicate collagen fibers are seen between the cells. × 6800.

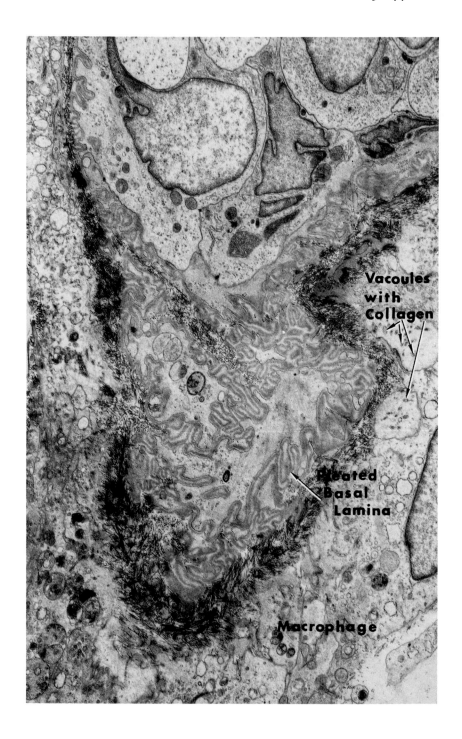

The hair germ and the club it formed during catagen are the characteristic structures of resting follicles (Montagna and Chase, 1956; Roth, 1965; Parakkal, 1970) (Figs. 8, 22, 27).

Since follicles in telogen are in a stable stage, such deleterious agents as x-rays and cholchicine, which in certain doses profoundly affect growing follicles, produce no visible changes. In fact, growing follicles are so sensitive to certain agents that when they are insulted they go into telogen and only later regenerate.

IV. Hair Growth in Man

Although the trichogram technique is well known and generally widely used clinically, only Pecoraro and his colleagues in Argentina have used it systematically for years to make valuable observations on this important subject. With this method, they have recorded the density of hair per square centimeter and the particular phase of growth cycles (Barman *et al.*, 1964), the thickness of individual hairs, the rate of growth, and the regeneration time after plucking. This brief account summarizes their major findings.

Following these markers, they found no apparent ethnic differences in the rate of hair growth in the scalp, axillae, and pubic regions. Individual differences, however, are apparent in the obviously different rates at which hair grows in each of the body areas. In 5-month-old human fetuses, for example, the hairs in the frontal, parietal, coronal, and occipital regions of the scalp all appear to be in anagen (Pecoraro *et al.*, 1964b). Later in fetal life, however, those in the frontal, and later in the parietal, areas change rapidly to catagen and then to telogen. In the occipital region, however, the follicles remain in anagen until near term, when they too change to telogen. In the frontal and parietal areas, a new growth cycle appears about 5 to 6 weeks before term, so that at birth the scalp follicles, even those in the same region, are in all three phases of the growth cycle (Barman *et al.*, 1964).

Except for individual variations, hair growth patterns in the scalp remain the same from infancy to puberty. From about 16 to 46 years of age, however, hair density and thickness decrease progressively in the scalp of both sexes, most noticeably in the vertex. The number of telogen follicles, which are smallest at puberty, increases with age in both sexes and is greater in the central than in the peripheral areas of

Fig. 25. Highly pleated basal lamina (hyalin membrane) of a follicle in catagen. The macrophages around the follicle contain phagocytosed collagen. × 6000.

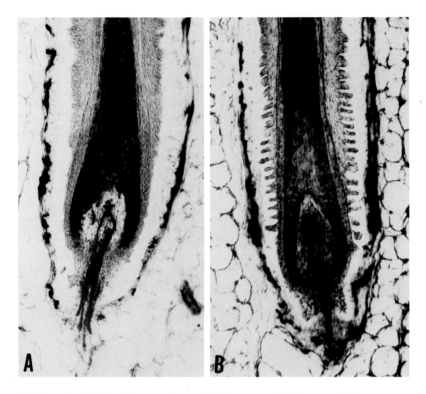

Fig. 26. Early (A) and more advanced (B) catagen. (B) shows a highly corrugated hyalin membrane. The outer root sheath is also highly plicated and degenerating rapidly.

the scalp. In both sexes, hair grows faster between the ages of 16 to 46 (Barman *et al.*, 1965); after 50, hair density decreases in both sexes, even in adults with no obvious signs of baldness. In this age group, the individual hairs are thicker and a higher proportion of follicles are in telogen.

Pregnancy does not affect the density of scalp hairs, but as it advances the percentage of thick hairs increases and the rate of growth apparently decreases. During pregnancy, most of the follicles remain in the growth phase. After delivery, most women undergo a temporary decrease in hair density which is a generalized loss of hair that is referred to as

Fig. 27. (A) is a scalp follicle in late catagen; the club hair is in the upper portion of the figure. (B) is a follicle in telogen. Note the spillage of alkaline phosphatase in (A). There is a bouquetlike arrangement of blood vessels under the dermal papilla of both figures.

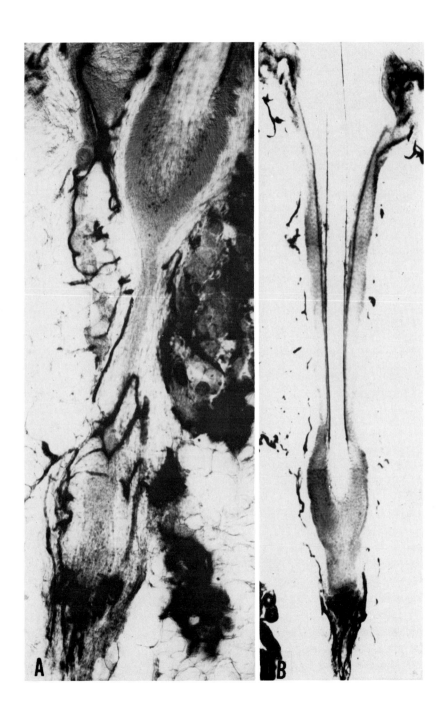

postpartum defluvium; in a short time, however, the rate of growth reverts to former values (Pecoraro *et al.*, 1964a). All such pregnancy-related changes decrease in intensity with successive pregnancies.

Like scalp hairs, axillary hairs show regional differences in growth patterns and are subject to the same factors of age, sex, race, and hormonal milieu. They are also affected by pregnancy and puerperium in the same way as scalp hairs: postpartum defluvium, decrease in growth rate, and return of normal growth rate. In both men and women, the density of hair is greater in the vault or central part of the axilla than in the surrounding region and is at a maximum during sexual maturity; these hairs also grow faster.

The growth of pubic hair has been studied in women—nonpregnant, pregnant, and postpartum. It has not been studied systematically in men. It appears at puberty, before axillary hair, and its rate of growth varies. In nonpregnant women, the growth rate is inversely correlated with age, with an estimated decline of ca. 15 mm/year. Hair density decreases with age at a rate of 0.31 hairs/cm^2/year, i.e., at a rate not significantly different from that in the axilla (0.38 hairs/cm^2/year) (V. Pecoraro, unpublished information). Pregnancy reduces this growth rate by 0.15 mm/week, a factor of 50. In the postpartum period, growth resumes its former rate and there is no loss of pubic hair.

Androgenic and anabolic hormones (even those advertised to be non-androgenic) induce some alopecia but stimulate the hairs in the chin and mustache (Puche *et al.*, 1971). These androgenic effects on scalp hair, which were observed in a group of normal men and women 20 to 25 years of age, were correlated with a high excretion of urinary androgens.

This resumé of Pecoraro and colleagues' work shows that both the cycles and the rate of hair growth in man are strikingly regional, and except for those in the pubic area, are controlled by the same internal factors.

V. Vascularity

Blood vessels have been demonstrated in macerated or cleared tissues of hair follicles that had been injected with opaque substances (Ryder, 1958; Durward and Rudall, 1958). Although these relatively crude preparations do not show all the small vessels, they are, nevertheless, instructive. Since the endothelium of the cutaneous arterial capillaries and arterioles can be stained quasi specifically with histochemical techniques for alkaline phosphatase (Montagna and Ellis, 1957b;

Klingmüller, 1957), thick frozen sections clearly show the vessels surrounding the hair follicles (Fig. 28).

The vascular patterns of follicles, which are fairly consistent, consist
of a dense and continuous plexus of arterioles and capillaries that
change in density in the transition from anagen to telogen. These
plexuses arise as branches of the dermal plexus or from more or less
direct vessels of the musculocutaneous arteries. ·In active follicles,
parallel, longitudinally oriented vessels extend from the base of the
bulb to the pilary canal. Cross-shunts and tortuous interconnections
between the parallel vessels outline a lattice pattern around the lower
part of follicles, just above the bulb (Fig. 28).

There are fewer cross-shunts above the keratogenous zone of the
follicle, the middle third of which is surrounded mostly by palisade
vessels. Cross-shunts also form a network at the level of the sebaceous
glands, which they envelop in tight nets that adhere to the sebaceous
acini. Above the entrance of the sebaceous duct, parallel vessels form

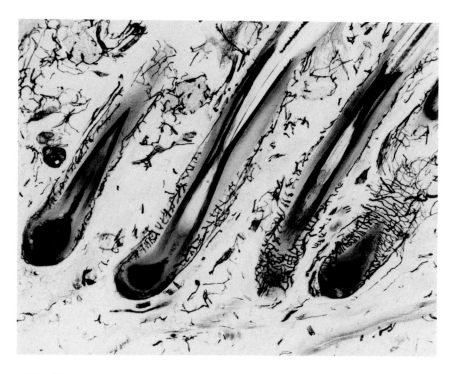

Fig. 28. Numerous capillary networks around eyebrow hair follicles demonstrated
with alkaline phosphatase.

a loose network that rises up to, and is continuous with, the loops of capillaries in the papillary body. The terminal part of the pilary canal is usually encircled by a vascular ring. Returning to the base of the follicle, some vessels penetrate the dermal papilla and form tufts that sometimes extend to the walls of the inner surface of the follicle (Fig. 29). The vascular system of each follicle and of its sebaceous glands forms a continuous unit.

Progressively smaller follicles are surrounded by relatively fewer vessels so that the lower part of vellus hair follicles has only a few capillaries and the dermal papilla has none. The large sebaceous glands that are often associated with these follicles, however, are richly vascularized. We have found that the strength of phosphatase reactivity in a dermal papilla is inversely related to the abundance of its capillaries.

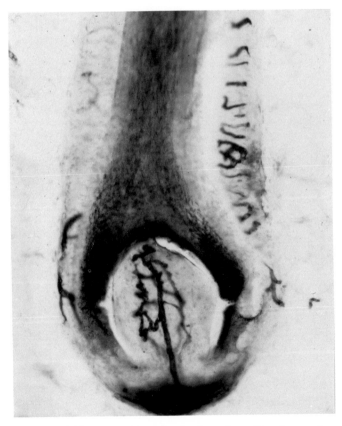

Fig. 29. A tuft of blood vessels inside the dermal papilla of an eyebrow follicle demonstrated with the alkaline phosphatase technique.

In the lower third of the follicle during early catagen, characteristic corrugations in the outer root sheath are accompanied by a thickening of the surrounding vitreous membrane and connective tissue sheath (Figs. 21, 26). But except for these and some later changes, the vascular system appears to remain intact (Ellis and Moretti, 1959). Later in catagen, during the resorption of the outer root sheath and the bulb matrix, the palisade vessels and the cross-shunts around the lower part of the follicles remain intact (Fig. 27). At this time, the follicle becomes shorter and the dermal papilla remains attached to its base; the path vacated by the degenerating follicle is marked by a trail of collapsed connective tissue sheaths, vessels, and nerves. Freed by this time from the bulb, the ball of dermal papilla cells nevertheless remains in contact with the retreating follicle. All of these changes take place inside the relatively intact vascular plexus of the follicle. Only when the lower third of the follicle is reduced to a thin strand of cells and phosphatase reactivity spills characteristically into the surrounding connective tissue (Figs. 26, 27) do some of the capillaries of the network degenerate. By telogen, those that once surrounded the lower part of the follicle have formed a collapsed bundle still crisply reactive for alkaline phosphatase; just under the dermal papilla, the vessels form a "bouquet" (Fig. 27). When a quiescent follicle becomes active again, the developing bulb advances through the collapsed vessels below the dermal papilla as a new vascular network regenerates. The most densely vascularized part of the follicle corresponds to the level of the keratogenous zone. When ^{35}S-labeled glucose is injected into the skin of sheep, the strongest accumulation of radioactive particles is recovered first around this general region (Ryder, 1956b). This and the basal cells of the bulb that face the dermal papilla are probably the most important sites of exchange.

The amount of vascular tissue in a papilla is related to the size of the follicle: the wider the diameter of a follicle, the larger the number of capillaries it contains (Ryder, 1956a). The wide papillae of large human follicles abound in capillaries which for the most part atrophy when the papillary cells become disengaged from them.

VI. Innervation

Hair follicles are richly supplied with nerves disposed in such a way as to resemble the dermal nerve networks; the two systems together represent nearly all the sensory nerves of the hairy skin (Winkelmann,

1959, 1960). Regardless of size, every follicle in man is surrounded by
nerve fibers from the base of the bulb to its junction with the epidermis.
Some nerves lie parallel to the follicle; other finer ones form a socklike
net around it. At the level of the bulge of large follicles and the bulb
of small, vellus follicles, a "basket" of blunt parallel nerve fibers forms
a collar or stockade (Figs. 30A,B,C); the upper parts of these nerves
often branch into plump, packed terminals like the tines of a fork
(Montagna and Giacometti, 1969) (Fig. 30D). The patterns of these
nerves, always more precise around vellus than around larger follicles,
are so well organized that we refer to them as follicle end-organs.
Despite their similarity, these recurrent patterns differ greatly even in
adjacent follicles of the same size. For example, the branched terminals
are sometimes missing and often the nerves that form the stockade are
loops rather than terminals. Silver impregnation techniques show only
the major nerves of these end-organs and the trunks toward which they
converge, but techniques for acetylcholinesterase also show other nerves
around the follicle (Figs. 31, 32). Underneath the thick parallel fibers
of the end-organs, clearly demonstrable finer fibers encircle the follicle
and surface to form a sparse network around the infundibulum; a few
are scattered here and there around the sebaceous glands. Some of
these fine nerves sometimes rise to the base of the epidermis, divide into
barely visible branches, and even penetrate it. Occasionally Häarscheiben
are seen, particularly in the eyebrows and glabella (Montagna and Ford,
1969). All of these nerves are clearly demonstrable in histochemical
preparations for acetylthiocholinesterase, but the end-organs are so
strongly reactive that the enzyme diffuses into the surrounding connective tissue. Since quiescent follicles are shorter than active ones,
their nerve end-organs are closer to their base. Furthermore, in quiescent
follicles the nerve plexus in the space vacated by the degenerated bulb
collapses in a train below the dermal papilla, but remains intact and
rich in cholinesterase (Montagna and Ellis, 1957a; Winkelmann, 1959).
    Recent electron microscopic observations have shown a perifollicular
nerve network around the hair follicles at the level of sebaceous glands
(Fig. 33). It consists of axons surrounded by Schwann cells, each nerve
arranged in a palisade around the follicle close to the outer root sheath.
The axon processes do not seem to penetrate the outer sheath. The basal
lamina separates the nerves from the outer root sheath cells. Each
Schwann cell is surrounded by connective tissue which contains delicate
collagenous fibers.
    Since follicles grow without nerves, the nerves just described must
for the most part be sensory. Their abundance clearly indicates that

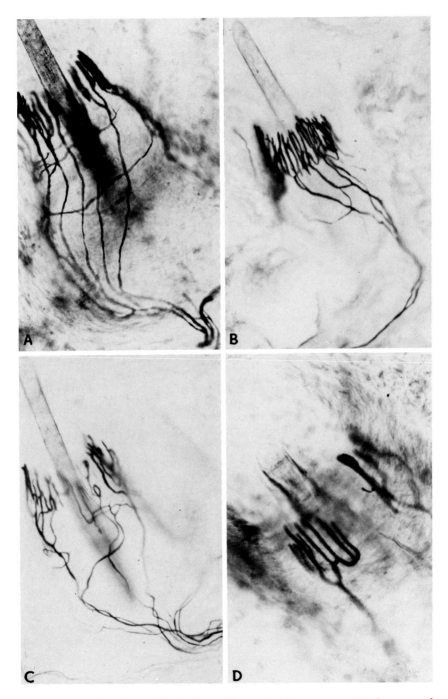

Fig. 30. In (A), (B), and (C) the similarities in the arrangement of nerve end-organs around vellus hair follicles are evident. (D) shows the tinelike terminals of the palisade nerves. Thick frozen sections prepared with Winkelmann's techniques.

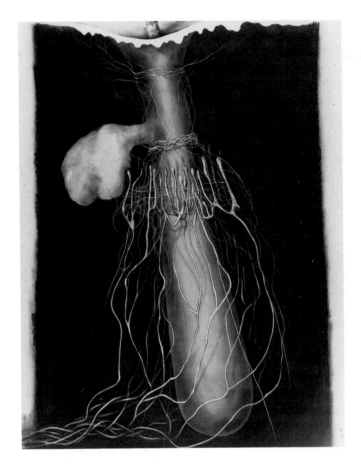

Fig. 31.   Diagram of the disposition of nerves around a small hair follicle.

they are among the most important elements of the sensory mechanism
(Winkelmann, 1959). This type of innervation is well adapted to tactile
sensations, the hair shaft acting as a lever that increases the range of
sensitivity to minute mechanical disturbances.

VII. Pigmentation

Whereas everyone is aware of the functional melanocytes in the upper
part of the bulb just above the dermal papilla of hair follicles, those in
the outer root sheaths of the follicles in adult skin have received little

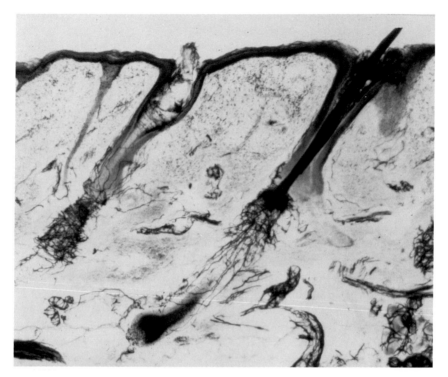

Fig. 32. Adjacent follicles from the scalp accompanied by numerous nerves even below the end-organ. Thick frozen section treated with actylcholinesterase technique.

attention. Scrupulous search has shown them in variable numbers there (Fig. 34), and even in the matrix, whether in man or other mammals. Melanocytes, then, active or not, are found throughout nearly the entire follicle. Even when amelanotic, melanocytes can be activated by such agents as x-rays and trauma (Montagna and Chase, 1956; Staricco, 1959, 1960, 1961).

Dendritic melanocytes in the upper part of the bulb provide melanin granules to the passing presumptive cells of the cortex and medulla of the hair, but not to those of the cuticle of the hair and the inner root sheath. When a follicle is nearing the end of its growth cycle, melanin formation and the formation of the medulla stop simultaneously. The last segment of hair that grows, therefore, is white and nonmedullated. During these preparatory changes, as we have seen, a club hair is formed in the keratogenous zone, after which the cells of the bulb degenerate and the follicle has no real matrix. At the end of the

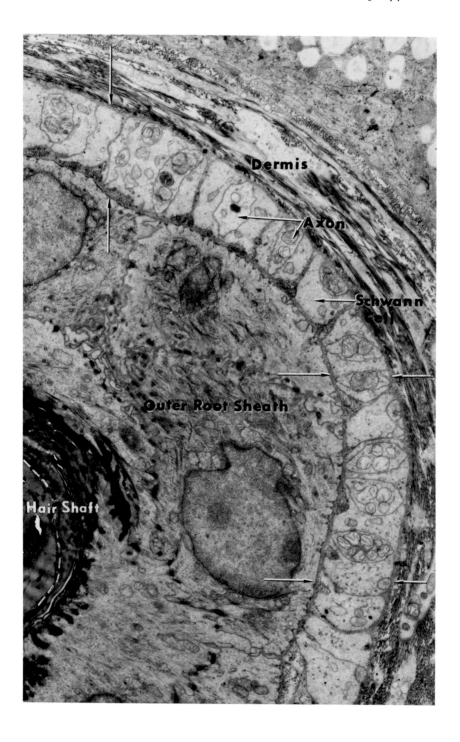

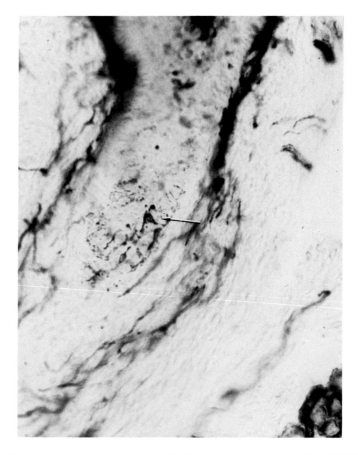

Fig. 34. Active melanocytes (arrow) in the outer root sheath of hair follicle from the eyelash in an unstained thick section.

quiescent period, the hair germ, consisting mostly of outer root sheath, rebuilds a new bulb which again manufactures hair and inner root sheath; at this point, melanocytes once again take their proper place in the follicle. According to some workers (Silver *et al.*, 1969), dormant melanocytes bide their time among the germ cells until the next growth phase when they become dendritic and again produce and transfer pigment.

Fig. 33. A ring of sensory nerves (between arrows) around a hair follicle. The axons are surrounded by Schwann cells. × 8000.

Hair color depends mostly on the kind and amount of pigment present in the medullary and cortical cells of the hair shaft. There is some structural basis for the differences between black, blonde, and red hairs. The melanosomes in black hair are large and elliptical, those in red and blonde hair are smaller and spherical. Moreover, ontogenetic differences affect both the formation of the membrane matrix and the deposition of pigment in the melanocytes of different colored hairs.

The effect of graying is produced by the admixture of black, white, and partially pigmented hairs. Completely white hair follicles have no melanocytes and in partially pigmented follicles, the melanosomes are incompletely melanized. Thus white hairs result from the progressive diminution in melanogenesis and the possible final loss of melanocytes. Whether melanocytes are completely lost or simply go into a dormant stage has not yet been determined (Fitzpatrick *et al.*, 1966). There is no basis for reports that stress results in graying of hair overnight. As a rule, graying starts in the beard and scalp during the fourth and fifth decades of life and later spreads to other regions. Despite occasional reports of repigmentation, graying is an irreversible process. Even if the melanocytes in most of the hair follicles are impaired and fail to produce pigment, it generally takes some time before this condition becomes apparent at the base of the hairs nearest the scalp. The section of already colored hair remains unaffected, since there is no physiological mechanism except bleaching or dyeing that can alter pigmentation in already formed hair.

VII. Keratinization

A. *The Medullary Protein*

The proteins in the medulla are so different from those in any other keratinized tissues that it is questionable whether they belong to the "keratin" proteins. Because they are resistant to most chemical treatment, they are difficult to isolate. The usual method of separating them is to treat them with concentrated solutions of sodium hydroxide which dissolves the cuticle and cortex and leaves the insoluble medulla as a residue (Blackburn, 1948; Matoltsy, 1953; Rogers, 1964b). However, some medullary material is also solubilized with this treatment. A less drastic method is to stain the hairs with gold and then shatter them in formic acid (Bradbury and O'Shea, 1969). Since gold is deposited preferentially on the medulla, it can be separated by density gradient centrifugation. Amino acid analysis of isolated medullary protein shows a very low cystine and a high glutamic acid and glutamine content. In addition, the medulla contains citrulline, which is not present in any of the other keratinized cells except the inner root sheath.

B. Cortical Proteins

Most of our knowledge about the molecular structure of keratinous proteins is derived from studies on wool and hair (Crewther *et al.*, 1965; Gillespie, 1965). About 85 to 90% of wool and hair consists of cortical cells. The keratinous protein in these cells is characterized mainly by insolubility and resistance to proteolytic enzymes. To solubilize the protein, the disulfide bonds must be severed by either reduction or oxidation. Proteins produced by reduction are called *kerateines*; those extracted after oxidation are called *keratoses* [S-carboxymethyl kerateines (SCMK)] (Goddard and Michaelis, 1934, 1935). With starch gel electrophoresis at alkaline pH values in the presence of 8 *M* urea, SCMK proteins obtained after reduction separate into a number of fast- and slow-moving bands (O'Donnell and Thompson, 1964). The five slow-moving bands are grouped into SCMKA or low-sulfur proteins and the five fast-moving bands into SCMKB or high-sulfur proteins. These low- and high-sulfur proteins constitute about 60 and 30% of the total proteins of hair and wool, respectively.

Significant differences distinguish the two classes of protein. For example, the molecular weight of the SCMKA proteins is about 45,000, that of SCMKB is 20,000. The amino acid composition of each differs substantially from the other. (For a detailed analysis of the amino acid composition of these fractions, see Fraser *et al.*, 1972.) After oxidation, the keratoses can be separated into two fractions, α-keratose and γ-keratose. In sulfur content and amino acid composition, α- and γ-keratoses are similar to SCMKA and SCMKB proteins, respectively (Crewther *et al.*, 1965).

In addition to the two groups of proteins mentioned above, crude preparations of low-sulfur proteins obtained from reduced wool contain small amounts of a protein rich in glycine and tyrosine (Gillespie, 1960; Harrap and Gillespie, 1963). This protein, which accounts for about 2 to 3% of the wool fiber, is extremely heterogenous (Zahn and Biela, 1968a,b).

Briefly, wool fiber contains more than 50 types of proteins extractable by different methods. Whether they are artifacts from the breakdown products of a single protein which constitutes the bulk of the fiber (Corfield *et al.*, 1968) is not known.

It has been suggested that the high- and low-sulfur proteins are derived from the matrix and the filaments of the cortical cells, respectively. A few minutes after ^{35}S-cystine is administered to sheep, the first uptake occurs in the bulb region. Two hours later, the radioactivity is much more intense in the keratogenous zone than it was earlier in the bulb (Downes *et al.*, 1967). A well-defined matrix around the filaments, first seen in the keratogenous zone by Mercer (1961), led him to believe that "keratin" is synthesized in two stages: (1) the filamentous part, consisting of low-sulfur protein, in the lower portion of the bulb; and (2) the sulfur-rich matrix in the keratogenous zone. When Downes *et al.* (1963) measured the radioactivity in the two groups of proteins after the administration of ^{35}S-cystine, their results concurred with but did not prove conclusively the two-stage hypothesis. The matrix-filament complex seen in the cortical cells of the bulb region when the filament bundles are first formed (Parakkal, 1969a) suggests that the matrix is formed comparatively early in keratinization. However, when the matrix later incorporates cystine at the keratogenous zone, it may be enriched by sulfur. For example, the fact that the content of the matrix, as well as the overall proportion of high-sulfur protein, increases in sheep maintained on a cystine-rich diet (Rogers, 1964b; Gillespie *et al.*, 1964) suggests a correspondence between the matrix and high-sulfur proteins.

C. Cuticular Proteins

The cuticle is divided into exocuticle and endocuticle according to its degree of solubility. Endocuticle can be digested with proteolytic enzymes, whereas exocuticle can be solubilized only after oxidation or reduction. Various methods of isolating cuticle cells have been used, but most preparations are impure or chemically degraded. Analyses of cuticle fragments isolated by being shaken in formic acid (Bradbury *et al.*, 1966) show appreciable amounts of cystine, proline, and serine. This would support the hypothesis that cuticular proteins are similar to cortical proteins (Mercer, 1961). Asquith and Parkinson (1966) even isolated an γ-keratose from cuticular material.

Because of its affinity with osmium tetroxide and silver nitrate, exocuticle is regarded as keratinous, whereas endocuticle consists of non-keratinous cellular debris. Cuticular proteins are amorphous and, unlike those of the cortex which have large quantities of filaments, have only a few filaments.

D. Proteins of the Inner Root Sheath

The proteins of the inner root sheath, like those of the other layers in the hair follicle, differ from those of the cortex. The keratinized cells of the inner root sheath contain a filamentous matrix complex similar to that in the cortex. Amino acid analyses of isolated inner root sheath filament yield no cystine, which is present in different amounts in both the low- and high-sulfur proteins of the cortex. On the other hand, citrulline is present in the filaments of the inner root sheath but not in the cortex. This amino acid is derived from arginine, but the enzymes responsible for the conversion have not been identified.

Keratinous proteins are stabilized by the disulfide bonds, but since there is no cystine in the inner root sheath, its proteins and those of the medulla may be stabilized by E-(γ-glutamyllysine) cross-links. Transglutaminase, the enzyme that catalyzes the cross-linking, has been identified in extracts of hair follicles from guinea pigs and rats. This cross-linking may be responsible for the extreme resistance of medullary and inner root sheath proteins to normal protein solvents.

In summary, the four layers of the hair follicle—medulla, cortex, cuticle, and inner root sheath—produce end products that are biochemically (sulfur content) and morphologically different (Fig. 35).

E. Proteins of the Cell Membrane Complex

During keratinization, modifications of the plasma membrane are probably as important as the formation and stabilization of intracellular proteins. In many instances, the cellular envelopes of keratinized cells are very resistant to keratinolytic agents. The most constant feature of these modifications is the deposition of material (50 to 150 Å thick) in the intercellular spaces and the resultant formation of the cell membrane complex. This intercellular substance is called the γ-layer.

Different extraction procedures result in the partial or complete loss of this layer. After the keratinous proteins are extracted, the residue consists mostly of cell envelopes, which account for about 3% of the total weight of the hair or wool. According to its amino acid composition, this residue is more or less rich in glycine and cystine (Bradbury et al., 1971). The cell envelopes of the stratum corneum also have high values of glycine and cystine (Matoltsy and Matoltsy, 1966).

The differences in the membrane complexes of the various keratinized layers become obvious in stained cuticle and cortex (Rhodin and Reith, 1962). However, more precise separation procedures have to be devised

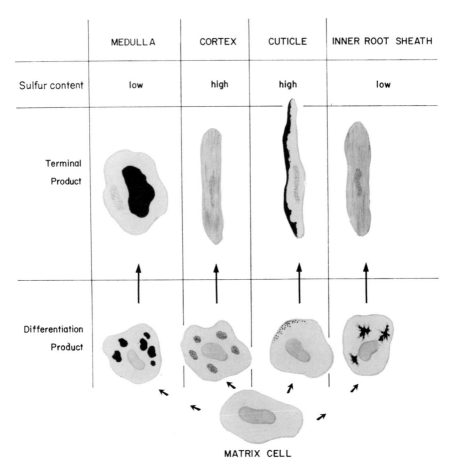

Fig. 35. Diagram showing sulfur content and differentiation products in the cells in the different layers of a hair follicle.

before this type of differentiation of membrane complex can be fully understood.

VIII. Factors That Influence Growth

A. *Hormonal*

A thorough knowledge of the normal cycles of follicular growth in the various species of animals is a prerequisite for determining how hormones affect the growth rate of hair. Different species sometimes respond differently to the same hormone, e.g., to adrenocorticosteroids which stimulate hair growth in certain alopecic diseases of the human scalp but suppress it in rats and other rodents.

Most male mammals, including man, have coarser hair than females. Spontaneous growth waves in female rats lag behind those of males, but the cycle of growth in each follicle is the same in both sexes. In animals gonadectomized at an early age, sex differences generally disappear. In animals gonadectomized, adrenalectomized, hypophysectomized, or rendered thyroid deficient, daily estrogen treatment retards the initiation and rate of both spontaneous and induced hair growth. However, estrogen induces the growth of fine, sparse hair in all animals except those that have been hypophysectomized. Daily androgen treatment has no apparent effect on hair growth except in hypophysectomized rats.

Spontaneous replacement of hair is noticeably retarded during pregnancy and lactation, even when the club hairs are plucked. These effects are not duplicated when pregnant or lactating females are treated with progesterone, but hair growth generally increases when nursing females are treated with luteotrophic hormone.

Adrenalectomy stimulates the growth and spread of follicles in rats but does not affect the growth rate of individual follicles. Daily treatment with small doses of cortisone inhibits spontaneous hair growth in intact, gonadectomized or adrenalectomized rats, but has no effect on growing follicles. Growth is inhibited in propylthiouracil-treated or hypophysectomized animals injected with small amounts of cortisone but resumes as soon as cortisol treatment is discontinued. Deoxycorticosterone has no effect on hair growth.

Continuous treatment with adrenaline inhibits spontaneous hair growth in intact rats and delays the response to plucking, but once it resumes, growth proceeds normally. Moreover, prolonged treatment with adrenaline inhibits both spontaneous and induced growth, neither of which is mediated or potentiated by thyroxin. Adrenaline, however, inhibits hair growth more effectively in cortisol-treated, adrenalectomized rats than in adrenalectomized animals without cortisol.

Spontaneous growth is markedly retarded in alloxan-diabetic animals, but once initiated again growth is normal. Treatment with phlorizin does not affect hair growth despite continued glycosuria and hypoglycemia. Insulin restores spontaneous replacement to normal in alloxandiabetic animals and increases growth in intact animals. Regrowth after plucking is normal in glucose-treated intact animals, but spontaneous growth is often retarded. Insulin, then, seems to be more directly involved in hair growth than glucose, perhaps by regulating the utilization of glucose from the blood during the early stages of follicular growth.

Amounts of propylthiouracil that produce a deficiency in thyroid hormone likewise inhibit spontaneous waves of hair growth; however, in these circumstances, induced growth by plucking is normal. Injections of thyroxine accelerate the spontaneous replacement of hair in both propylthiouracil-treated and normal rats, in both of which the rate of growth, once stimulated, remains normal. Thyroxine and cortisol have antagonistic effects on hair growth, one hormone offsetting the effects of the other, but thyroxine and gonadal hormones do not act antagonistically.

Hypophysectomy accelerates the initiation and spread of spontaneous growth waves but has no effect on the rate of hair growth in rats except to make the pelage infantile. ACTH inhibits hair growth in intact, gonadectomized, and hypophysectomized, but not in adrenalectomized, rats.

In hypophysectomized rats, implants of pituitary tissue and injections of growth hormone both restore the pelage to adult texture whereas even after treatment with gonadal hormones their hair remains infantile. Sex hormones, then, modify the type of hair produced only if growth hormone is present.

About 15% of the follicles in the scalp of nonpregnant young women are quiescent (Lynfield, 1960), whereas in pregnant women the percentage is appreciably less. Clinical observations indicate that during the postpartum period, women experience a diffuse loss of scalp hair.

The beards of Caucasian men grow somewhat faster than those of Japanese (Hamilton et al., 1969). By the same token, the incidence of facial hirsutism in Caucasian women is relatively high but extremely low in Japanese women. In both ethnic groups, men have similar values for the mean diameter of coarse hairs and for the percentage of gray hairs with advancing age. Growth of axillary hair is more pronounced in Caucasian than in Japanese men and women. Likewise, Caucasian men tend to develop coarse hairs on the external ears and sternal area and to become bald more than Japanese men.

B. Nutritional

Matrix cells have no diurnal rhythms. Not even starvation depresses their mitotic activity which seems to be affected only by full shock.

Malnutrition affects hair growth variously, depending on the species (Ryder, 1958; Bradfield, 1968). The weight and quality of wool that a sheep produces, for example, is controlled by the amount and quality of the food he eats. A poor diet reduces the breaking strength and the length and diameter of wool fibers; reinforcing the diet with both protein and carbohydrate corrects these defects and increases wool production. Carbohydrate provides the energy to utilize protein, releases protein for keratin formation, and to a lesser extent, maintains mitotic activity in the matrix. Cystine or methionine are essential for hair growth in many animals, but not for sheep, which synthesize cystine from sulfate. The B vitamins are necessary agents for hair growth, and pantothenic acid is probably essential for copper utilization. Copper deficiency results in a loss of pigment in the hair and a loss of crimp in the wool. Trace amounts of copper appear to catalyze the oxidation of SH– to –S–S– groups, although this has not been detected. Follicles easily utilize ^{35}S-labeled cystine, the radioactive particles of which appear almost at once after being injected into the outer root sheath above the bulb (Ryder, 1958). Injections of ^{14}C-labeled glucose are immediately followed by the appearance of radioactive particles in the bulb, but not above it.

The effects of protein malnutrition will be treated in greater detail in the next chapter.

C. Metabolic

The metabolism of the hair follicle still defies exact analysis, and as a result very few biochemical data are available. This lack of information is understandable when we consider the limited samplings of human hair follicle to be had and the inadequate methods of analyzing them. The metabolic studies that have been reported have been carried out mostly on rodent skin after the hair cycles had been synchronized by plucking. The chief drawbacks of this system are that the samples include the whole skin, not just the hair follicles, and that information obtained from rodents does not necessarily apply to man.

At the present time, only a few studies comply with the following requirements for an adequate analysis: (1) that hair follicles be obtained either by plucking or by microdissection and be studied without other tissue contaminants; and (2) that studies be made on human follicles or on such satisfactory substitutes as those of stump-tailed macaques which also become bald.

1. ENERGY METABOLISM

Microtechniques have been devised to study glucose metabolism in human follicles plucked from the nonbald scalp (Adachi and Uno, 1968, 1969). The general metabolic pattern is extremely "glycolytic," i.e., more than 85% of the glucose consumed is reduced to lactate. When the various glucose pathways are studied in growing and resting follicles, the metabolic activities are found to be much higher during the growing phase. For example, in growing follicles glucose utilization increases 200%, glycolysis 200%, the activity of pentose cycle 800%, metabolism by other pathways 150%, and ATP production via the respiratory chains 270%. Every hour 1 gm of growing follicles produces 57.6 μmoles of ATP and 2.3 μmoles of TPNH, but 1 gm of resting follicles produces only 23.8 μmoles of ATP and 0.21 μmoles of TPNH. Clearly, by activating the pentose cycle, glucose assimilation in growing hair follicles produces not only sufficient energy for itself but also essential substances (TPNH, ribose, etc.) for the follicle to metabolize fatty acid, nucleic acids, and steroid hormones. When the enzymes of carbohydrate metabolism were assayed with the microtechnique of Lowry (1953), the results agreed with those obtained above. For example, in the bulb portion of growing follicles, the activity of glucose-6-phosphate dehydrogenase, a key enzyme of the pentose cycle, increased 350% over that in resting follicles.

During anagen, the activities of the enzymes involved in carbohydrate metabolism are generally higher in the bulb portion than in the sheath of follicles except for phosphorylase activity which is much higher in the external sheath than in the bulb (Adachi and Uno, 1968; Uno et al., 1968a). Glycogen concentrations were correspondingly elevated, i.e., the outer sheath of anagen follicles contained 2.8 gm glycogen per 100 gm dry weight, the bulb 0.44 (Uno et al., 1968b). Thus, an inverse relationship exists between glycogen metabolism and biological activity. Furthermore, the parallelism between phosphorylase activity and glycogen concentration remains unexplained, since phosphorylase is not a synthetic but a degradative enzyme for glycogen. Glycogen accumulates in skin when the metabolism is suppressed (Montagna and Chase, 1956), but how this functions physiologically in the external sheath is not known. Practically no information has been reported about control mechanisms, except that of phosphofructokinase which plays a significant role in controlling energy metabolism in human hair follicles (Kondo and Adachi, 1971).

2. ANDROGEN METABOLISM

Studies of the metabolism of steroid hormones by hair follicles have been limited to androgen. Northcutt *et al.* (1969) demonstrated that pubic hair follicles can convert testosterone to dihydrotestosterone (DHT). In freshly plucked scalp hair follicles incubated with ^{14}C-testosterone and glucose as a carbon source, Adachi *et al.* (1970) showed the major metabolites to be, in decreasing order, androstenedione, DHT, and androstanedione. Fazekas and Lanthier (1971), on the other hand, found that beard hair follicles metabolize ^{3}H-testosterone to, again in decreasing order, androstenedione, androstanedione, DHT, and androsterone. The minor discrepancy in the rate of DHT formation reported in these two studies may be due to different incubation conditions rather than to differences in the metabolic capacity of scalp and beard follicles.

Because DHT is a more potent androgen than testosterone (Dorfman and Dorfman, 1962), attention has been focused on its formation from testosterone. Moreover, it is the predominant androgen in such target organs as the prostate and seminal vesicles (Bruchovsky and Wilson, 1968; Anderson and Liao, 1968). The conversion of testosterone to DHT is catalyzed by 5α-reductase with TPNH as a cofactor. One wonders, then, whether 5α-reductase is a key enzyme in the production of the "tissue-active androgen" DHT. Since the independent works just cited cannot be compared for their 5α-reductase activities, Takayasu and Adachi (1972) devised a microtechnique for measuring optimal reductase activities and when they compared these activities in various hair follicles from different regions of the scalp, they found that they are three to eight times higher in growing follicles than in resting ones. Hair follicles from the beard, axilla, chest, pubis, arms, and legs showed essentially the same activities as those of the scalp. Thus 5α-reductase cannot be attributed to the differential hormonal effects in relation to patterns of hair distribution. Therefore, the key control mechanisms should be sought in factors other than 5α-reductase.

3. HORMONAL CONTROL OF HAIR FOLLICLE METABOLISM

In general, basic hormone action can be classified into two groups: (1) a hormone attaches to a specific binding protein which carries it to the nuclei of a target tissue to trigger new protein (enzyme) syntheses; and (2) a hormone stimulates or inhibits adenyl cyclase, which accordingly increases or decreases cyclic AMP. Cyclic AMP activates protein kinase which in turn phosphorylates (activates) various enzymes, particles, and membranes. The action of steroid hormones is usually mediated by specific binding proteins, that of nonsteroidal hormones by the cyclic-AMP system.

Whether or not human hair follicles contain specific binding proteins is not known. Androgen and estrogen receptors are present in rat skin (Eppenberger and Hsia, 1972) which varies in its binding capacity, depending on which stage of hair cycle it is in. However, since these authors analyzed whole skin and not isolated hair follicles, the specific localization of these receptors has yet to be determined.

Adachi and Kano (1970), who investigated adenyl cyclase activities in 8000 μg of a particulate fraction from freshly plucked scalp follicles, found that these activities are inhibited by DHT and activated by estrone. This concurs with the fact that androgens initiate baldness but estrogens do not. If the same diverse biochemical consequences that are seen in other tissue are assumed to occur in hair follicles after the changes in cyclic AMP levels, then perhaps DHT inhibits energy production by keeping phosphofructokinase relatively inactive, and suppresses various protein syntheses. Thus, a relatively low cyclic AMP level may prematurely terminate follicular growth; the frequent repetition of these processes could transform terminal follicles as in pattern baldness. Perhaps in that case, estrone could help maintain a high cyclic AMP level in hair follicles and by keeping various metabolic processes active maintain vigorous hair growth. However, such a tentative hypothesis for the molecular genesis of common baldness (Adachi, 1973) must be supported by direct evidence, such as the measurement of intrafollicular levels of cyclic AMP in relation to various hormonal states.

With the discovery that some subhuman primates also develop baldness (Montagna, 1963), a condition long considered peculiar to man, these animals have been used in a limited way in current studies of human alopecia. Histological examination, i.e., analysis of a spectrum of enzyme activities and clinical (genetical) observations, has shown that baldness in stump-tailed macaques is the same as common baldness in man (Uno *et al.*, 1967, 1968a,b; Takashima *et al.*, 1970; Takashima and Montagna, 1971). Using microchemical methods, Takashima *et al.* (1970) studied *in vitro* the uptake rates of labeled testosterone and its catabolism in follicles from bald and nonbald areas of stump-tailed macaques and compared them with rates from the hairy areas. They found that in both male and female animals the metabolism of testosterone was less active in follicles in the bald area than in those in the hairy area, whereas the production rate of androstenedione was much faster in follicles in the bald area. Specifically, the ratio was less than one in hair follicles from the hairy scalp region or from the back region, but more than one in hair follicles from the bald or presumptively bald scalp. These data appear to be the first demonstration that follicles in bald and hairy regions differ biochemically.

Despite these biochemical differences, however, the results seem to be paradoxical. Since common baldness is triggered by an elevated androgen level, hair follicles transformed to vellus in the frontal bald scalp should retain more testosterone than those in nonbald areas. However, data showed that the content of the presumably active androgen (testosterone) was less abundant and that of the inactive one (androstenedione) was more abundant in vellus than in terminal hair follicles. A plausible explanation is that an extremely active male sex hormone, such as dihydrotestosterone, is formed from testosterone by vellus hair follicles (or by terminal hair follicles before transformation). Thus the increased production of the metabolites may indicate a corresponding increase in the production of the tissue-active androgen. Reviewing the previous experiments, Takashima and Montagna (1971) found that the hair follicles and epidermis in the frontal scalp region of a juvenile stumptail produced much larger amounts of dihydrotestosterone than those in the occipital and other nonbalding scalp regions.

Thus available data suggest that the initial step in the process of baldness is the accumulation of the "tissue-active androgen," DHT. However, further molecular mechanisms subsequent to DHT accumulation need to be explored.

X. Alopecia

Baldness is a human attribute still so widely misunderstood that it is advisable to explain a few facts about it here. The phenomenon is one of biological interest; although its cosmetic implications may loom as catastrophic to a balding young man who sees his crowning glory thin and wane, they are of no great consequence to the biologist.

Surprisingly few dermatologists realize that other mammals also become bald as part of the normal process. Some nonhuman primates, like the uakari (*Cacajao,* sp.), the stump-tailed macaque (*Macaca speciosa*), the orangutan (*Pongo pygmaeus*), and the chimpanzee (*Pan troglodytes*), become bald in different degrees. In almost all primates except man, the scalp hair extends visibly down to the eyebrows. For example, the entire scalps of preadolescent uakaris from South America are covered from the brows up with long, thin, reddish hair. As the animals mature, very short hairs first appear on the forehead and then progressively cover the entire scalp.

Since despite its naked appearance the human forehead is *not hairless,* the so-called hair line does not, as is commonly believed, actually represent a demarcation between two different kinds of tissue. If we are ever to understand baldness, we must first realize that the forehead and scalp are one continuous and indistinguishable tissue and that they must be studied as such. For example, in early fetal life, the lanugo hairs that cover the forehead are as long as those on the rest of the scalp; thus there is no real line of demarcation between the skin of the forehead and that of the scalp. After the fifth fetal month, the follicles on the forehead as well as those elsewhere on the body undergo a slight, gradual involution, whereas those in the rest of the scalp grow larger. Newborn infants are often more hirsute than they are months or even years later. How long the hair of a newborn infant will be is unpredictable; some are born so hirsute that even the forehead is covered with coarse hairs, others are nearly bald. In some infants, most of the hair is lost during the first two postnatal years, after which the scalp hairs are completely restored. In others, an initially low hair line gradually recedes during childhood.

By the time of birth, the follicles on the forehead are so small and the hairs so almost colorless that the area appears naked. And yet, the number of follicles at that site is only slightly less than that of the "hairy" scalp. During successive postnatal years, these forehead hairs become more and more diminutive and invisible. The low hair line found in some infants retreats during childhood and finally establishes an individual pattern that is probably familial. All of these patterns of shifting hair line and the establishment of a naked forehead are inseparably tied in with the process of baldness. When pattern baldness is first noticed in the twenties or earlier, the follicles are behaving exactly like those on the forehead of fetuses and infants.

Why is it that the human scalp, once the site of the vigorous growth of hair follicles in the young, becomes progressively bald? The phenomenon does not, as often believed, represent a catastrophic destruction of hair follicles but a systematic involution which culminates in organs similar to follicles in the fetus and involves no appreciable diminution in absolute numbers.

Despite the loss of hirsute adornment it entails, baldness is a natural tendency of the follicles in the frontal, parietal, and upper occipital areas of the scalp to become very small as individuals mature. Although there is nothing extraordinary about these sequences of events, practically nothing is known about the mechanisms that control them. Similar events occur in other parts of the body. The shoulders and backs of infants are often covered with long, coarse hairs which later fall out and are replaced by either small terminal or vellus hairs. The visible pubic hair frequently seen in infant girls is later replaced by vellus hairs which become coarse at adolescence. Baldness occurs in both men and in women, but since it is more precisely patterned and precipitous in men, it is more obvious in them.

During the many years that we have studied the scalps of young and old people, we have never failed to find vellus-type follicles in the scalps of children (Fig. 36). These disappear, or at least are rarely found, in late puberty, but they reappear in the twenties. The origin of these follicles in children is incomprehensible and their fate unknown. Their presence on the scalps of young adults suggests that they are converted terminal follicles as the scalp becomes bald. Alopecia consistently occurs at the hair line, and yet, whether the signs of baldness are visible or not, the hair line always has a population of follicles in every conceivable transitional stage, from perfect terminal hairs to fully converted vellus hairs.

A society that is early indoctrinated into believing that scalp hair is indispensable for beauty is rarely receptive to the facts of the matter. If, however, it could rid itself of its prejudices and learn a lesson from what happens in other animals, it might ultimately accept the fact that these ornamental hairs are replaced by a different kind of adornment, total nakedness. Normally all human beings become progressively balder; the fact that in some the more advanced signs of baldness appear at an earlier age is only of genetic significance. No one contemplating the opposite phenomena of one man whose trunk is covered with a coarse blanket of long hairs and of another whose trunk is practically "naked" would seriously enter a judgment of "normal" for either. Why, then, do we still refer to baldness as a disease? Since we all have alopecia in some degree or other, it can only be regarded as a human characteristic.

Alopecia in old people is a related point. Although it is similar to

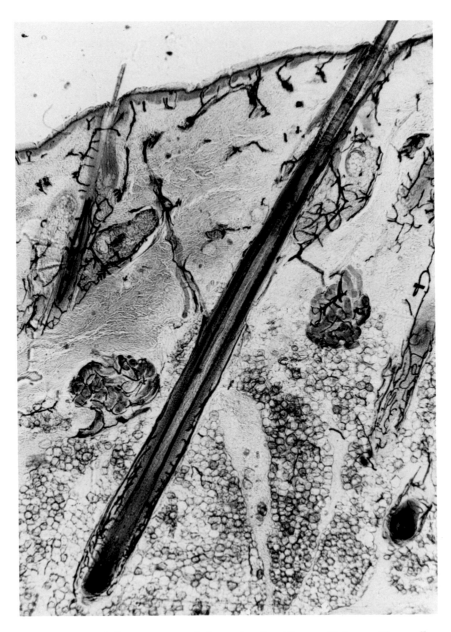

Fig. 36. Thick frozen section of the scalp of a 12-year-old boy, with a vellus follicle on the left and a terminal follicle on the right. The preparation has been treated with the technique for alkaline phosphatase to show blood vessels.

that which occurs in younger people, senile baldness also involves the actual destruction of follicles. Senile alopecia, then, is always superimposed upon that which occurs normally during the prolonged ontogeny of human skin.

Another point is rarely, if ever, appreciated. Regardless of the totality of apparent baldness in a young scalp, numbers of quiescent guard hair follicles are still to be found in these areas. These follicles have very long quiescent periods but occasionally can be seen growing singly or in scattered groups even on the forehead. Split-skin preparations often show considerable numbers of them (Figs. 37 and 38). If these scattered quiescent guard hair follicles are stimulated through chemical or physical agents, observers get the false impression that hair has "regrown" in such areas. Several such instances have been published; in one, topical applications of testosterone propionate were said to have restored hair "growth" and thus to have partially cured baldness (Papa and Kligman, 1965). What really happened was that the quiescent guard hair follicles in the bald areas were stimulated to grow all at once. No doubt other agents could have performed the same service.

The progressive diminution in the size of scalp follicles during balding takes place over several successive generations of hair cycles (Fig. 39). At each cycle, the size of the anagen follicle becomes smaller until it reaches that of the anagen vellus follicle. A follicle, then, becomes smaller only when it regrows after telogen. Once this stage is reached, vellus follicles have increasing periods of rest which far exceed the periods of growth. Baldness thus occurs with marvelous uniformity over a long period of time.

The hair follicles of the human scalp are unique in that they can grow uninterruptedly for years and produce hairs of considerable length. With some exceptions, the scalp hairs of most other mammals are only slightly longer than those of the body. A few baboons, the lion-tailed macaque (*M. silenus*), the Celebes ape (*M. nigra*), some marmosets, and some maned mammals have longer scalp than body hairs.

The etiology of baldness is still conjectural. Those who attributed baldness to faulty scalp tissue have been proved wrong by the success of hair transplants. Furthermore, the apparently degenerative changes in the bald scalp occur after the follicles have decreased in size. No one can seriously regard diminished vascularity in the scalp as a cause of baldness since the great vascular bed of the tissue is of little use to the skin itself, and nearly ischemic skin can support good hair growth.

Fig. 37. Split skin preparations from the scalp of a 77-year-old man. (A) The nonbald temporal area; the follicles are largely normal although some are smaller than others. (B) The upper occipital area which had a fuzz of hair with vellus and some quiescent terminal follicles. The underside of the epidermis is beginning to flatten out.

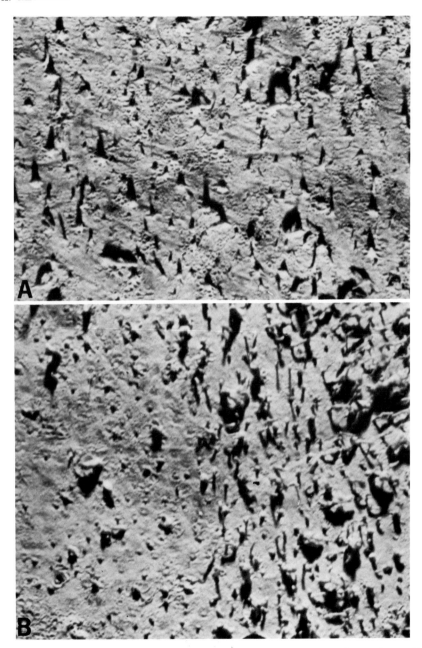

Fig. 38. Split skin preparations from the same scalp as in Fig. 37. (A) The totally bald frontal area still has numerous vellus follicles and occasional quiescent terminal follicles. (B) The transition between temporal (right) and frontal (left), area has a mixed population of vellus and terminal follicles. The underside of the epidermis is almost completely flattened out.

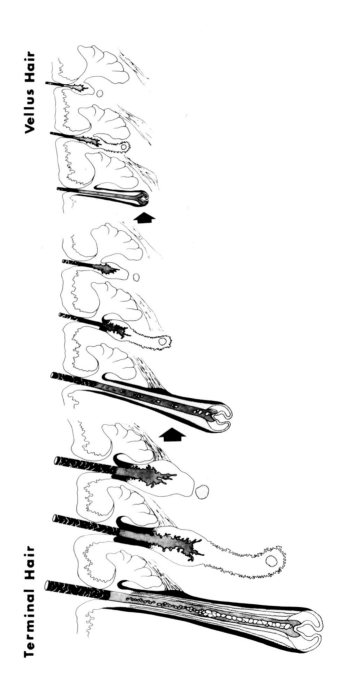

Fig. 39. The diminution of follicle size in succeeding generations of hair cycles which culminates in a vellus type.

Everyone will agree with Hamilton (1942) that a certain level of androgen is necessary for alopecia to develop. But *how* it develops no one knows. The only clues so far have come from the chemists who are beginning to show first, that scalp hair follicles metabolize testosterone at a high rate (p. 241) and second, that the follicles of the frontal and parietal areas metabolize it at an even greater rate than those of the temporal, lower parietal, and occipital areas. Thus, it is the follicles themselves, not the environment, which are programmed to respond to androgens in such a way as to induce baldness.

References

Adachi, K. 1973. The metabolism and control mechanism of human hair follicles. *In* "Current Problems in Dermatology," Vol. 5, pp. 37–78. Karger, Basel.

Adachi, K., and M. Kano. 1970. Adenyl cyclase in human hair follicles: Its inhibition by dihydrotestosterone. *Biochem. Biophys. Res. Commun.* 41: 884–890.

Adachi, K., and H. Uno. 1968. Glucose metabolism of growing and resting human hair follicles. *Amer. J. Physiol.* 215: 1234–1239.

Adachi, K., and H. Uno. 1969. Some metabolic profiles of human hair follicles. *In* "Advances in Biology of Skin. Hair Growth" (W. Montagna and R. L. Dobson, eds.), Vol. 9, pp. 511–534. Pergamon, Oxford.

Adachi, K., S. Takayasu, I. Takashima, M. Kano, and S. Kondo. 1970. Human hair follicles: Metabolism and control mechanisms. *J. Soc. Cosmet. Chem.* 21: 901–924.

Anderson, K. M., and S. Liao. 1968. Selective retention of dihydrotestosterone by prostatic nuclei. *Nature (London)* 219: 277–279.

Asquith, R. S., and D. C. Parkinson. 1966. The morphological origin and reactions of some keratin fractions. *Text. Res. J.* 36: 1064–1071.

Auber, L. 1952. The anatomy of follicles producing wool-fibres with special reference to keratinization. *Trans. Roy. Soc. Edinburgh* 62: 191–254.

Barman, J. M., V. Pecoraro, and I. Astore. 1964. Method, technic and computations in the study of the trophic state of the human scalp hair. *J. Invest. Dermatol.* 42: 421–425.

Barman, J. M., I. Astore, and V. Pecoraro. 1965. The normal trichogram of the adult. *J. Invest. Dermatol.* 44: 233–236.

Bell, M. 1969. The ultrastructure of differentiating hair follicles in fetal rhesus monkeys *(Macaca mulatta)*. *In* "Advances in Biology of Skin. Hair Growth" (W. Montagna and R. L. Dobson, eds.), Vol. 9, pp. 61–81. Pergamon, Oxford.

Birbeck, M. S. C., and E. H. Mercer. 1956. Cell membranes and morphogenesis. Nature 178: 985–986.

Birbeck, M. S. C., and E. H. Mercer. 1957a. The electron microscopy of the human hair follicle. 1. Introduction and the hair cortex. *J. Biochem. Biophys. Cytol.* 3: 203–214.

Birbeck, M. S. C., and E. H. Mercer. 1957b. The electron microscopy of the human hair follicle. 2. The hair cuticle. *J. Biochem. Biophys. Cytol.* 3: 215–222.

Birbeck, M. S. C., and E. H. Mercer. 1957c. The electron microscopy of the human hair follicle. 3. The inner root sheath and trichohyaline. *J. Biochem. Biophys. Cytol.* **3**: 223–230.

Blackburn, S. 1948. The composition and reactivity of medullated keratins. *Biochem. J.* **43**: 114–117.

Bonnet, R. 1892. Über Hypotrichosis congenita universalis. *Anat. Hefte Abt.* **1**: 233–273.

Bradbury, J. H., and J. M. O'Shea. 1969. Keratin fibres II. Separation and analysis of medullary cells. *Aust. J. Biol. Sci.* **22**: 1205–1215.

Bradbury, J. H., J. D. Leeder, and I. C. Watt. 1971. The cell membrane complex of wool. *Appl. Poly. Symp.* No. **18**: 227.

Bradbury, J. H., G. V. Chapman, A. N. Hambly, and N. L. R. King. 1966. Separation of chemically unmodified histological components of keratin fibres and analyses of cuticles. *Nature (London)* **210**: 1333–1334.

Bradfield, R. B. 1968. Changes in hair associated with protein-calorie malnutrition. *In* "Calorie Deficiencies and Protein Deficiencies" (R. A. McCance and E. M. Widdowson, eds.), p. 213. Churchill, London.

Breathnach, A. S., and J. Smith. 1968. Fine structure of the early hair germ and dermal papilla in the human foetus. *J. Anat.* **102**: 511–526.

Bruchovsky, N., and J. D. Wilson. 1968. The intranuclear binding of testosterone and 5α-androstan-17β-ol-3-one by rat prostate. *J. Biol. Chem.* **243**: 5953–5960.

Bullough, W. S., and E. B. Laurence. 1958. The mitotic activity of the follicles. *In* "The Biology of Hair Growth" (W. Montagna and R. A. Ellis, eds.), pp. 171–186. Academic Press, New York.

Butcher, E. O. 1934. The hair cycles in the albino rat. *Anat. Rec.* **61**: 5–19.

Butcher, E. O. 1951. Development of the pilary system and the replacement of hair in mammals. *Ann. N. Y. Acad. Sci.* **53**: 508–516.

Chapman, R. E., and R. T. Gemmell. 1971. Stages in the formation and keratinization of the cortex of the wool fiber. *J. Ultrastruct. Res.* **36**: 342–354.

Chase, H. B. 1954. Growth of the hair. *Physiol. Rev.* **34**: 113–126.

Chase, H. B., and G. J. Eaton. 1952. The growth of hair follicles in waves. *Ann. N. Y. Acad. Sci.* **83**: 365–368.

Corfield, M. C., J. C. Fletcher, and A. Robson, 1968. Recent work on the chemical structure of wool proteins. *In* "Symposium on Fibrous Proteins" (W. G. Crewther, ed.), pp. 289. Butterworths, Sydney.

Crewther, W. G., R. D. B. Fraser, F. G. Lennox, and H. Lindley. 1965. The chemistry of keratins. *In* "Advances in Protein Chemistry" (C. B. Anfinsen, M. L. Anson, J. T. Edsall, and F. M. Richards, eds.), Vol. 20, pp. 191–346. Academic Press, New York.

Danforth, C. 1925. Hair in its relation to questions of homology and phylogeny. *Amer. J. Anat.* **36**: 47–68.

Dobb, M. G., and J. Sikorski. 1961. La spécificite de l'interaction métalkératine et l'interprétation quantitative des images fournies par le microscope électronique des coupes ultra-minces des fibres de laine. *In* "Structure de la Laine," p. 37–50. Institut Textile de France, Paris.

Dorfman, R. I., and A. S. Dorfman. 1962. Assay of subcutaneously administered androgens on the chick's comb. *Acta Endocrinol. (Copenhagen)* **41**: 101–106.

Downes, A. M., L. F. Sharry, and G. E. Rogers. 1963. Separate synthesis of fibrillar and matrix proteins in the formation of keratin. *Nature (London)* **199**: 1059–1061.

Downes, A. M., W. H. Clark, and T. C. Dagg. 1967. Use of radioisotopes in the measurement of wool growth. *At. Energy (Sydney)* **10**: 2–7.

Durward, A., and K. M. Rudall. 1949. Studies on hair-growth in the rat. *J. Anat.* **83**: 325–335.

Durward, A., and K. M. Rudall. 1958. The vascularity and patterns of growth of hair follicles. *In* "The Biology of Hair Growth" (W. Montagna and R. A. Ellis, eds.), pp. 189–218. Academic Press, New York.

Ellis, R. A., and G. Moretti. 1959. Vascular patterns associated with catagen hair follicles in the human scalp. *Ann. N. Y. Acad. Sci.* **83**: 448–457.

Eppenberger, E., and S. L. Hsia. 1972. Binding of steroid hormones by the 105,000 X g supernatant fraction from homogenates of rat skin and variations during the hair cycle. *J. Biol. Chem.* **247**: 5463–5469.

Epstein, W. L., and H. L. Maibach. 1969. Cell proliferation and movement in human hair bulbs. *In* "Advances in Biology of Skin. Hair Growth" (W. Montagna and R. L. Dobson, eds.), Vol. 9, pp. 83–98. Pergamon, Oxford.

Fazekas, A. G., and A. Lanthier. 1971. Metabolism of androgens by isolated beard hair follicles. *Steroids* **18**: 367–379.

Fitzpatrick, T. B., W. C. Quevedo, Jr., A. L. Levene, V. J. McGovern, Y. Mishima, and A. G. Oettle. 1966. Terminology of vertebrate melanin-containing cells: 1965. *Science* **152**: 88–89.

Fraser, R. D. B., and T. P. MacRae. 1956. The distribution of ortho- and para-cortical cells in wool and mohair. *Text. Res. J.* **26**: 618.

Fraser, R. D. B., and G. E. Rogers. 1953. Microscopic observations of the alkaline thioglycolate extraction of wool. *Biochim. Biophys. Acta* **12**: 484–485.

Fraser, R. D. B., and G. E. Rogers. 1965. The bilateral structure of wool cortex and its relation to crimp. *Aust. J. Biol. Sci.* **8**: 288–299.

Fraser, R. D. B., T. P. MacRae, and G. E. Rogers. 1972. "Keratins. Their Composition, Structure and Biosynthesis." Thomas, Springfield, Illinois.

Gemmell, R. T., and R. E. Chapman. 1971. Formation and breakdown of the inner root sheath and features of the pilary canal epithelium in the wool follicle. *J. Ultrastruct. Res.* **36**: 355–366.

Gillespie, J. M. 1960. The isolation and properties of some soluble proteins from wool I. The isolation of a low-sulphur protein. *Aust. J. Biol. Sci.* **13**: 81–103.

Gillespie, J. M. 1965. The high-sulphur proteins of normal and aberrant keratins. *In* "Biology of Skin and Hair Growth" (A. G. Lyne and B. F. Short, eds.), pp. 377–398. Amer. Elsevier, New York.

Gillespie, J. M., P. J. Reis, and P. G. Schinckel. 1964. The isolation and properties of some soluble proteins from wool I. The proteins in wools of increased sulphur content. *Aust. J. Biol. Sci.* **17**: 548–560.

Goddard, D. R., and L. Michaelis. 1934. A study on keratin. *J. Biol. Chem.* **106**: 605–614.

Goddard, D. R., and L. Michaelis. 1935. Derivatives of keratin. *J. Biol. Chem.* **112**: 361–371.

Hamilton, J. B. 1942. Male hormone stimulation is prerequisite and an incitant in common baldness. *Amer. J. Anat.* **71**: 451–480.

Hamilton, J. B., and W. Montagna. 1950. The sebaceous glands of the hamster. I. Morphological effects of androgens on integumentary structures. *Amer. J. Anat.* **86**: 191–233.

Hamilton, J. B., H. Terada, G. E. Mestler, and W. Tirman. 1969. I. Coarse sternal hairs, a male secondary sex character that can be measured quantitatively: The influence of sex, age and genetic factors. II. Other sex-differing characters: Relationship to age, to one another, and to values for coarse sternal hairs. *In* "Advances in Biology of Skin. Hair Growth" (W. Montagna and R. L. Dobson), Vol. 9, pp. 129–151. Pergamon Press, Oxford.

Happey, F., and A. G. Johnson. 1962. Some electron microscope observations on hardening in the human hair follicle. *J. Ultrastruct. Res.* **7**: 316–327.

Harding, H. W. J., and G. E. Rogers. 1971. The E- (γ-glutamyl) lysine cross-linkage in citrulline containing protein fractions from hair. *Biochemistry* **10**: 624–630.

Harrap, B. S., and J. M. Gillespie. 1963. A further study on the extraction of reduced proteins from wool. *Aust. J. Biol. Sci.* **16**: 542.

Hashimoto, K. 1970. The ultrastructure of the skin of human embryos. IX. Formation of the hair cone and intraepidermal hair canal. *Arch. Klin. Exp. Dermatol.* **238**: 333–345.

Hausman, L. A. 1932. The cortical fusi of mammalian hair shafts. *Amer. Natur.* **66**: 461–470.

Hausman, L. A. 1944. Applied microscopy of hair. *Sci. Monthly* **59**: 195–202.

Horio, M., and T. Kondo. 1953. Crimping of wool fibers. *Text. Res. J.* **23**: 373–386.

Kassenbeck, P. 1961. Le polymorphisme de fibres keratiniques. *In* "Structure de la Laine," pp. 51–74. Institut Textile de France, Paris.

Kligman, A. M. 1959. The human hair cycle. *J. Invest. Dermatol.* **33**: 307–316.

Kligman, A. M. 1961. Pathologic dynamics of human hair loss. I. Telogen effluvium. *Arch. Dermatol.* **83**: 175–198.

Klingmüller, G. 1957. Kapillardarstellungen mittels Phosphatasefärbungen in gefriergetrocknetem Hautgewebe. Proc. 11th Intern. Congr. Dermatol., *Acta Dermato-Venereol. Suppl.* **3**: 433–436.

Knutson, D. D. 1973. Electron microscopic observations of acne. *In* "Advances in Biology of Skin. Sebaceous Glands and Acne vulgaris," (W. Montagna, M. Bell, and J. Strauss, guest eds.). Vol. 14. *J. Invest. Dermatol.* Special Issue. (In press.)

Kondo, S., and K. Adachi. 1971. Phosphofructokinase (PFK) regulation of glycolysis in skin. *J. Invest. Dermatol.* **57**: 175–179.

Lagermalm, G. 1954. Structural details of the surface layers of wool. *Text. Res. J.* **24**:17–25.

Ling, J. F. 1965. Hair growth and moulting in the Southern elephant seal *(Mirounga leonina)* (Linn.). *In* "Biology of the Skin and Hair Growth" (A. G. Lyne and B. F. Short, eds.), pp. 525–544. American Elsevier Publishing Co., New York.

Lowry, O. H. 1953. The quantitative histochemistry of the brain. Histological sampling. *J. Histochem. Cytochem.* **1**: 420–428.

Lundgren, H. P., and W. H. Ward. 1963. The keratins. *In* "Ultrastructure of Protein Fibers" (R. Borasky, ed.), pp. 39–121. Academic Press, New York.

Lynfield, Y. L. 1960. Effect of pregnancy on the human hair cycle. *J. Invest. Dermatol.* **55:** 323–327.

Matoltsy, A. G. 1953. A study of the medullary cells of hair. *Exp. Cell Res.* **5:** 98–109.

Matoltsy, A. G., and M. N. Matoltsy. 1966. The membrane protein of horny cells. *J. Invest. Dermatol.* **46:** 127–129.

Mercer, E. H. 1949. Some experiments on the orientation and hardening of keratin in the hair follicle. *Biochim. Biophys. Acta* **3:** 161–169.

Mercer, E. H. 1953. The heterogeneity of the keratin fibres. *Text. Res. J.* **23:** 388–397.

Mercer, E. H. 1961. "Keratin and Keratinization." Pergamon, Oxford.

Montagna, W. 1962. "The Structure and Function of Skin," 2nd ed. Academic Press, New York.

Montagna, W. 1963. Considerazioni sulla filogenesi del cuoico capelluto. *Minerva Dermatol.* **38** (Suppl.): 202–208.

Montagna, W., and H. B. Chase. 1956. Histology and cytochemistry of human skin. X. X-irradiation of the scalp. *Amer. J. Anat.* **99:** 415–446.

Montagna, W., and R. A. Ellis. 1957a. Histology and cytochemistry of human skin. XII. Cholinesterases in hair follicles of the scalp. *J. Invest. Dermatol.* **29:** 151–157.

Montagna, W., and R. A. Ellis. 1957b. Histology and cytochemistry of human skin. XIII. The blood supply of the hair follicles. *J. Nat. Cancer Inst.* **19:** 451–456.

Montagna, W., and D. M. Ford. 1969. Histology and cytochemistry of human skin. XXXIII. The eyelid. *Arch. Dermatol.* **100:** 328–335.

Montagna, W., and L. Giacometti. 1969. Histology and cytochemistry of human skin. XXXII. The external ear. *Arch. Dermatol.* **99:** 757–767.

Montagna, W., H. Machida, and E. Perkins. 1966. The skin of primates. XXVIII. The stump-tail macaque *(Macaca speciosa). Amer. J. Phys. Anthropol.* **24:** 71–85.

Myers, R. J., and J. B. Hamilton. 1951. Regeneration and rate of growth of hairs in man. *Ann. N. Y. Acad. Sci.* **53:** 562–568.

Northcutt, R. C., D. P. Island, and G. W. Liddle. 1969. An explanation for the target organ unresponsiveness to testosterone in the testicular feminization syndrome. *J. Clin. Endocrinol. Metab.* **29:** 422–425.

O'Donnell, I. J., and E. O. P. Thompson. 1964. Studies on reduced wool IV. The isolation of a major component. *Aust. J. Biol. Sci.* **17:** 973–989.

Oliver, R. F. 1967a. Ectopic regeneration of whiskers in the hooded rat from implanted lengths of vibrissa follicle wall. *J. Embryol. Exp. Morphol.* **17:** 27–34.

Oliver, R. F. 1967b. The experimental induction of whisker growth in the hooded rat by implantation of dermal papillae. *J. Embryol. Exp. Morph.* **16:** 231–244.

Orentreich, N. 1969. Scalp hair replacement in man. *In* "Advances in Biology of Skin. Hair Growth" (W. Montagna and R. L. Dobson, eds.), Vol. 9, pp. 99–108. Pergamon, Oxford.

Orfanos, C., and H. Ruska. 1970. Die keratine der Haut and des Haares. *Der Hautarzt* **21:** 343–351.

Papa, C. M., and A. M. Kligman. 1965. Stimulation of hair growth by topical application of androgens. *J. Amer. Med. Assoc.* **191:** 521–525.

Parakkal, P. F. 1969a. The fine structure of anagen hair follicle of the mouse. *In* "Advances in Biology of Skin. Hair Growth" (W. Montagna and R. L. Dobson, eds.), Vol. 9, pp. 441–469. Pergamon, Oxford.

Parakkal, P. F. 1969b. Ultrastructural changes of the basal lamina during the hair growth cycle. *J. Cell Biol.* **40**: 561–564.

Parakkal, P. F. 1969c. Role of macrophages in collagen resorption during hair growth cycle. *J. Ultrastruct. Res.* **29**: 201–217.

Parakkal, P. F. 1970. Morphogenesis of the hair follicle during catagen. *Z. Zellforsch.* **107**: 174–186.

Pecoraro, V., I. Astore, and J. M. Barman. 1964a. Cycle of the scalp hair of the newborn child. *J. Invest. Dermatol.* **43**: 145–147.

Pecoraro, V., I. Astore, J. M. Barman, and C. I. Araujo. 1964b. The normal trichogram in the child before the age of puberty. *J. Invest. Dermatol.* **42**: 427–430.

Pinkus, F. 1927. Die normale Anatomie der Haut. *In* "Handbuch der Haut-und Geschlechtskrankheiten" (J. Jadassohn, ed.), pp. 1–378, Vol. 1, Part 1. Springer-Verlag, Berlin, New York.

Pinkus, H. 1958. Embryology of hair. *In* "The Biology of Hair Growth" (W. Montagna and R. A. Ellis, eds.), pp. 1–32. Academic Press, New York.

Puccinelli, V. A., R. Caputo, and B. Ceccarelli. 1967. The structure of human hair follicle and hair shaft: an electron microscope study. *G. Ital. Dermatol.* **108**: 453–498.

Puche, R. C., V. Pecoraro, I. Astore, and J. M. Barman. 1971. Relationships between the urinary 17-ketosteroides and some characteristics of the human scalp hair. *Steroidologia* **2**: 121–127.

Rhodin, J. A. G., and E. J. Reith. 1962. Ultrastructure of keratin in oral mucosa, skin, esophagus, claw, and hair. *In* "Fundamentals of Keratinization" (E. O. Butcher, and R. F. Sognnaes, eds.), pp. 61–94. Amer. Assoc. Advance. of Sci., Washington, D. C.

Robins, E. G., and A. S. Breathnach. 1969. Fine structure of the human foetal hair follicle at hair-peg and early bulbous-peg stages of development. *J. Anat.* **104**: 553–569.

Robins, E. G., and A. S. Breathnach. 1970. Fine structure of bulbar end of human foetal hair follicle at stage of differentiation of inner root sheath. *J. Anat.* **107**: 131–146.

Rogers, G. E. 1957. Electron microscope observations on the glassy layer of the hair follicle. *Exp. Cell Res.* **13**: 521–528.

Rogers, G. E. 1959. Electron microscope studies of hair and wool. *Ann. N. Y. Acad. Sci.* **83**: 378–399.

Rogers, G. E. 1963. The localization and significance of arginine and citrulline in proteins of the hair follicle. *Histochem. Cytochem.* **11**: 700–705.

Rogers, G. E. 1964a. Structural and biochemical features of the hair follicle. *In* "The Epidermis" (W. Montagna and W. C. Lobitz, Jr., eds.), pp. 179–232. Academic Press, New York.

Rogers, G. E. 1964b. Isolation and properties of inner sheath cells of hair follicles. *Exp. Cell Res.* **33**: 264–276.

Rogers, G. E. 1969. The structure and biochemistry of keratin. *In* "The Biological Basis of Medicine" (E. E. Bittar and N. Bittar, eds.), Vol. 6, pp. 21–57. Academic Press, New York.

Roth, S. I. 1965. The cytology of the murine resting (telogen) hair follicle. *In* "Biology of the Skin and Hair Growth" (A. G. Lyne and B. F. Short, eds.), pp. 233–250. Amer. Elsevier, New York.

Roth, S. I. 1967. Hair and nail. *In* "A. S. Zelickson's Ultrastructure of Normal and Abnormal Skin," pp. 105–131. Lea & Febiger, Philadelphia, Pennsylvania.

Roth, S. I., and E. B. Helwig. 1964. The cytology of the cuticle, the cortex and the medulla of the mouse hair. *J. Ultrastruct. Res.* **11**: 52–67.

Ryder, M. L. 1956a. Blood supply of the wool follicle. *Wool Indust. Res. Assoc. Bull.* **18**: 142–147.

Ryder, M. L. 1956b. Use of radioisotopes in the study of wool growth and fibre composition. *Nature (London)* **178**: 1409–1410.

Ryder, M. L. 1958. Investigations into the distribution of thiol groups in the skin follicles of mice and sheep and the entry of labelled sulphur compounds. *Proc. Roy. Soc. Edinburgh* B**67**: 65–82.

Saitoh, M., M. Uzuka, M. Sakamoto, and T. Kobori. 1969. Rate of hair growth. *In* "Advances in Biology of Skin. Hair Growth" (W. Montagna and R. L. Dobson, eds.), Vol. 9, pp. 183–201. Pergamon, Oxford.

Schinckel, P. G. 1961. Mitotic activity in wool follicle bulbs. *Aust. J. Biol. Sci.* **14**: 659–663.

Silver, A. F., H. B. Chase, and C. T. Arsenault. 1969. Early anagen initiated by plucking compared with early spontaneous anagen. *In* "Advances in Biology of Skin. Hair Growth" (W. Montagna and R. L. Dobson, eds.), Vol. 9, pp. 265–286. Pergamon, Oxford.

Staricco, R. G. 1959. Amelanotic melanocytes in the outer sheath of the human hair follicle. A preliminary report. *J. Invest. Dermatol.* **33**: 295–297.

Staricco, R. G. 1960. The melanocytes and the hair follicle. *J. Invest. Dermatol.* **35**: 185–194.

Staricco, R. G. 1961. Mechanism of migration of the melanocytes from the hair follicle into the epidermis following dermabrasion. *J. Invest. Dermatol.* **36**: 99–104.

Steinert, P. M., P. Y. Dyer, and G. E. Rogers. 1971. The isolation of nonkeratin protein filaments from inner root sheath cells of the hair follicle. *J. Invest. Dermatol.* **56**: 49–54.

Straile, W. E. 1962. Possible functions of the external root sheath during growth of the hair follicle. *J. Exp. Zool.* **150**: 207–224.

Takashima, I., and W. Montagna. 1971. Studies of common baldness of the stump-tailed macaque *(Macaca speciosa)*. VI. The effect of testosterone on common baldness. *Arch. Dermatol.* **103**: 527–534.

Takashima, I., K. Adachi, and W. Montagna. 1970. Studies of common baldness in stump-tailed macaque. IV. *In vitro* metabolism of testosterone in the hair follicles. *J. Invest. Dermatol.* **55**: 329–334.

Takayasu, S., and K. Adachi. 1972. The conversion of testosterone to 17β-hydroxy-5α-androstan-3-one (dihydrotestosterone) by human hair follicles. *J. Clin. Endocrinol. Metab.* **34**: 1098–1101.

Trotter, M. 1930. The form, size, and color of head hair in American whites. *Amer. J. Phys. Anthropol.* **14**: 433–445.

Trotter, M., and O. H. Duggins. 1950. Age changes in head hair from birth to maturity. III. Cuticular scale counts of hair of children. *Amer. J. Phys. Anthropol.* **8**: 467–484.

Uno, H., K. Adachi, F. Allegra, and W. Montagna. 1968a. Studies of common baldness of the stumptailed macaque. II. Enzyme activities of carbohydrate metabolism in the hair follicles. *J. Invest. Dermatol.* **51**: 11–18.

Uno, H., K. Adachi, and W. Montagna. 1968b. Glycogen contents of primate hair follicles. *J. Invest. Dermatol.* **51**: 197–199.

Uno, H., F. Allegra, K. Adachi, and W. Montagna. 1967. Studies on common baldness of the stumptailed macaque. I. Distribution of the hair follicles. *J. Invest. Dermatol.* **49**: 288–296.

Van Scott, E. J., and T. Ekel. 1958. Geometric relationships between the matrix of the hair-bulb and its dermal papilla in normal and alopecic scalp. *J. Invest. Dermatol.* **31**: 281–287.

Winkelmann, R. K. 1959. The innervation of a hair follicle. *Ann. N. Y. Acad. Sci.* **83**: 400–407.

Winkelmann, R. K. 1960. "Nerve Endings in Normal and Pathologic Skin: Contributions to the Anatomy of Sensation." Thomas, Springfield, Illinois.

Wolbach, S. B. 1951. The hair cycle of the mouse and its importance in the study of sequence of experimental carcinogenesis. *Ann. N. Y. Acad. Sci.* **53**: 517–536.

Zahn, H., and M. Biela. 1968a. Über die isolierung tyrosinreicher proteine aus wolle. *Textil. Praxis* **23**: 103–106.

Zahn, H., and M. Biela. 1968b. Tyrosin reiche proteine im ameisensaureextrakt von reduzierter wolle. *Europ. J. Biochem.* **5**: 567–573.

8

Effects of Malnutrition on the Morphology of Hair Roots

ROBERT B. BRADFIELD[*]

I. Introduction

Today there is an increasing need to develop age-independent techniques that will enable us to assess rapidly the incidence and severity of protein-calorie malnutrition (PCM) during field studies. Such techniques are important in the tropics where routine laboratory techniques are not adaptable to high temperatures and humidity and are rendered useless by irregular or nonexistent electrical power.

Hairs are good models for diagnosing protein-calorie malnutrition. For example, of the eleven indicators of PCM listed by the World Health Organization, four deal with hair (Jelliffe, 1955). Moreover, changes in hair color have already been used successfully to diagnose protein deprivation. On the other hand, during dietary protein deprivation, protein levels are maintained in the blood at the expense of other tissue. Therefore, protein-rich tissue provides a better test for early protein deprivation than blood.

---

[*] University of California, Berkeley.

But in tropical areas, hair probably provides the best gauge for the early recognition of PCM. It can be transported in hot, humid areas without special precautions for chemical stabilization or refrigeration. The samples can be obtained and examined in ambient temperatures with a low-power microscope even in laboratories without electricity. To many tropical hospital laboratories that are forced to operate under such handicaps, these advantages provide cogent reasons for selecting hair as a model for determining PCM. Moreover, the use of the root rather than the shaft eliminates the 3-week lag between the time it takes for protein deficiency to affect growing hair and the time it takes for hair to reach a suitable length to be studied. Moreover, hair root preparations are unaffected by some ethnic influences on hair color and texture.

II. Protein Synthesis in the Follicle

The synthesis of nuclear DNA is an essential prelude to cell division, but experiments with radioactive thymidine, a specific precursor of DNA, have shown that DNA synthesis takes place only in the nuclei in the matrices of the bulb and the external root sheath. Studies of autoradiographs show that cells labeled with ^{3}H-thymidine originate in the matrix and move up to form all the layers of the follicle except the external root sheath, which is largely self-sustaining. From the matrix the cells rise towards the skin surface as a unit.

The rate at which the matrix cells divide has been studied mostly in rodents injected with a pulse dose of ^{3}H-thymidine. For about 2 hours after the injection, some nuclei are labeled, but there are no mitotic figures. After an abrupt increase to 100% of labeled cells, mitosis falls to less than 10% over the next few hours and is followed by another sudden rise between 14 and 15 hours after the injection. Since the interval between the two increases indicates the length of the mitotic cycle, the matrix cells in rodents divide about once every 12 hours. At the same time, they must double their mass, which to remain constant has to move out of the bulb. As the cells from the matrix rise to the lower keratogenous zone, their volume increases about four times.

The synthesis of ribonucleic acids has been investigated by observing the incorporation of ^{3}H-cytidine, a precursor of both DNA and RNA; since DNA synthesis is restricted to the matrix and external root sheath, ^{3}H-cytidine is a specific precursor of RNA in the rest of the follicle.

Up to 3 hours after the injection of ^{3}H-cytidine into rats, radioactivity gradually decreases in cortical cells from the matrix to the top of the keratogenous zone. This indicates that the rate of RNA synthesis in the keratogenous zone is less than that in the bulb. The statistically significant correlation between the rate of RNA synthesis and the size of the nucleoli of cortical cells is compatible with current theories that the nucleolus manufactures most of the cellular RNA. The high rate of RNA synthesis by the cells in the bulb is followed by a high rate of protein synthesis by the same cells when they reach the keratogenous zone 12 to 24 hours later.

Studies on the incorporation of radioactive amino acids into growing follicles show that the cortical cells of the keratogenous zone have a higher rate of protein synthesis than the matrix. For example, in rats the rate of protein synthesis per unit volume of tissue is about three times greater in the keratogenous zone than in the matrix. The cells of the keratogenous zone are about four times larger than those of the matrix, and the ratio of the rates of protein synthesis per cell is 12:1. Since the matrix cells double their mass and divide once every 12 hours, the rate of protein synthesis in the matrix equals that required to produce the protein mass of one matrix cell every 12 hours. The rate of protein synthesis, then, is about twelve times greater in the cells of the keratogenous zone, and the protein equivalent of one matrix cell is added to each of these cells every hour. These rough estimates emphasize the magnitude of the metabolic activity of the cells that form the hair cortex. The intense anabolic activity of hair follicles in terms of protein synthesis and the fact that no measurable breakdown of protein occurs in these cells during differentiation make them valuable organs in which to investigate protein synthesis. If any factor reduces the rate of protein synthesis in the body, organs with a high rate of activity like hair follicles can be expected to be sensitive indicators of such a reduction (Sims, 1970).

III. Changes in the Hair Shaft

During PCM, and particularly in persons with kwashiorkor, the color and texture of human hair change. Hypochromotrichia, loss of natural curl, brittleness, and sparseness are so evident and universal that they are used to diagnose the disorder (Jelliffe, 1955; Jelliffe and Welbourn, 1963).

Between the acute and recovery phases of both marasmus and marasmic-kwashiorkor, significant changes occur in the diameter of the hair of Indian children in the Andes (Bradfield, 1967; Bradfield *et al.*, 1966). These changes have also been reported in experimental animals whose diets were modified to induce marasmiclike and kwashiorkorlike conditions (Pond *et al.*, 1965). Because the reduction of hair shaft diameter was closely associated with decreased serum albumin values (Bradfield, 1967; Bradfield *et al.*, 1966), the stress-strain characteristics of hair were examined in children suffering from marasmus and marasmic-kwashiorkor and in experimental animals in which the condition had been simulated.

The hair of the pigs with PCM showed a much reduced maximum supercontraction in 8 M LiBr at 96°C for both the kwashiorkorlike syndrome (20.8%) and the marasmiclike syndrome (22.8%) compared with the hair of littermate controls (29.0%). The fibers of protein-deficient animals also contracted at a slower rate.

Mechanical properties were studied in a controlled atmosphere at 70°F and 65% relative humidity. Extension at break was significantly lower in the protein-deficient animals, particularly in those with the kwashiorkorlike syndrome (control 42%, kwashiorkor 10%, marasmus 30%). In addition, less than half as much force was required to break the fibers of protein-deficient animals than those of littermate controls. Mechanically, the fibers of protein-deficient pigs were weaker than and behaved differently from normal hairs of the same diameter.

The same techniques were applied to examine the hair of children suffering from marasmus and marasmic-kwashiorkor. Hair samples from children with acute marasmic-kwashiorkor exhibited significantly lower values in the modulus, stress at yield, and stress at break than hair samples from the same children after they had recovered. Both the molecular orientation and the lateral order appeared to have increased considerably with clinical recovery (Rebenfeld, 1965). To a lesser extent, this was also true in marasmus.

Less complicated methods of studying the mechanical properties of hair are being tried in an attempt to develop a system easily adapted to tropical conditions. These methods include break at 223° to 225°F as a measure of cross-linking, density gradient in chloral hydrate (1.495 to 1.415 sg), rotation, torsional modulus, gross physical reaction to solvents (particularly benzene followed by HNO_3), and swelling in water and alcohols.

The levels of cystine in hair under PCM conditions appear to vary. Bigwood and Robazza (1955) found less cystine in the hair of African children with kwashiorkor than in normal children from the same area. According to Close (1958), in the acute phases of kwashiorkor, the cystine content in the hair of African children was reduced whereas under similar clinical conditions straighthaired Guatemalan Indian children did not show cystine changes even when the color had changed. A recent Japanese study demonstrated significant cystine changes in the hair of elementary school children after an experimental school lunch program which lasted several months (Koyanagi et al., 1965). To some degree, the disagreement between these groups is due to ethnic differences, to analytical procedures, and in some cases to a lack of adequate controls.

Using the method of Schram and co-workers (1954) in studies with Andean Indians, Bradfield (1967) determined cystine as cysteic acid after oxidation with performic acid. Cysteic acid was isolated from the hydrolysate by passage through a 15 cm Dowex 50 column (Moore and Stein, 1951). During the acute stages of both marasmic-kwashiorkor and marasmus, neither nitrogen nor cystine was lower than the recovery values for the same child (Sanda and Bradfield, 1967). Microbiological cystine determinations confirmed these results. Studies in South Africa demonstrate that linear growths of scalp hair are drastically reduced by 90% in hospitalized children with kwashiorkor (Sims, 1970).

The soft texture of scalp hair and its lack of curl in Zulu patients with kwashiorkor suggests a reduction in protein density. Microanalysis of x-ray emission in cryostat sections of the hair of kwashiorkor patients did not reveal any changes in density but did show that it absorbed more water. Since the amount of water absorbed is limited by the ability of the cross-links to resist deformation, perhaps the cross-links are weak. The hair of Andean Indians with acute kwashiorkor demonstrates significant deviations from normal in the modulus, stress at yield, and stress at break. These changes suggest reduced efficiency of the cross-links betwen polypeptide chains.

IV. Changes in Hair Roots

On the assumption that the cyclic activity of scalp follicles is independent, each type of follicle must be distributed at random. In the steady state, the ratio of empty, anagen, and telogen follicles is correlated with the average time each follicle spends at each stage of the cycle.

Since counts of empty, anagen, and telogen follicles can be made only on biopsy specimens, a second best index of hair cycles is the proportions of anagen, telogen, and broken shafts in samples of plucked hair; these broken shafts must be regarded as follicles in anagen that produced hair of such narrow diameter that it broke when plucked.

Studies have been made on changes in hair roots in (1) marasmus and kwashiorkor patients, (2) young adults who volunteered for experimental protein deprivation, and (3) nutritional status surveys of moderate malnutrition.

The morphology of hair roots has been studied in 26 Andean Indian children, 13 with kwashiorkor, and 13 healthy ones of the same age. The hair bulbs from the healthy children were about the size of those of Caucasian children of comparable age. In all the samples studied, the number of well-formed and heavily pigmented growing hairs was normal ($66 \pm 6\%$); the bulbs were not atrophied, and the mean diameter was $18 \pm 0.7 \times 10^{-2}$ mm. The rest of the follicles were in telogen ($10 \pm 3\%$); the number of dysplastic hairs was $25 \pm 5\%$ of the total sample and the inner and outer root sheaths were fairly intact (100% and $60\% \pm 10\%$, respectively).

The hair of children with kwashiorkor, however, had strikingly different physical characteristics. Instead of the usual coarse, lustrous black color typical of Andean Indians, the hair was lusterless, fine-textured, and dry, and there were numerous dysplastic hairs, many broken at a constriction in the shaft.

The hair roots differed noticeably from those of normal children, i.e., the number of anagen follicles was significantly less than normal ($26 \pm 6\%$) ($p < 0.01$), and those that were present were abnormally formed with severe atrophy and shaft constriction immediately above the bulb. In most cases, anagen follicles showed a marked pigment depletion and many appeared speckled. The extent of atrophy was indicated by the mean bulb diameter of $7 \pm 0.4 \times 10^{-2}$ mm, one-third that of normal values ($p < 0.01$). In the telogen phase, the number of clubs increased significantly to $45\% \pm 5\%$, and the number of dysplastic hairs was about equal to that of the normal samples ($29 \pm 6\%$). There was a significant loss of both internal and external sheaths ($53 \pm 8\%$ and $36 \pm 7\%$, respectively).

Hair root samples were also taken during several stages of recovery. Both clinical and hair root response to treatment varied greatly with the individual according to the severity of protein deprivation and the relative virulence of concurrent gastrointestinal and respiratory infection. From successive samples in each child, however, several observations could be made that indicated the sequence of events during protein repletion. The shift to the growth phase appeared to be rapid, since after 6 weeks no sample exceeded 4% telogens. During the first 2 months after admission, a number of atrophied anagens were found and there was still fraying of the bulbs. As the latter regained normal size and form, the pigmentation remained speckled instead of regaining the normal solid dark appearance. After about 3 months of dietary and medical treatment, the bulbs were well-formed and heavily pigmented and the sheaths complete. However, the bulbs were not yet normal; the number of dysplastic hairs was double that of normal, and stretched and hooked shafts continued to appear. In several cases, hypochromotrichia persisted for 6 to 12 months after the children were discharged from the hospital. Little relationship between hypochromotrichia and serum albumin levels was found during either acute or recovery stages (Bradfield, 1968; Bradfield *et al.*, 1969).

The morphology of hair roots in cases of classical marasmus has also been studied (Bradfield *et al.*, 1969). Although kwashiorkor is the more spectacular form of PCM in terms of incidence and severity, marasmus is a more significant public health problem in many tropical developing areas, and its relative importance is probably grossly underestimated. The most striking change in the hair roots of children with marasmus was the almost complete absence of bulbs in the anagen phase; in 8 of the 15 cases studied, no anagen follicles were present in the samples. In four of the remaining cases, less than 1% of the follicles were in the growing phase, and these were abnormal, with shaft constriction and almost complete atrophy. The extent of the atrophy was reflected by the mean bulb diameter of $6 \pm 1 \times 10^{-2}$ mm. All cases showed marked pigment depletion in anagen, and many were completely dyspigmented; there was a further significant shift to telogen ($60 \pm 7\%$). About one-half of the total sample (46%) consisted of dysplastic hairs, many of which were broken at some constriction in the shaft. In 14 of the 15 cases, inner and outer root sheaths were incomplete.

Unlike the kwashiorkor children studied previously, the marasmic children exhibited a striking and highly significant shift to the resting phase of hair growth, demonstrated both by the almost complete absence of bulbs in the growing phase and by the large increase in the number of clubs in the resting phase. The number of broken hairs increased 50% over that in children with kwashiorkor. The absence of complete sheaths in 14 of the 15 marasmic children contrasted with the findings in the kwashiorkor children, in whom more than one-third of the bulbs had complete inner sheaths and more than one-half had complete outer sheaths.

These results suggest a physiological adaptation to chronic insufficient calorie and protein intake by a complete shift to the resting phase of hair growth. This reestablishment of priorities reduces the amount of nitrogen loss that would otherwise occur if the follicles remained in the growing phase. The morphological differences in the hair roots of marasmic and kwashiorkor children are no doubt due more to comparative differences in chronicity than to specific differences in relative protein–calorie density. In classical marasmus, which is a severe chronic undernutrition, the child adapts to the stress by failing to grow. The long-term effect on the hair follicle is a shift to the resting phase and a consequent conservation of nitrogen. Classic kwashiorkor represents a period of normal growth that has been interrupted by an acute condition. Sometimes linear growth continues as hair follicles adapt to this stress by diminution in the size of the bulbs already in the growing phase and by a partial shift to the resting phase. The defensive physiological adaptations to the calorie and protein stress of marasmus in hair roots provide a convenient clinical model since hair is easily accessible (Bradfield *et al.*, 1969).

Even granted that these reports were limited to classic cases of kwashiorkor and marasmus, studies of malnourished patients admitted to hospitals leave much to be desired from the standpoint of controlling and isolating the specific effects of malnutrition because (1) the children are studied on admission and differences in staff and treatment are inevitable, (2) differences in the type and amount of concurrent infection and infestation affect both the treatment and recovery rate, and (3) sometimes other nutritional deficiencies further complicate the situation.

To better control the conditions under which initial nutritional status is determined, studies were carried out in a metabolic ward. Healthy men 24 to 29 years old were fed a liquid diet that contained all known nutrients. The protein deprivation diet was prepared by the isocaloric substitution of dextromaltose for egg albumin.

In the first study, eight volunteers were fed the protein deprived diet for 15 consecutive days. During this period the hair roots underwent severe atrophy and decreased pigmentation, and there was a significant reduction in bulb diameter ($p < 0.05$); external root sheaths were consistently absent in the atrophied bulbs. During the anagen phase, marked atrophy occurred in the bulb in one-half the total sample. When protein was restored to the diet (Bradfield and Margen, 1967; Bradfield and Bailey, 1969), the changes were reversible to some degree. A second study of six young men confirmed these findings. At 11 days of protein depletion, consistent morphological changes were evident in the hair roots, including reduction in bulb diameter, atrophy, dyspigmentation, and absence of sheaths, but the growth phase of the hair did not change. Urinary nitrogen values reached minimal values, but serum protein and albumin levels remained normal. When protein was added to the diet, the hair root changes were partially reversed. The sequence of root changes, which commenced with bulb depigmentation, occurred as soon as bulb diameters began to reduce, but there was wide individual variation. Next, when there was substantial reduction in root diameter, sheath abnormalities occurred, but before atrophy. However, since sheath loss is also found in otherwise normal bulbs as a result of epilation techniques, the significance of this change was limited. Bulb atrophy proved to be a useful morphological sign because it was uniformly progressive, easy to assess, and showed less individual variation than other measures. Early in protein repletion, the mean bulb diameters tended to return to normal size, an indication that the bulb is a sensitive indicator of protein status in the body (Bradfield, 1971).

Bulb diameter is a more sensitive index of protein deprivation than the percentage of atrophy. The diameter of a well-pigmented bulb is reduced before much atrophy occurs. Nevertheless, the percentage of gross bulb atrophy can be assessed more rapidly than mean bulb diameter and occurs to such an extent that it has been a useful field index in work with malnourished children. In general, bulb diameters respond to protein deprivation more quickly and with less individual variation than either sheath or shaft parameters.

The changes in the morphology of the hair root described here show not only that they are due to protein deprivation, but also that they occur while serum albumin levels are still normal. The next step was to find out whether such changes occur early enough and regularly enough in preschool children with mild to moderate malnutrition.

The most commonly used criterion for ascertaining malnutrition in preschool children is a reduction in the normal weight-for-age relationship. Therefore in 72 West Indian children of African descent, the depression of this ratio was compared with the degree of change from normal hair root morphology. The sample included positive and negative controls, 16 healthy controls (with normal weight) and 12 clinical cases of PCM with a weight-for-age ratio less than 60% of normal. Both the control groups were the same age and came from the same ethnic group in the same Caribbean island.

Of several morphological characteristics, the most obvious was the reduction in mean root diameter. The analysis of variance revealed a very significant ($p < 0.01$) reduction in mean root diameter with reduction in weight-for-age. The samples were then grouped by weight-for-age, and the method of least significant differences was used to ascertain at which weight-for-age class the reduction in mean bulb diameter differed most from normal values. There was a sizable reduction in mean root diameter ($p < 0.01$) from the normal controls in the 81 to 90% weight-for-age group (Bradfield and Jelliffe, 1970; Bradfield et al., 1970).

The morphological changes occurred early enough and regularly enough to be used to differentiate between normal and moderately malnourished children. Under the conditions of this study, objective measurements of hair root diameter alone could not be used to classify the various levels of weight depression suggested by Jelliffe (1955). However, subjective classification of the incidence of certain abnormalities (bulb fraying, bulb dyspigmentation, incomplete outer root sheaths) helped to assess the relative severity of PCM. The correlation between reduction in weight-for-age and mean root diameter was high (r = 0.96) (Bradfield et al., 1972).

In another public health evaluation, the biochemical, hairs, and anthropometric methods used for early recognition of malnutrition were compared simultaneously in 170 preschool children living in the Guatemalan highlands. Decreases in hair root diameter and urinary urea/creatinine ratio were closely related as early indicators of inadequate protein intake. As later indicators of PCM, increased hair-root atrophy was related to changes in the ratio of nonessential/essential amino acids in sera and also to depressed weight-for-height (Nammacher et al., 1972).

The soluble protein and deoxyribonucleic acid (DNA) contents of hair roots have been studied in relation to protein stress. The DNA/hair root was found to correlate well with the protein content, and hair root volume and protein content were directly proportional (r = 0.9) (Crounse et al., 1970a,b). The results of these studies suggest that the response of hair root to PCM is one of reduction in size and cell number, a phenomenon that also occurs in such other tissues as intestinal mucosa. Biochemical evidence that hair root DNA and soluble protein correlate well with cell size corroborates the morphological findings, but morphological measurements are easier to perform in the field.

Comparisons of tests of PCM suffer from the lack of an accepted predictor of malnutrition against which each one can be compared. The interpretation of prevalence of cross-sectional surveys is further complicated by a number of factors. Sometimes the tests show varied responses to different types and severity of nutritional stress, and some reflect nutritional changes more rapidly than others. Moreover, the nutritional insult may be chronic or acute and its timing in relation to the time of the survey will influence test response. Because of homeostatic mechanisms, many of the tests probably change in a curvilinear rather than in a linear fashion. Infection and infestation synergistically affect nutrition status; therefore interpretations of cross-sectional studies are complicated. Age-independent measurements are useful because in many areas birth dates have little social significance and cannot be verified by documentation.

Like other methods used to assess nutritional status, the hair root method is not really a quantitative assessment. However, the ease and speed of sampling, the simplicity of examination, and the possibility of using this method under adverse climatic and other physical conditions make it more attractive for less well-developed tropical areas.

This method is primarily intended for assessing the nutritional status of groups, rather than of individuals. As in all such tests, there is a range of variation both in normal and abnormal sets of individuals. In public health practice, this test could be helpful in locating those segments of the population that are most likely to be malnourished.

References

Bigwood, E. J., and F. Robazza. 1955. Amino acid and sulfur content of hair in normal African natives and in kwashiorkor. Voeding 87: 251. (Abstr.)

Bradfield, R. B. 1967. Changes in hair associated with protein-calorie malnutrition. In "Proc. Colloquium on Protein Deficiencies and Calorie Deficiencies" (E. M. Widdowson, ed.). Churchill, London.

Bradfield, R. B. 1968. Changes in hair root morphology and hair diameter associated with protein-calorie malnutrition. In "Protein Deficiencies and Calorie Deficiencies" (R. McCance and E. M. Widdowson, eds.). Churchill, London.

Bradfield, R. B. 1971. Protein deprivation: comparative response of hair roots, serum albumin, and urinary nitrogen. Amer. J. Clin. Nutr. 24: 405–410.

Bradfield, R. B., and M. A. Bailey. 1969. Hair root response to protein undernutrition. In "Advances in Biology of Skin. Hair Growth" (W. Montagna and R. L. Dobson, eds.), Vol. 9, pp. 109–119. Pergamon, Oxford.

Bradfield, R. B., and E. F. P. Jelliffe. 1970. Early assessment of malnutrition. Nature (London) 225: 283–284.

Bradfield, R. B., A. Cordano, and G. G. Graham. 1969. Hair-root adaptation to marasmus in Andean Indian children. Lancet 2: 1395–1397.

Bradfield, R. B., and S. Margen. 1967. Morphological changes in human scalp hair roots during protein deprivation. *Science* 157: 438–439.

Bradfield, R. B., E. F. P. Jelliffe, and D. B. Jelliffe. 1972. Assessment of marginal malnutrition. *Nature (London)* 235: 112.

Bradfield, R. B., E. F. P. Jelliffe, and R. Neil. 1970. A comparison of hair root morphology and arm circumference as field tests of protein-calorie malnutrition. *J. Trop. Pediat.* 16: 196.

Bradfield, R. B., B. Poresky, and A. Cordano. 1966. Hair diameter changes during protein-calorie malnutrition. *In* "Proc. 11th Pacific Science Congress," Tokyo, Japan. (Abst.)

Close, J. 1958. Les modifications chimiques et morphologiques des cheveux, accompagnant le kwashiorkor. *Ann. Soc. Belg. Med. Trop.* 38: 95.

Crounse, R. G., A. J. Bollet, and S. Owens. 1970a. Quantitative tissue assay of human malnutrition using scalp hair roots. *Nature (London)* 228: 465–466.

Crounse, R. G., A. J. Bollet, and S. Owens. 1970b. Tissue assay of human protein malnutrition using scalp hair roots. *Trans. Ass. Amer. Physicians* 83: 185.

Jelliffe, D. B. 1955. "Infant Nutrition in the Tropics and Subtropics." World Health Organization, Monograph Ser. No. 29.

Jelliffe, D. B., and H. F. Welbourn. 1963. Clinical signs of mild-moderate protein-calorie malnutrition. *In* "Mild-Moderate Forms of Protein-Calorie Malnutrition" (G. Blix, ed.). Almqvist & Wiksells, Sweden.

Koyanagi, T., S. Hareyama, and T. Takanohashi. 1965. Effect of supplementation of vitamin, phosphorus, methionine or skim milk on the cystine content of hair, dark adaptation, creatine-creatinine excretion and growth of undernourished. *Tohoku J. Exp. Med.* 85: 108–114.

Moore, S., and W. H. Stein. 1951. Chromatography of amino acids on sulfonated polystyrene resins. *J. Biol. Chem.* 192: 663–681.

Nammacher, M. A., R. B. Bradfield, and G. Arroyave. 1972. Comparing nutritional status methods in a Guatemalan survey. *Amer. J. Clin. Nutr.* 25: 871.

Pond, W. G., R. H. Barnes, R. B. Bradfield, E. Kwong, and L. Krook. 1965. Effect of dietary energy intake on protein deficiency symptoms and body composition of baby pigs fed equalized but suboptimal amounts of protein. *J. Nutr.* 85: 57–66.

Rebenfeld, L. 1965. Personal communication.

Sanda, M., and R. B. Bradfield. 1967. Hair cystine changes in protein-calorie malnutrition. *Fed. Proc., Fed. Amer. Soc. Exp. Biol.* 26: 630. (Abstr.)

Schram, E., S. Moore, and E. J. Bigwood. 1954. Chromatographic determination of cysteic acid. *Biochem. J.* 57: 33–37.

Sims, R. T. 1970. Condensed and adopted from: Hair as an indicator of incipient and developed malnutrition and response to therapy—principles and practice. *In* "An Introduction to the Biology of the Skin (R. H. Champion, T. Gillman, A. J. Rook, and R. T. Sims, eds.). Blackwell, Oxford.

9

Nails

I. Introduction

The tips of the last digital phalanges of reptiles, birds, and most mammals terminate in keratinous claws or hoofs; only those of primates have nails. These protective plates at the end of the primate digits can be used for many purposes, but their primary function is to grasp and manipulate small objects. Despite the many tools fashioned by his genius, man still finds his nails nearly indispensable. Furthermore, like other cutaneous structures, nails have attained cosmetic value, especially among women, who have been extravagantly wooed by diverse media to improve the quality, shape, color, and luster of their nails.

Phylogenetically, it is apparent that nails have evolved from claws (Clark, 1936), which also protect the tips and sides of toes. Romer (quoted in Thorndike, 1968) has stated that nails are flattened claws whose matrices are restricted to their proximal ends. However, unlike avian and reptilian claws, whose keratin gives a β-diffraction pattern (Baden, 1970), all nails and the single claw possessed by some prosimians give an α-type diffraction pattern. This indicates that all nails are basically similar and that they are both structurally and chemically different from claws.

Man's knowledge of his nails is incomplete and even fragmentary. Perhaps because of the difficulty in obtaining adequate material for study, less is known about them than about any other cutaneous appendage. The bony phalanx to which the proximal edge of the nail matrix is attached and the toughness of the nail plate itself constitute the principal obstacles to good histological preparations. Because little progress has been made in recent years on the structure and function of nails, this chapter is limited to the basic nature of these structures. The reader interested in the minutiae of morphology should consult Horstmann's (1957) and Krantz's (1939) exhaustive accounts.

II. Structure

In spite of their apparent simplicity, nails have a complex architecture; to facilitate communication in describing them, we use the standard terminology that follows (Figs. 1 and 2). The *nail plate,* composed of tightly packed keratinized cells, is a roughly rectangular, convex horny shield set into a depression on the dorsal surface of the last phalanx of the digits. On the surface of the plate are fine longitudinal parallel striations, more conspicuous in the aged than in the young. The underside of the plate is grooved by more or less parallel longitudinal ridges. That portion of the nail just proximal to its free, distal edge has a characteristic underside, strikingly different from the longitudinal ridges (Fig. 3). The plate rests on the *nail bed,* a thickened stratified epidermis whose surface mirrors the underside of the nail plate and which is continuous with the variably bulging *hyponychium,* over which emerges the free edge of the nail. The hyponychium is continuous with the epidermis that covers the ventral surface of the digits. The proximal end of the nail bed is confluent with the *nail matrix,* an oblique wedge of tissue that extends down to the terminal interphalangeal joint. The upper and lower parts of the wedge are the *dorsal and ventral matrix,* respectively. At the proximal end of the nail plate, a fold of skin extends over it as the *proximal nail fold.* The horny epidermal extension of the tip of this fold is the *eponychium* or cuticle. The fold is continuous laterally as the *lateral nail fold.* At the point where it emerges underneath the proximal nail fold, the nail has a whitish crescent-shaped area called *lunula* (L. *luna* = moon), found predictably in the thumbs; it may or may not be present in the other digits. Its whitish color is due to the scattering of light by the thick matrix cells. The pink color of the

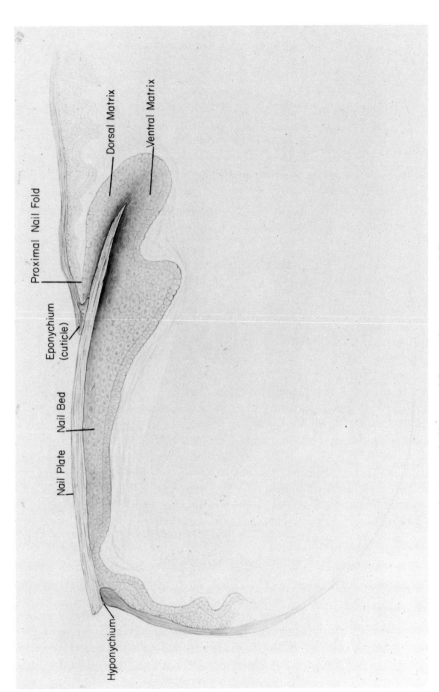

Fig. 1. Diagram of a longitudinal view of fingernail showing the different anatomical structures.

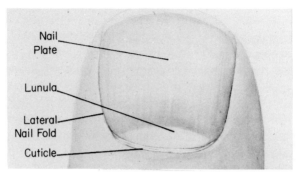

Fig. 2. Surface view of a nail.

rest of the nail is due to the extensive superficial bood vessels under the nail bed showing through the translucent nail plate.

III. Development

Nail anlagen appear in human embryos at about 9 weeks. By 10 weeks, "nails" consist of undifferentiated cells that mark the primary nail field (Zaias, 1963; Zaias and Baden, 1971). At this stage, the epithelium is 2 to 3 layers thick and like the rest of the epidermis is covered by a periderm. A tongue of cells growing into the dermis at the proximal end of the nail field establishes a primordial matrix. That portion of the skin that extends over the matrix is the future proximal nail fold. A nail plate, extending distally from underneath the proximal nail fold, is not discernible until 14 weeks of gestation. The plate is confluent with the stratum corneum of the nail bed upon which it rests. At this time, the epidermis of the nail bed has a well-developed granular layer. In 16-week-old fetuses, the nail plate covers about half the area of the nail bed, and the matrix and the basal portion of the nail bed have many ridges extending into the dermis. At about 17 weeks, the nail plate is fully grown and completely covers the nail bed.

IV. Formation of the Nail Plate

The origin of the nail plate remains controversial. The earlier theory that it is produced entirely by the matrix has now been challenged.

Fig. 3. (A) Scanning electron micrograph of the longitudinal grooves of the dermis after the nail plate is removed. At the hyponychial end, the grooves become honeycombed. (B) Enlarged view of longitudinal grooves. (Courtesy of Dr. W. H. Fahrenbach.)

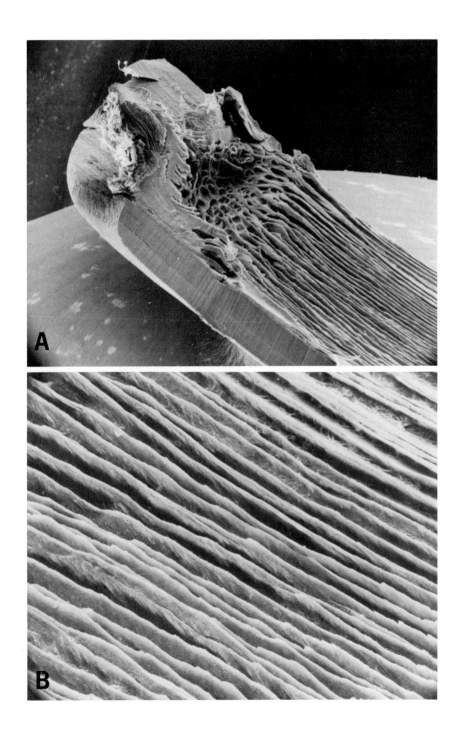

Several investigators (Lewis, 1954; Achten, 1963; Jarret and Spearman, 1966) have suggested that the plate is formed mainly by the matrix but also receives contributions from the nail bed and the proximal nail fold. Zaias and Alvarez (1968), tracing the course of labeled glycine in the nail matrix and plate of squirrel monkeys, concluded that the matrix alone gives rise to the nail plate. Norton (1971), however, experimenting with the toenails of human volunteers, arrived at somewhat different conclusions. One hour after the administration of tritiated thymidine, he found labeled cells in the nail matrix but not in the nail bed; one and two weeks later, however, there were labeled cells both in the matrix and in the nail bed. But one hour after he had injected labeled glycine, he found cells in both the matrix and the bed, with much greater uptake by the matrix cells than by those of the nail bed. In spite of their limitations and contrary to general belief, these experiments indicate that even though most of the nail plate is formed by the matrix, the nail bed is not inert but contributes to the formation of the nail plate.

V. Keratinization

When first formed in the fetus, the nail plate is wedged between the dorsal and ventral matrices, both of which, at this stage of development, add keratinized cells to the proximal edge of the nail. At this time, both the dorsal and the ventral side of this part of the nail plate have a granular layer (Hashimoto et al., 1966). The ventral part of the nail is attached to and apparently is continuous with the nail bed, which at this stage of development also has a granular layer. Once the nail plate is completely formed, the granular layer disappears from both the matrices and the nail bed. In the fetal nail bed, keratinization is similar to that in the epidermis, but in the adult nail bed it resembles that in the hair cortex. There are no keratohyalin granules in the cells during the differentiation of the nail plate in adults, and the keratinized cells are held tightly together like those in the hair cortex. Unlike the events that occur in the hair cortex, the membrane-coating granules (MCG) of the forming nail are discharged before the horny cells are formed. The cytoplasmic filaments are grouped into bundles as they are during the differentiation of the cortical cells of the hair.

X-ray diffraction studies (Baden, 1970) have shown that nails, like hair and epidermis, contain α-keratin. The water-binding capacity of nail is similar to that of hair but different from that of the stratum corneum. Both hair and nail plate are more permeable to water than the stratum corneum. The flux of water across the nail plate is ten times that across epidermis even though nail is much thicker and more compact that the stratum corneum (Burch and Winson, 1946).

VI. Other Cell Types in the Nail

Melanocytes are often present in the nail matrix. Negroes have pigmented nails but Caucasians rarely do (Monash, 1932). Embryologically, melanocytes migrate into the primordium of the nail matrix and the epidermis at the same time. Both Langerhans and Merkel cells have been reported in the nail matrix.

VII. Regeneration and Growth

If a nail plate is lost or surgically evulsed, a new one is soon formed from the matrix. When the nail plate is removed, most of the nail bed adheres to it and is later replaced by the cells that apparently migrate from the neighboring epidermis of the nail folds. The nail matrix is firmly attached to the underlying dermis and is therefore extremely difficult to remove completely. The thickness of the nail plate is determined by the size of the germinative cell population of the nail matrix.

How fast the nail grows is a matter of some disagreement. The rate depends on many factors, including systemic disorders that have nothing to do with nails (Pinkus, 1927). Pregnancy, trauma, and nail biting are said to increase the growth rate (Hamilton et al., 1955).

According to Bean (1963), the rate of nail growth is about 0.10 to 0.12 mm per day; earlier authors (Pinkus, 1927; Knobloch, 1951; Mörike, 1954), however, record nails grow about 1 mm per day. Growth is different in each finger, and this may explain the disparate reports of different authors (Knobloch, 1951). Moreover, the fluctuations in the growth rate of a single nail or the nails of a single individual should caution one against errors. In an extended study of 180 persons of different age groups, Knobloch (1951) found an increase in growth during the first twenty years, a higher increase during the twenties, and the highest in the thirties; from there on, the growth rate slowly decreased until the eighties. Silvestri (1955) also observed slower nail growth in the aged. Regardless of age, nails appear to grow faster on the right hand than on the left, perhaps a reflection of handedness,

20% more during the summer than during the winter, and twice as fast during the day than at night.

Whereas it is generally believed that injuries to nerves slow down nail growth (Head and Sherren, 1905; Pinkus, 1927), no one has done a systematic quantitative observation on the trophic influence of nerves on nail growth.

Physicians use such nail aberrations as "drummer" fingers, hourglass-shaped nails, leukonychy, and the cross furrows of Beau's as diagnostic aids. The histology of these alterations has been studied but little is known about their etiology.

References

Achten, G. 1963. L'ongle normal et pathologique. *Dermatologica* **126**: 229–245.

Baden, H. P. 1970. The physical properties of nail. *J. Invest. Dermatol.* **55**: 115–122.

Bean, W. B. 1963. Nail growth—a twenty-year study. *Arch. Intern. Med.* **111**: 476–482.

Burch, G. E., and T. Winson. 1946. Diffusion of water through dead plantar, palmar and torsal human skin and through toenails. *Arch. Dermatol. Syph.* **53**: 39–41.

Clark, W. E. LeGros. 1936. The problem of the claw in primates. *Proc. Zool. Soc. London,* pp. 1–24.

Hamilton, J. B., H. Terado, and G. E. Mestler. 1955. Studies of growth throughout the life span in Japanese: Growth and size of nails and their relationship to age, sex, heredity and other factors. *J. Gerontol.* **10**: 401–415.

Hashimoto, K., B. G. Gross, R. Nelson, and W. Lever. 1966. The ultrastructure of the skin of human embryos. III. The formation of the nail in 16–18 week old embryos. *J. Invest. Dermatol.* **47**: 205–217.

Head, H., and J. Sherren. 1905. The consequences of injury of the peripheral nerves in man. *Brain* **28**: 116–338.

Horstmann, E. 1957. Der nagel. *In* "Die Haut (Handbuch der Mikroskopischen Anatomie des Menschen)" (W. v. Mollendorff, ed.), pp. 176–196. Springer-Verlag, Berlin and New York.

Jarret, A., and R. Spearman. 1966. Histochemistry of the human nail. *Arch. Dermatol.* **94**: 652–657.

Knobloch, H. 1951. Fingernagelwachstum und Alter. *Z. Alternsforsch.* **5**: 357–362.

Krantz, W. 1939. Beitrag zur Anatomie des Nagels. *Dermat. Z.* **64**: 239–242.

Lewis, B. L. 1954. Microscopic studies of fetal and mature nail and surrounding soft tissue. *Arch. Dermatol.* **70**: 732–747.

Mörike, K. D. 1954. Das Verhalten des Hyponychiums beim normalen Nagelwachstum. *Anat. Anz.* **101**: 289–293.

Monash, S. 1932. Normal pigmentation in the nails of the Negro. *Arch. Dermatol. Syph.* **25**: 876–881.

Norton, L. A. 1971. Incorporation of thymidine-methyl H^3 and glycine-2-H^3 in the nail matrix and bed of humans. *J. Invest. Dermatol.* **56:** 61–68.

Pinkus, F. 1927. Die normale anatomie der Haut. *In* "Handbuch der Haut und Geschlechtskrankheiten" (J. Jadassohn, ed.), Bd. 1/1, S. 1–378. Springer-Verlag, Berlin and New York.

Silvestri, U. 1955. Su le unghie dei vecchi. *G. Gerontol. Suppl.* **5:** 277–287.

Thorndike, E. E. 1968. A microscopic study of the Marmoset claw and nail. *Amer. J. Phys. Anthropol.* **28:** 247–262.

Zaias, N. 1963. Embryology of the human nail. *Arch. Dermatol.* **87:** 37–53.

Zaias, N., and H. P. Baden. 1971. Disorders of the nail. *In* "Dermatology in General Medicine" (K. A. Arndt, W. H. Clark, Jr., A. Z. Eisen, E. J. Van Scott, and J. H. Vaughan, eds.), pp. 331–353. McGraw-Hill, New York.

Zaias, N., and J. Alvarez. 1968. The formation of the primate nail plate: an autoradiographic study in squirrel monkey. *J. Invest. Dermatol.* **51:** 120–136.

10

Sebaceous Glands

I. Introduction

First described by Eichorn in 1826, holocrine sebaceous glands, like hairs, are unique to mammalian skin. Despite their similarity, the glands of other classes of vertebrates (e.g., some amphibians and reptiles and the preen glands of birds) are not really sebaceous. All human sebaceous glands are structurally, grossly, and microscopically similar but they vary in size, activity, and response to trophic agents according to their location in the body. In this chapter, the emphasis will be on human glands, and reference will be made to the glands of other mammals only to elucidate through comparison and contrast.

Man has more sebaceous glands than any other known mammal. They are largest and most numerous in the scalp, forehead, face, and ano-genital area where from 400 to 900 or more occupy each square centimeter of skin surface. Fewer than 100 are found per square centimeter in the rest of the body (Benfenati and Brillanti, 1939). In some regions, such as the anconal surface of the hands and feet, wrists and ankles, the number varies from none to 50 per square centimeter of skin (Johnsen and Kirk, 1952), and there are none on the palms, soles, and proximal nail folds. However, such figures must be interpreted with reservation since the number and size of the glands vary greatly among individuals and ethnic groups and according to sex and age. They are larger in men than in women and larger and more numerous on the dorsal midline of the trunk than on its ventral surface.

In man their size varies inversely with that of the hair follicles to which they are attached; but in the scalp, eyebrows, and eyelashes, both follicles and glands are large. In the face, scrotum, forehead, and anogenital areas the glands associated with vellus hairs reach an enormous size and empty into the greatly dilated pilary canals, some of which contain brushlike tufts of vellus hairs that remain in the follicle when a new hair is formed (Fig. 1). To distinguish these gigantic glands from smaller ones, Horner (1846) called them "sebaceous follicles," a name later adopted by Kligman and Shelley (1958) and now in general use. Some glands open directly onto the surface: the meibomian glands in the palpebrae, large rosettes of glands in the oral and buccal mucosa, the vermilion surface of the lips (Miles, 1958a,b), the nipples and areolae (Perkins and Miller, 1926; Montagna, 1970), the prepuce (Tyson's glands), and the labia minora (Machado de Sousa, 1931). The little-known single-cell glands of Wolff (Wolff, 1951a,b), which embody the special properties of both sebaceous cells and melanocytes (Pelfini *et al.*, 1970), are found on the surface of the palpebrae.

Regardless of variability in size, human sebaceous glands consist of multiple aggregates of acini that empty into a duct (Figs. 1, 4, and 5). Their general configuration is determined by their relative abundance (hence by the amount of crowding they are subjected to) and by the nature of the dermis in which they grow (Clara, 1929). Despite differences in size, shape, or position, their cellular morphology and sequences of differentiation are similar but not necessarily identical. As used here, sebaceous differentiation denotes the orderly synthesis, segregation, and accumulation of lipid droplets which culminate in enlarged, misshapen cells that fragment to form sebum.

Sebaceous glands, particularly the larger ones, are richly supplied with blood vessels. Since the larger acini are compressed masses of lobular units separated only by very delicate connective tissue trabeculae (Fig. 2), blood vessels may appear to be, but actually are not, *inside* the acini (Fig. 3A, B).

For the most part, intact sebaceous glands are structured alike. Each acinus is attached to a common excretory duct (Figs. 1, 4, and 5) composed of cornifying, stratified squamous epithelium continuous with the

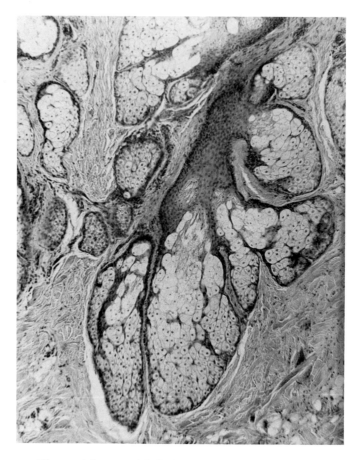

Fig. 1. Sebaceous follicle from frontal area of the scalp.

wall of the pilary canal or with the surface epidermis. As the acini develop, the cells enlarge centripetally. In the center, they are large, misshapen, or undergoing fragmentation; at the outer periphery, the undifferentiated cells resemble those of the epidermis. The acini in a glandular unit, like the individual cells within each acinus, vary in differentiation and maturity. Occasional acini are completely undifferentiated showing little or no lipid accumulation in any of their cells; others have lipid droplets around the nuclei only in the center cells, and some are full of lipid-laden cells that extend to the outer periphery of the acinus and rest against greatly attenuated undifferentiated

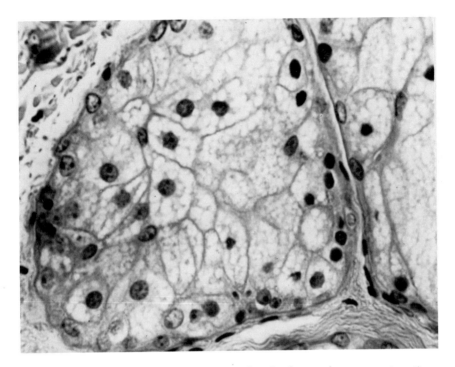

Fig. 2. Fused sebaceous acini separated only by a thin connective tissue trabecula.

peripheral cells. The culmination in the life cycle of sebaceous cells is the accumulation of so much lipid in their cytoplasm that they eventually die and fragment. The center of mature acini and the ducts contain a melange of lipids, cell detritus, flakes of keratinized cells, and microorganisms such as *Corynebacterium acnes, Pityrosporum ovale,* and others. Mites are occasionally found in the ducts of facial sebaceous follicles.

Sebaceous cells have bimodal differentiating potentialities. Even differentiating cells contain many cytoplasmic filaments identical with those in epidermal cells. Bimodality is especially evident in the ducts that enter into the pilary canals. Sebaceous ducts are lined with stratified squamous epithelium, the superficial cells of which are keratinized and often contain lipid droplets.

In other mammals, sebaceous glands are, with some exceptions, particularly numerous in the external auditory meatus and always largest

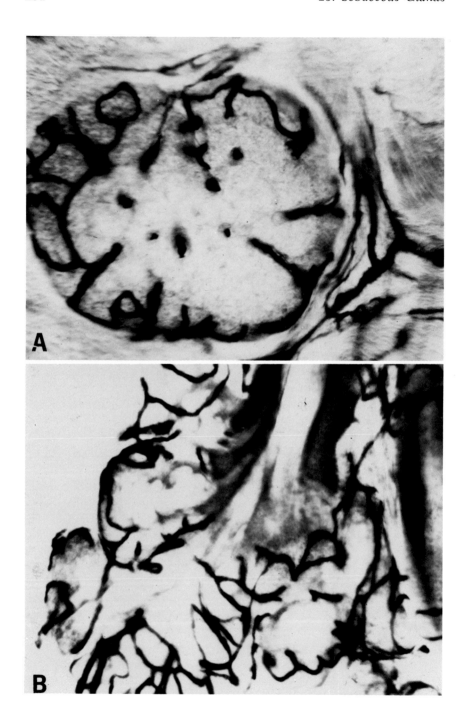

around the facial area and perineum. The rich fields of glands in the ears of rabbits and in the costovertebral spot of hamsters, which respond readily to steroid hormones, make these animals excellent experimental models. Some mammals are relatively free of sebaceous glands, and whales and porpoises apparently lack them altogether. Of the many mammals studied, only lemurs have glands about as numerous and large as man's.

Sebaceous glands are appendages of hair follicles and in man generally open inside the pilary canals. In adult lemurs, however, the glands open directly onto the surface of the hairy skin and not inside the pilary canal (Montagna and Yun, 1962); they originate from hair follicles but become separated postnatally (Yun and Montagna, 1964). In some animals, aggregates of glands, like the preputial and inguinal glands in rats, mice, rabbits, and gerbils, become encapsulated to form specialized organs. Marmosets and shrews have large fields of abdominal sebaceous glands; in man and most mammals the meibomian gland in the palpebrae represents such glands.

II. Development

Since most sebaceous glands arise from hair follicles, they appear asynchronously in a cephalocaudal sequence. In the scalp and face, where some hair follicles differentiate first, the glands are well formed in fetuses 3½ months old, although elsewhere on the body there are none (Serri and Huber, 1963). Even before the embryonic hair follicles become slanted or have begun to form a hair, they develop two humps, one above the other from the outer root sheath. The lower is the "bulge" proper, to which the fibers of the arrectores pilorum muscles will attach when they differentiate from the mesenchyme; the upper is the anlage of a sebaceous gland. When the follicles become slanted, these bulges are always oriented on the follicle margin facing the obtuse angle with the surface epidermis.

The cells in these anlagen are indistinguishable from those of the basal (germinative) layer of the epidermis and from those of the pilary canal with which they are continuous. Primordial sebaceous cells have a high nucleo:cytoplasmic ratio and the cytoplasm stains well with most

---

Fig. 3. (A) Thick section of a sebaceous acinus from the areola, surrounded by alkaline phosphatase reactive blood vessels. (B) A large gland from the same specimen showing all of the acini accompanied by alkaline phosphatase reactive vessels.

basic dyes; since this basophilia is abolished by crystalline ribonuclease, it is probably RNA. Seen under the electron microscope (Fig. 6), the cytoplasm of these cells contains numerous aggregates of glycogen, tonofilaments, a high concentration of free ribosomes, a few profiles of granular endoplasmic reticulum (RER) and agranular endoplasmic reticulum (SER), and only small Golgi zones. The apposing membranes of adjacent cells are attached by desmosomes and the basal cells have hemidesmosomes against the basal lamina (Breathnach, 1971).

Under the light microscope sebaceous cells appear to differentiate first in the center, where barely visible perinuclear vesicles gradually accumulate and fill the cytoplasm; as lipid accumulation is completed, the cytoplasm is reduced to flimsy, nonbasophilic, intervesicular strands. As individual cells accumulate lipid droplets, they attain many times their former volume and cause the entire glandular fundus to enlarge. The nucleus remains about the same size, round, and relatively intact until the later stages of cell disintegration when it shrinks and disappears. By the fourth month of gestation, most of the glands on the head are nearly mature; elsewhere on the body they appear at different times after the hair follicles are formed. When the large mature cells in the center of primordial glands are fully extended and misshapen, the cell membrane ruptures. A similar kind of differentiation occurs also in the cells in the center of the prospective duct, which at first is a solid cord. These cells differentiate linearly in a column that extends through the epidermis to the surface and proceeds for a distance parallel to it (Fig. 7). When these cells are mature, i.e., full of sebum vesicles, they lose their integrity, rupture, and form a sebum channel that represents the first pilosebaceous canal. Even fetal sebaceous glands are relatively large and apparently functional, and presumably some of the vernix caseosa is, at least in part, composed of sebum. Sebaceous glands are large in newborn infants, regress noticeably shortly after birth, remain relatively small throughout infancy and most of childhood, and develop fully at puberty. They are larger in boys than in girls. In prepubertal children the glands have many of the features of newly formed fetal anlagen, with high concentrations of free ribosomes, tonofilaments, and large Golgi zones, but virtually agranular ER.

In the labia minora of infant girls, where hair follicles develop and later mostly disappear, sebaceous glands consist of nests of lobules sometimes arranged around a minute vellus hair follicle, which later

Fig. 4.   (A) Intact sebaceous glands from a labium minor, in a split-skin preparation. (B) A large intact sebaceous gland from the scalp in a split-skin preparation.

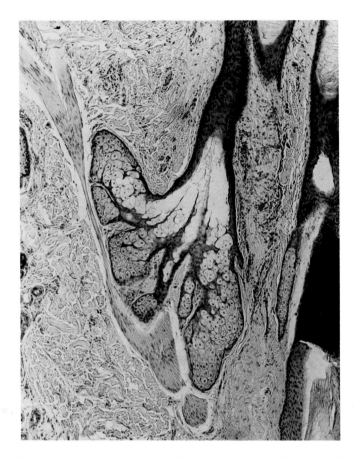

Fig. 5.   Tangential section through a sebaceous gland from the scalp.

disappears. Aborted or disoriented hair follicles are sometimes found
even in the labial sebaceous glands of mature women. Miles (1963)
found one hair follicle associated with the sebaceous glands in the
buccal mucosa.

Most adult human beings normally have sebaceous glands on the
buccal, oral, and even gingival mucosa (Fordyce spots). Glands also
develop after adolescence in the glabrous vermilion border of the lips,
inexplicably more numerous in the upper than in the lower lip. Their
presence in these areas underscores the relative unimportance of a hair
follicle in the genesis of these glands and their long ontogeny in man.

Sebaceous glands are common, even numerous, in the vermilion surface of the lips and in the cheek, buccal, and gingival mucosa of adult

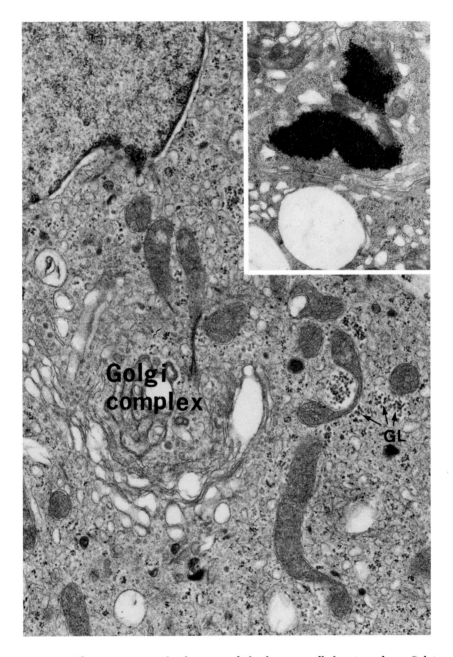

Fig. 6. Electron micrograph of a primordial sebaceous cell showing a large Golgi complex, mitochondria, and clusters of glycogen (GL) granules. × 24,000. The inset shows aggregates of glycogen granules. × 16,450. (Courtesy of Dr. M. Bell.)

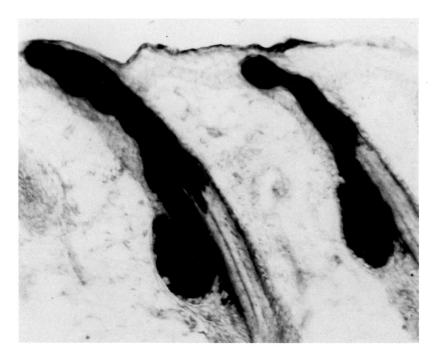

Fig. 7. Two sebaceous glands developing in association with hair follicles in the scalp of a 3½-month-old fetus; colored with sudan black.

human beings (Miles, 1958a,b; 1963). Some appear at the onset of puberty, and continue to increase in number particularly after the age of 35. This suggests that the glands differentiate directly from the epithelium of the lips and of the buccal and oral cavities. No histological studies of these areas in subadolescent subjects have ever shown anlagen of sebaceous glands.

Sebaceous glands can also differentiate in other organs. They are often found in the parotid and submaxillary glands of men and women (Hartz, 1946; Andrew, 1952; Meza Chavez, 1949). Sebaceous metaplasias have been found in the cervix uteri (Nicholson, 1918–1919; Dougherty, 1948; Donnelly and Navidi, 1950), and numerous large glands are always present between the lactiferous ducts at the tip of the nipple of mammary glands in men and in women (Montagna, 1970). The areolae of women always have large randomly distributed sebaceous tubercles associated with the lactiferous ducts of the glands of Montgomery

(Montagna and Yun, 1972). Sebaceous metaplasia in salivary glands is surprising only because it occurs so often. However, sebaceous differentiation is unexpected in the epithelium of the cervix, which derives its origin from the entoderm of the urogenital sinus, and in the larynx (Geipel, 1949) and esophagus (de La Pava and Pickren, 1962), which originate from the entoderm of the pharynx. The lesson to be learned here is that sebaceous glands can and do develop from a number of assorted tissues and that *de novo* formation occurs normally.

III. The Structure

Because much has been written about the structure of sebaceous glands and the differentiation of their cells, this section focuses only on some salient cytological details and electron microscopic findings. Additional information can be obtained in Montagna (1962); Montagna *et al.,* (1963); Strauss and Pochi (1969); and Ellis (1967, 1968).

The many descriptions of the ultrastructure of sebaceous glands have been uneven, the quality of the illustrations has been erratic, and the interpretations of structure have been variant (Rogers, 1957; Palay, 1958; Charles, 1960; Kurosumi *et al.,* 1960; Hibbs, 1962; Ellis and Henrikson, 1963; Brandes *et al.,* 1965; Henrikson, 1965; Ellis, 1967). Since sebaceous glands contain much lipid, special care must be taken in fixing and subsequently preparing them for electron microscopy; the glands of even closely related species must sometimes be treated differently. Furthermore, some of the best fixatives for preserving structural integrity in most other tissues, including other cutaneous organs, prove less than adequate for sebaceous glands. Only the relevant details of ultrastructure are given here as they pertain to the elaboration of lipid vesicles. Although cell differentiation is a continuous process with no discernible stages or phases, to avoid confusion certain arbitrary "stages" will be imposed. Moreover, no two differentiating sebaceous cells, even adjacent ones, are likely to be at exactly the same state of differentiation. Cells *do not actively move* centripetally in the acinus. What appears to be a centripetal migration of the differentiating cells from the periphery of the acinus is in reality a displacement of these cells by the undifferentiated cells which come to lie under them against the basement membrane and thus to displace them so that by the time they are mature, they are more or less in the center of the acinus. This is a process not of active cell migration but of peripheral accretion synchronous with differentiation. Having alerted the reader to the true process, let us now, for the sake of convenience, describe differentiation as if it occurs in well-defined stages: (1) *undifferentiated sebaceous cells* not yet undergoing transformation; (2) *differentiating cells* synthesizing and accumulating lipid, and (3) *mature cells* about to fragment into sebum.

A. *Undifferentiated Cells*

These cells usually rest against the basement membrane of an acinus but occasionally are found among partially differentiated cells (Figs. 8 and 9). Although their surfaces are generally smooth, cytoplasmic interdigitations occur at intervals along the cell membrane where 0.2 μm-wide folds project against adjacent cells and make mirror-image indentations. Ellis (1967, 1968), who calls these extensions microvilli, finds them most common where three or more cells meet. Regardless of the "stage" of differentiation, all sebaceous cells have desmosomal attachments as in the epidermis; the cell surfaces that rest against the basla lamina have hemidesmosomes. Ellis (1967) suggests that cellular "microvilli" protrude into alleged intercellular spaces to participate in pinocytotic activity. Specialized junctional areas between cells consist of highly electron-opaque membranes between which are insinuated an electron-opaque amorphous material and a centrally located electron-opaque line. Unlike desmosomes, these junctional areas have no tonofilaments associated with them; like the fascia occludens-type junctions (Fawcett, 1966), they often appear circular and can be pinched off and incorporated into sebaceous cells.

The cytoplasm of undifferentiated cells, whether at the periphery or elsewhere, is full of free ribosomes, some of which are attached to the membranes of the RER, which is sometimes organized into sparse parallel lamellae. Glycogen particles, too, are often prominent in these cells. One or more small Golgi complexes, usually seen around the oval nucleus, are composed of stacks of smooth membranes with some dilated cisternae. Centrioles consisting of a centrosphere crowned by radiating microtubules are commonly found. The numerous pleomorphic, dense mitochondria scattered throughout the cytoplasm of potential sebaceous cells have loosely packed, ill-defined cristae mostly slanted perpendicular to the long axis.

In the cytoplasm of presumptive sebaceous cells are tonofilaments (Fig. 9) which form fine meshes or bundles of tonofibrils similar to those in undifferentiated epidermal cells.

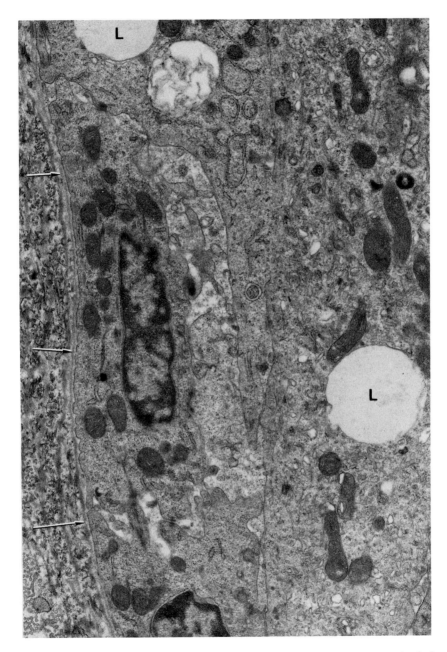

Fig. 8. An undifferentiated sebaceous cell at the periphery of an acinus. The dark arrows indicate the basal lamina. Note the two large lipid vacuoles (L) in the adjacent cell. × 15,700. (Courtesy of Dr. M. Bell.)

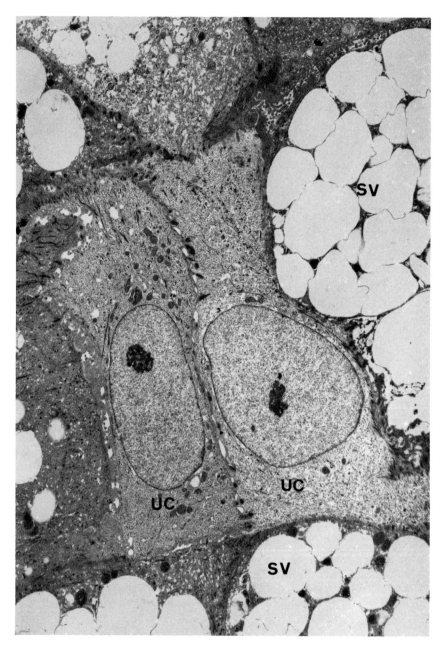

Fig. 9. Undifferentiated sebaceous cells (UC) in the center of an acinus. Numerous tonofibrils are seen in the cytoplasm. The surrounding cells have numerous sebum vesicles (SV). × 4300. (Courtesy of Dr. M. Bell.)

B. *Differentiating Cells*

The plasma membrane of differentiating cells, regardless of how much lipid they contain, is mostly smooth; at intervals it is thrown into short folds that are either pressed against the surface of another cell or interdigitated with it. Adjacent cell surfaces are connected by small desmosomes and occasional fascia occludens tight junctions. During the early stages of lipid synthesis, differentiating cells are very different from those replete with lipid vesicles.

At the beginning of maturation, differentiating cells have prominent Golgi membranes (Fig. 11), numerous free ribosomes, glycogen particles, many small smooth-surfaced vesicles that constitute the AER (Ellis, 1967), and variable numbers of different-sized sebum vesicles. The many small, dense mitochondria are scattered at random in the cytoplasm. Cells with few lipid vacuoles have numerous free ribosomes in their cytoplasm and large glycogen particles. The tubules of the smooth endoplasmic reticulum are often connected with the nearly ubiquitous Golgi membranes around the nucleus (Fig. 10). As the cells accumulate more lipid vacuoles in their cytoplasm, they become larger and acquire more small vesicles with ribosomes and glycogen particles interspersed between them.

Every differentiating cell has sebum vesicles of different sizes. The smaller vesicles contain somewhat uniformly dense material, but the larger ones appear irregular and mottled depending on the fixative used. Small vesicles of the SER are often continuous with the smaller lipid vesicles, and regardless of size, all lipid droplets are surrounded by smooth membranes. It is not clear, however, whether the vesicles are membrane bounded, or whether there is merely an interphase between the lipid and the surrounding Golgi and AR membranes.

Some controversy also persists about the respective roles of agranular reticulum and Golgi zones in the formation of vesicles. When fetal sebaceous anlagen begin to form, agranular reticulum is absent even in cells with vesicles that are sometimes as large as adult ones (Bell, 1971a,b; Breathnach, 1971). However, large Golgi zones are often clearly related to the circumferences of the vesicles. When in later development the agranular reticulum becomes abundant, it too becomes associated with the sebum vesicles.

Tonofilaments, numerous in undifferentiated cells, are less prominent in the partly differentiated ones and difficult to identify in mature cells. Mitochondria are relatively as numerous as in the undifferentiated cells but have allost no intramitochondrial granules (Ellis, 1967).

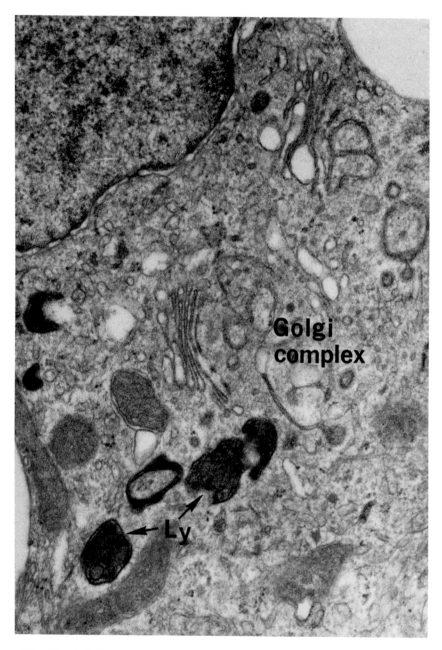

Fig. 10. A differentiating cell showing diffuse Golgi membranes. Some lysosome-like (Ly) particles are seen in the cytoplasm. × 41,250. (Courtesy of Dr. M. Bell.)

Cells in all stages of maturity contain somewhat rounded, electron-dense lysosomes. The ultrastructure of lysosomes in human sebaceous cells shows some structural elements embedded in a homogeneous opaque matrix. They are usually round, but vary in shape. The inner portion of the masses is usually less conspicuous than the surrounding matrix. Lysosomes originate in the Golgi zones and appear first as single, membrane-bounded vesicles that eventually become clumped in groups of three to seven inside another limiting membrane. As soon as a primordial cluster forms, acid phosphatase can be demonstrated in the space between the membranes of each vesicle and the membrane limiting the cluster (Fig. 11). As the clusters mature, the materials inside them become electron opaque. During the early formation of the clusters, the vesicles contain free ribosomes and glycogen (M. Bell, unpublished data).

The nearly spherical nucleus of partially differentiated cells is surrounded by sebum vacuoles. Inside the nucleus, large ovoid bodies, some of which are as large as the nucleolus, consist of concentric whorls of fibrils around a core of dense particles. Nuclear chromatin and the nucleolus apparently remain unchanged throughout sebacous maturation.

C. Mature Cells

These large, often deformed cells with a misshapen nucleus are bulging with sebum vacuoles separated only by wisps of cytoplasm. The surface is irregular, with microvilli extending into the surrounding spaces (Ellis, 1967). Small desmosomes attach adjacent cells, and what remains of the cytoplasm contains remnants of small smooth-surfaced vesicles, some ribosomes, barely recognizable tonofilaments packed against the plasma membrane, and small dense mitochondria. There are practically no membrane systems and only fragments of Golgi zones. Most sebum vacuoles have a nearly uniform size, but with the fusion of adjacent ones some can be as large as the nucleus. All vesicles contain a dense, cloudy substance, probably squalene, which is always present in sebum and which becomes blackened when exposed to osmium (Ellis, 1967). The clear material in vacuoles may represent saturated lipids or other substances extracted by the solvents used in the preparation of the material (Ellis, 1967).

Even though nearly all cytoplasmic inclusions fade away, the small, widely separated, electron-opaque mitochondria remain relatively intact and insinuated between the large sebum vesicles. Since the cells are

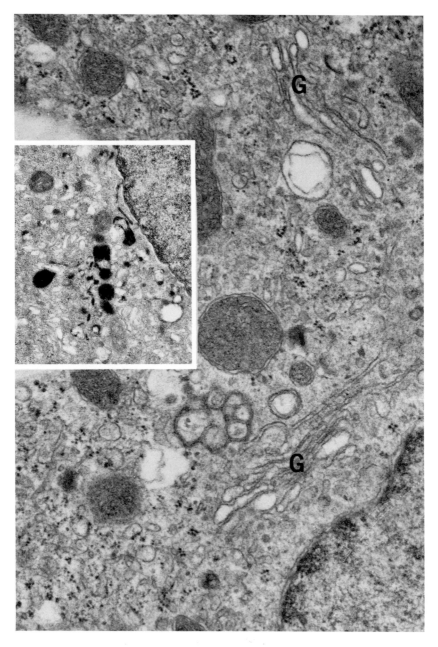

Fig. 11.   Differentiating sebaceous cell showing Golgi zones (G), and forming lysosomes. × 41,350. The inset shows localization of acid phosphatase reaction products. × 24,750. (Courtesy of Dr. M. Bell.)

now many times their original volume, the number of mitochondria is only relatively diminished.

The nucleus, compressed by the sebum vesicles around it, is distorted; the nuclear chromatin is clumped but the nucleolus is intact. When these cells finally rupture and spill their contents, the nucleus also degenerates.

Ultrastructural differences of glands in various parts of the body have not been studied sufficiently to be included here, but since such differences occur in other animals they can be expected in man as well. Particularly interesting are the differences seen in the prosimians and other nonhuman primates (Bell, 1970; 1971a,b). For example, in several species of *Macaca,* all lysosomes have a crystalline architecture not seen in man (Fig. 12).

D. The Duct

Sebaceous glands are connected to the pilosebaceous canal by a duct of stratified squamous epithelium. The part of the duct nearest the acini produces keratin as well as sebum; thus among its cellular components are membrane-coating and keratohyalin granules and lipid droplets. These droplets are smaller than those in the mature sebaceous cells and do not coalesce. The resemblance of the duct epithelium to that of the epidermis increases as the duct fuses with the pilosebaceous canal; however, instead of forming a compact epidermal horny layer, the epithelium produces a fragmented, thin, wispy material that is constantly sloughed into the lumen.

Recent ultrastructural investigations of pilosebaceous units in both normal and involved skin show little difference in the differentiation of sebaceous acini, but the keratinized portion that lines the pilosebaceous canal in acne lesions, especially in the intradermal portion, contains various numbers of characteristic lipid droplets (Fig. 13). The thick, compact horny layer is composed of cells engorged with lipids (Knutson, 1973). When lipids, keratinous cells, cell debris, and bacteria accumulate in these sites, the resulting impaction forms the comedo. These blocked deposits of sebaceous debris often become highly inflamed and form papules. However, it has not yet been determined whether comedones are caused by this specific aberration in the keratinization of the pilary canal.

E. Pigmentation

Of all the authors who have studied sebaceous glands, whether in man or other mammals, surprisingly only Montagna and his colleagues

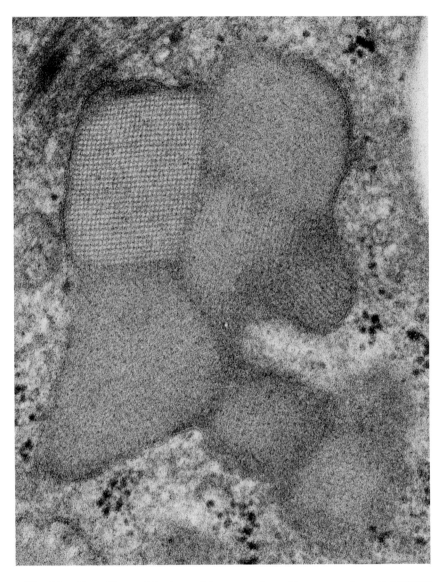

Fig. 12.   Crystalline architecture of lysosomes from *Macaca speciosa*. × 107,000.
(Courtesy of Dr. M. Bell.)

have found active melanocytes. In 1957, Montagna and Harrison first
reported melanotic melanocytes in the sebaceous glands of the Atlantic
seal. Later studies (Montagna *et al.*, 1961; Montagna, 1962; Montagna

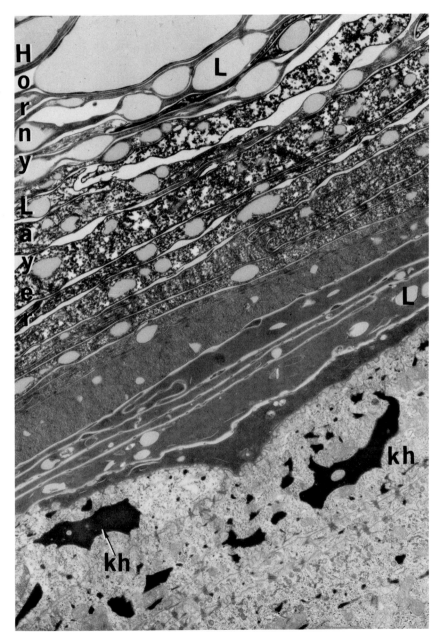

Fig. 13. The granular and horny layers from the pilosebaceous canal of a comedo. Note the lipid-laden horny layer. The granular layer is prominent. L, lipid; kh, keratohyalin. × 9000. (Courtesy of Dr. D. Knutson.)

and Yun, 1962) showed melanocytes in the glands of lemurs and other primates. Later still, the large sebaceous glands in the nipples and areolae of women's breasts were shown predictably to have melanocytes (Montagna, 1970; Montagna and Yun, 1972), usually intermingled with groups of undifferentiated cells with long dendrites extending for some distance between differentiating cells (Fig. 14). The mature cells in these glands contain only a few melanin granules, which have not as yet been studied under the electron microscope. In black lemurs, melanin is so copious that even the sebum is dark brown. The role of melanocytes in sebaceous glands or other organs, e.g., the meninges, brain, and viscera, is unknown.

IV. Growth and Proliferation

Once they attain maturity sebaceous cells die and disintegrate; to balance this loss they are replaced by dividing undifferentiated cells (Fig. 15). How this is accomplished is still debatable. From Bizzozero and Vassale (1887) through Stamm (1914), Kyrle (1925), and Schaffer (1927) to Epstein and Epstein (1966) mitotic activity was assigned to the cells at the periphery of the acini. But others (Bab, 1904; Brink-

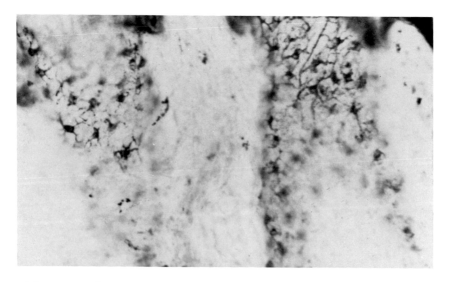

Fig. 14. Melanocytes in sebaceous glands from the areola of a 23-year-old woman.

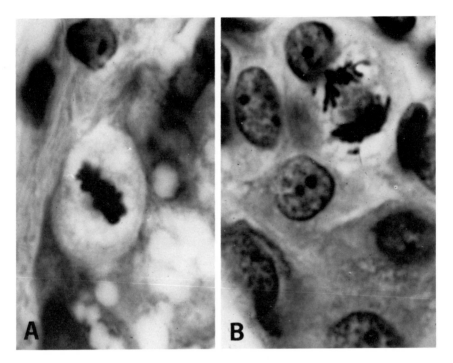

Fig. 15. Undifferentiated cells at the periphery of two acini undergoing mitosis: (A) metaphase, (B) telophase.

mann, 1912; Clara, 1929) described it as occurring mainly in the epithelium of the ducts at their junction with the acini, where cells are believed to flow down into the fundus of the acini. Both theories agree that new cells *migrate* toward the center of the glands. Arguing for the dynamics of growth solely on the location of mitotic activity is an exercise in futility. In thus concentrating on the foci of mitotic activity, most investigators have failed to investigate what is probably a more patent feature of growth, the shape or configuration of the glands. If the complex shape of the glands is ignored, arguments based on the locus of mitotic activity can yield only faulty or uninterpretable data. Epstein and Epstein (1966) partially avoided this pitfall by studying mitosis in scalp and arm skin injected with ^{3}H-thymidine.

The overall features of sebaceous glands can best be appreciated in split skin preparations where they maintain a nearly intact relationship with epidermis and hair follicles (Fig. 4). Such preparations are helpful in interpreting what is seen in histological preparations. The gross appearance and histology of sebaceous glands differ widely according to their size, the amount of crowding of the acini, and the various stages of maturity. Small to medium-sized glands in uncrowded areas consist of a few rounded or sacculated acini that clearly converge toward a duct more or less in the center of the aggregate. Histologically such glands differ from large, normally crowded or compressed and distorted glands, in which many of the acini are separated by barely visible, very delicate connective tissue septa. When intact, these large glands resemble heads of cauliflower (Fig. 4). In them, undifferentiated cells, which is undistorted glands appear mostly at the periphery of the individual acini, are often found even between the sometimes fused, adjacent acini, or, for that matter, almost anywhere in the glandular mass (Fig. 9). Hence the location of mitotic cells in neatly defined glandular acini differs from that in deformed ones. Sebaceous differentiation can and does occur wherever undifferentiated potential sebaceous cells are present, whether at the periphery of an acinus or elsewhere. For that matter, epithelial buds that grow from the walls of the ducts differentiate, grow into sebaceous acini, expand, and in a crowded environment, encroach upon nearby acini, sometimes fusing with them and becoming part of larger units. The boundaries of fused acini can still be traced by the small, elongated undifferentiated sebaceous cells and by the traces of connective tissue and fibrocytes that adhere to them. Such trabeculae can extend into or across the body of large and medium-sized sebaceous units. Sebaceous acini can also develop from the tabs of undifferentiated cells at the periphery of existing acini, as appendages of these parent acini.

In any consideration of glandular growth, it must be remembered that although in acne lesions and other pathological conditions the glands are often obliterated, the ductal tree appears relatively undamaged. When the lesion heals, the gland regrows, probably from ductal cells. Glands easily regenerate from the cells of the outer root sheath of hair follicles after they have been destroyed by dermabrasions (Staricco, 1960).

Epstein and Epstein (1966) determined that the steady state replenishment from the periphery of the acini continues for 2 to 4 weeks and that the average renewal time is about 7½ days.

Even cells of the epidermis can undergo sebaceous transformation when unduly irritated. Eisen *et al.* (1955), who studied shallow wounds that had destroyed the proximal portion of pilosebaceous systems together with the surrounding epidermis, found that during the reparative process, after an initial burst of mitosis in the basal cells of pilary canals and in the remaining basal undifferentiated glandular and ductal cells, the cells flowed to the surface of the wound where they differentiated mostly into keratinocytes; some, however, underwent sebaceous transformation on the surface.

Bullough and Laurence (1970) found a water-soluble substance in skin that can be precipitated by ethanol and activated by adrenaline and hydrocortisone and that inhibits mitotic activity in sebaceous glands both *in vitro* and *in vivo*. This substance is believed to be a sebaceous gland chalone.

V. Sebaceous Secretion

Although there are no special mechanisms that express sebum from the glands, a number of factors contribute to the extrusion. The contraction of the arrectores pilorum muscles of hair follicles may have some minor effect (Kligman and Shelley, 1958; Pontén, 1960) and skin temperature surely affects the viscosity of the sebum and, therefore, its spread on the surface. Except in the meibomian glands, there is no evidence that nerves control sebaceous secretion (Montagna and Ford, 1969); furthermore, as with hair growth and the differentiation of epidermal cells, complete denervation of a skin area does not affect sebaceous formation and differentiation (Doupe and Sharp, 1943; Pontén, 1960).

In agreement with Emanuel (1936, 1938), Kligman and Shelley (1958) suggest that the flow of sebaceous glands is continuous and that the quantity of sebum released in a given time per unit area is proportional to the total glandular volume. This is a function of the total number of differentiated cells. A large part of the sebum on the forehead and face is apparently produced by the sebaceous follicles, which secrete a clear, fluid sebum. Sebum is also brought to the surface by the stratum corneum which sucks it up from the ducts like a wick.

Since the secretion from individual glands has not been satisfactorily observed, some controversy remains about the rate of sebaceous flow onto the skin. Temperature, humidity, and altitude may influence the rate of sebaceous secretion, but no definitive work has been done on them. However, the amount of sebum produced is certainly related to the number and size of the sebaceous glands in any given area.

Both the size and the rate of maturation of sebaceous glands are regulated by hormones (Hamilton, 1941; Rony and Zakon, 1943;

Bullough and Laurence, 1960; Strauss *et al.*, 1962; Takayasu and Adachi, 1970) and in childhood are subject to topographic and individual differences.

VI. Hormonal Control

Hamilton (1941) was the first to show the influence of androgen on human sebaceous glands. Since then investigation of the hormonal control of sebaceous glands, especially those of the face, has accelerated. Histological methods have been widely used to determine their activity, but a more accurate method is to measure the amount of sebum secreted by them on the surface of the skin. Table I is a compounded estimate of the amount of sebum secreted by normal human subjects.

TABLE I

Production of Sebum

Age	mg Lipid/10 cm^2/3 hours	
	Males	Females
4–12	0.38	—
13–15	0.69	—
16–19	1.78	1.96
20–29	2.45	1.88
30–39	2.49	1.84
40–49	2.21	1.83
50–69	2.39	0.96
70–90	1.69	0.89

The significant drop in sebum production in women after age 49 is related to the menopause. Sebaceous glands are large at birth, smaller in infancy and childhood, and large at puberty and adulthood.

Studies of the influence of androgen on sebum production showed that as little as 5 mg methyl testosterone administered orally to a prepubertal child appreciably increased the size of sebaceous glands and the amount of sebum production. Since this dosage is probably equivalent to 1 mg testosterone injected intravenously and since methyl testosterone undergoes considerable degradation in the liver, the results indicate that the sebaceous glands are extremely sensitive to androgens.

Studies of sebum production show that the increased response to testosterone occurred about one week after the hormone was administered; moreover, the fact that under appropriate conditions the sebaceous glands enlarged only in the area where testosterone was applied topically indicates that the hormone had a direct effect.

After the administration of testosterone, however, sebum production increases significantly in eunuchs who normally produce very little. Precisely because sebaceous glands are highly sensitive to androgen, their activity is not a good indicator of plasma testosterone levels; under normal conditions, the glands are maximally stimulated by low but normal circulating levels of testosterone. The situation is somewhat more complex in women. The rather common occurrence of premenstrual flares of acne and the influence of progesterone on the sebaceous glands of laboratory animals at first suggested that progesterone is the prime stimulatory hormone in women. However, glandular enlargement in animals occurs only after large doses of progesterone, and physiological amounts of progesterone administered (about 50 mg/day) intramuscularly to women produced no response in the sebaceous glands. Furthermore, there is no rise in sebum production during the luteal phase of the cycle. Considering the length of time it takes a sebaceous gland to respond to androgens, the luteal phase seems to be too short to stimulate sebaceous glands functionally.

Only when the effects of estrogens on sebaceous glands are understood can the enlargement of these glands in women be explained. When large doses are administered over several weeks (for example, 100 μg ethinyl estradiol/day), sebum secretion decreases. However, unlike the effects of androgen in men, estrogen does not have a direct local effect. When it is applied to one side of the forehead, sebum production decreases on both sides, at the same time and to the same degree, undoubtedly because of a systemic effect. Estrogens then have a suppressive effect on sebum production. Furthermore, after sebaceous secretion has been suppressed with estrogen, it can be stimulated again by androgen regardless of the amounts of estrogen previously administered. Sebaceous glands are maximally stimulated in normal men and additional exogenous androgens do not cause measurable increases in sebum production. But once the glands have been inhibited with estrogen, the administration of methyl testosterone, testosterone propionate, or other potent androgens causes increased sebum production. Progesterone, however, has no effect. Apparently, then, the two hormones exert different or antagonistic effects on the cells.

This somewhat lengthy description of the effect of estrogens on the sebaceous glands is necessary if the data collected by the assay method that provides an alternate explanation are to be understood. Since physiological amounts of the adrenal androgen dehydroepiandrosterone, another catabolic product of testosterone metabolism, stimulate sebaceous glands in women, it could be the source of androgens. Furthermore, the ovary produces such androgens as Δ^4-androstenedione, a known stimulator of sebaceous glands. Ovarian and adrenal androgens combined are probably responsible for sebum production in women. Glucocorticoids, once thought to stimulate sebaceous glands directly, are apparently ineffective. Nor does prednisone have any effect on sebaceous glands; patients with Cushing's syndrome, who have elevated levels of glucocorticoids but no increase in androgenic metabolites, have likewise no increase in sebum secretion. Prednisone causes a 20% decrease in sebum production in women but not in men. The decreased production is roughly equivalent to the amount of sebum produced by eunuchs after prednisone treatment. Perhaps in men the effect of testicular testosterone is so overwhelming that adrenal hormones have no effect, whereas in women the relatively weak ovarian androgens are probably suppressed by adrenal hormones.

According to most experimental evidence, although androgens are the key stimulants of sebaceous glands, other hormones also affect their growth. The rate of sebum secretion, lowered after hypophysectomy in both young (Nikkari and Valavaara, 1969) and adult (Ebling et al., 1969a,b) rats, was increased by testosterone but not by estradiol or progesterone. The fact that adrenalectomy depressed sebum secretion in male rats (Thody and Shuster, 1971a,b,c) suggests that adrenal androgens are significant in the pituitary-adrenal, androgen-sebaceous gland stimulatory cycle. In hypophysectomized, castrated rats, thyrotrophic hormones boosted the action of testosterone (Ebling et al., 1970a,b) and ACTH increased sebum secretion.

In summary, the primary controlling factor of sebaceous glands in men and other male animals is testicular testosterone; in women it is a combination of ovarian and adrenal androgens. The decline in sebum production in menopausal women is due to a decrease or cessation of ovarian androgens. There is some evidence that dihydrotestosterone and not testosterone is the trophic hormone of sebaceous glands. This compound is produced by the catalytic action of α-reductase in its conversion from testosterone and other androgens.

Several studies corroborate the theories reported here. For example, the administration of testosterone or progesterone to normal rats induces an enlargement of the sebaceous glands (Haskin *et al.*, 1953; Ebling, 1957). When the pituitary and the adrenal glands are removed from ovariectomized rats, the ensuing atrophy of the sebaceous glands is not counteracted by the administration of progesterone and only slightly by testosterone (Lasher *et al.*, 1954). Hypophysectomy alone is also followed by a reduction in the size of the sebaceous glands which is only partially overcome by the administration of progesterone and testosterone. The pituitary, then, is necessary for the proper maintenance of the sebaceous glands. A "sebotropic" hormone, perhaps associated with specific foci in the molecules of prolactin and growth hormones, may directly affect the growth in mass and the differentiation of the glands (Ebling *et al.*, 1969a,b). Moreover, the glands apparently remain functional, though at reduced levels, in castrated juvenile animals.

The mechanism of androgen action at the local site has been studied best in the costovertebral glands of hamsters, which yield enough glandular material for microanalyses. Whether the data from these studies can be directly applied to the sebaceous glands of man is not yet known. A number of workers have studied tissue-active androgen in sebaceous glands (Adachi and Takayasu, 1972). Dihydrotestosterone has been shown to be much more "androgenic" than testosterone (Dorfman and Dorfman, 1962). When labeled testosterone is administered to animals, the predominant androgen in the prostate is dihydrotestosterone (Bruchovsky and Wilson, 1968a,b,c; Anderson and Liao, 1968), which is bound to receptor proteins in the nuclear fraction and, therefore, is not catabolized like testosterone. Since the levels of dihydrotestosterone in plasma are low, the tissue-active androgen is likely to be produced from testosterone in the peripheral target tissue. When ^{3}H-testosterone is injected intraperitoneally (Takayasu and Adachi, 1972), dihydrotestosterone (DHT) is the major metabolite formed in the sebaceous glands of hamsters as well as in the prostate and seminal vesicles. More than 90% of the steroid recovered from the nuclear fractions of these target tissues is dihydrotestosterone. These findings suggest that dihydrotestosterone is the active androgen at the target tissues—prostate, seminal vesicles, and sebaceous glands. Practically no traces of dihydrotestosterone are found in the cells of nontarget organs. Yet, liver contains both the enzyme 5α-reductase and the cofactor TPNH to convert testosterone to dihydrotestosterone. Hence one would expect this important metabolite to accumulate in it, but liver tissue contains practically no dihydrotesterone. Therefore, in this type of experiment, the amount of dihydrotestosterone detected reflects both the formation of dihydrotestosterone and its retention by the tissues. These data are consistent with the "receptor hypothesis" for hormone action, i.e., the target tissue possesses a specific receptor for a specific hormone.

After castration, sebaceous glands rapidly lose their ability to produce and retain dihydrotestosterone, unlike the seminal vesicle and prostate, which continue for six weeks to retain more of it than of testosterone. The dihydrotestosterone level in sebaceous glands falls rapidly to less than $\frac{1}{7}$ of the normal value 2 to 3 weeks after castration, probably because of either a reduction in the conversion of testosterone to dihydrotestosterone (caused by a decrease in a 5α-reductase or cofactor TPNH activity) or a decrease in the specific receptor protein for dihydrotestosterone.

To ascertain which was the cause, Takayasu and Adachi (1972) first assayed 5α-reductase activity in sebaceous glands and found that it increased steadily and remained high even 3 weeks after castration. The 5-fold increase in activity was probably due not to enzyme synthesis but to the shrinkage of cytoplasm and the remarkably long half-life of this enzyme, which therefore could not have been responsible for the reduced DHT level in sebaceous glands. Furthermore, in normal glands, DHT formation *in vitro* (a maximal rate) far exceeds that *in vivo* (a physiological rate). Since data indicate an "enzyme excess" even in normal sebaceous glands, the enzyme itself is not rate limiting.

Takayasu and Adachi (1972) also investigated the possibility that changes in DHT in the sebaceous gland after castration are due to the availability of the cofactor TPNH. In normal glands, the reduced form represents most of the total TPN, and this favors the conversion of testosterone to DHT by 5α-reductase. After castration, the endogenous level of the total TPN decreases to 30% and that of the reduced TPN to almost 50% of the normal. This further supports the theory that TPNH, not the enzyme itself, is the controlling factor. However, the decrease in TPNH is limited to only 50% of the normal value and does not appear to be the only cause for the rapid reduction in DHT after castration.

VII. Receptor Proteins for Dihydrotestosterone in Sebaceous Glands

Jensen and Jacobson (1962) gave the first clear evidence for a receptor by demonstrating the specific uptake of ^{3}H-estradiol in the uterus and vagina, but the nature of these receptors has been characterized only within the last few years. For example, in a radioactive estradiol-receptor complex, Toft and Gorski (1966) found a supernatant fraction with a sedimentation coefficient of 9.5 S.* Later Jensen *et al.* (1968), in addition to the 9 S cytosol receptor, extracted 5 S receptor protein from uterine nuclei. These authors then proposed a two-step mechanism of estradiol action, i.e., specific uptake by the 9.5 S (cytosol) and retention by the 5 S receptor (nuclear). DeSombre *et al.* (1969) have purified 2500 times the cytosol 8 S receptor.

There have been fewer studies on androgen than on estrogen receptors, mostly because of the additional mechanisms needed to form tissue-active hormone DHT from testosterone and because of the limited availability of samples (prostate and seminal vesicles). Fang *et al.* (1969) and Fang and Liao (1971), whose work describes the uptake of DHT by the 7 S cytosol receptor protein of the prostate and its subsequent retention by the 3 S nuclear receptor, support the two-step mechanism of hormone action in the prostate.

Data on the *in vivo* metabolism of testosterone, already described, presaged a DHT receptor in the sebaceous glands, and *in vitro* binding experiments demonstrated it directly. Adachi *et al.* (1972) have consistently demonstrated the ^{3}H-dihydrotestosterone-macromolecular complex after incubating sebaceous gland homogenates with labeled dihydrotestosterone at 3°C. These macromolecular receptors appear to be proteins, since incubating the ^{3}H-dihydrotestosterone complex with proteolytic enzymes, but not with nucleases, released the bound ^{3}H-dihydrotestosterone. The sedimentation coefficient of the cytosol receptor protein is 7 S and that of the nuclear receptor 3 S. When the cytosol and nuclear fractions were first separated and each fraction was incubated with ^{3}H-dihydrotestosterone at 3°C for the same period, only the cytosol receptor was bound to dihydrotestosterone. The nuclear fraction was labeled only after the cytosol fraction had been added. Thus, both cytosol and nuclear fractions contain their own specific dihydrotestosterone receptors, and the cytosol receptor complex transfers dihydrotestosterone to the nuclear receptor which interacts with the gene.

Adachi and Takayasu (1972) conducted further *in vitro* experiments with hamster sebaceous glands to determine whether the receptor proteins have a specific role in regulating androgen action and whether

---

* "S" value differs according to the experimental conditions.

changes occur in the receptor protein contents in animals up to 11 days after castration. Since the ability of the sebaceous glands to bind ^{3}H-dihydrotestosterone decreases dramatically after castration, the turnover of the receptor protein seems to be the major controlling factor in androgen action.

The fact that the receptor protein of sebaceous glands does not bind estrogen coincides with the clinical observations, i.e., the inability of estradiol to compete with androgen in sebaceous secretion. The binding of DHT to the receptor protein is markedly inhibited by such antiandrogens as cyproterone.

According to the receptor hypothesis of hormone action, specific target organs must contain specific macromolecules to interact with specific proteins, which, in turn, facilitate the retention of DHT to serve as a "positive feedback" mechanism. DHT also stimulates the synthesis of glucose-6-phosphate dehydrogenase, a key enzyme of the pentose cycle (within 24 hours) and that of a series of key glycolytic enzymes (1 to 2 weeks) which may belong to the same operon in the gene. By increasing the level of TPNH, the activation of the pentose cycle also serves the positive feedback mechanism. Finally, the activation of certain regulatory enzymes causes increases in metabolic reactions and energy formation and further leads to a hypertrophy of sebaceous glands.

VIII. Lipids

Sebum contains a number of unusual lipids not found elsewhere in the body. Cholesterol and free fatty acids are the only substances that have been consistently found in the sebum of other mammals (Wheatley, 1952, 1953). Cyclic triterpene alcohols occur in the wool fat of sheep and goats but not in the skin of dogs and rabbits. The sebum of man and some other mammals contains the acyclic triperpene, squalene, in high concentration.

The amount of surface lipids on a particular area of skin depends partly on the number, the size, and the rate of secretion of sebaceous glands, the thickness of the stratum corneum, and perhaps on the wetness of the skin, since sweating aids in spreading sebum, which is extremely hydrophilic (Rothman, 1954). The age and sex of the individual and a variety of physiological and physical factors no doubt play important but still unknown roles in controlling the amounts of skin surface lipids.

There are at least five different classes of lipids in sebum, several of them with complex mixtures. Since most investigations on cutaneous lipids have been carried out on samples of skin-surface lipids, these analyses must include lipid from the epidermis, trace amounts produced by cutaneous microorganisms (Saito and Asada, 1967; Moss *et al.*, 1967; Kellum and Strangfeld, 1970), those contained in the secretions of apocrine and possibly eccrine sweat glands, and those from exogenous contamination.

Table II summarizes the classes of lipids believed to occur in sebum and their estimated percentages by weight (Greene *et al.*, 1970).

TABLE II

Approximate Composition of Sebum [a]

Constituents	Weight %
Triglycerides	57.5
Wax esters	26.0
Squalene	12.0
Cholesterol esters	3.0
Cholesterol	1.5

[a] From Greene *et al.* (1970).

Two major biosynthetic pathways probably occur in human sebaceous glands. The first leads to the "fatty acid series" of sebaceous lipids, i.e., triglycerides, wax esters, and the fatty acids esterified to cholesterol (Weitkamp *et al.*, 1947; James and Wheatley, 1956; Haahti, 1961; Nicolaides *et al.*, 1964; Nicolaides, 1965; Kellum, 1967a,b). Fatty acids ranging from 5 to 30 carbons in length are probably synthesized by repeated condensation with activated 3-carbon malonyl-acyl carrier protein (ACP) molecules outside the mitochondria in sebaceous cells (Wakil *et al.*, 1964). The precursors probably include acetyl-CoA (leading to straight chain fatty acids with an even number of carbon atoms), propionyl-CoA (leading to straight chain fatty acids with an odd number of carbon atoms), isovaleryl-CoA (leading to isobranched fatty acids with an even number of carbon atoms), or α-methylbutyryl-CoA (leading to anteisobranched fatty acids), even though the latter two precursors are not really known to be present (Wheatley *et al.*, 1961). (Isobranched refers to a methyl branch on the second carbon from the terminal end of the fatty acid chain; anteisobranched denotes the methyl branch found on the third carbon from the terminal end of the fatty acid chain.) Unlike the fatty acids of internal lipid metabolisms, which have long chains with *even* numbers of carbon atoms, either saturated or unsaturated between the ninth and tenth carbon atoms from the carboxyl group, the fatty acid chains in sebum have both *even* and *odd* numbers of carbon atoms, both *straight* and *branched* chain configurations of several types, and a pattern of unsaturation predominantly between the sixth and seventh carbon atoms from the carboxyl group. The fatty acid molecules of sebaceous glands are unique (Weitkamp *et al.*, 1947; Nicolaides *et al.*, 1964). Fatty acid chains with two or more branches are sometimes found.

The long fatty acid chains in sebaceous glands can go in one of three directions (Nicolaides, 1965). First, the esterification of three molecules of fatty acids with a molecule of glycerol forms the triglycerides. Next, other fatty acid chains are reduced to form wax alcohols with a high degree of unsaturation or branching, and chain lengths that range from C_{14} to C_{24} (Brown *et al.*, 1954; Hougen, 1955; Nicolaides and Foster, 1956; Boughton and Wheatley, 1959; Haahti, 1961; Nicolaides, 1961; Haahti and Horning, 1961, 1963). These long chain wax alcohols are esterified with another molecule of fatty acid to yield wax esters. The wax esters (fatty acid plus wax alcohol) range in total chain length from C_{26} to C_{42} with an average of 36 carbon atoms.

The wax esters also show branching and unsaturation in all combinations. Nicolaides (1963) suggested that the structural features in the wax ester chain lower the melting point sufficiently to cause the wax ester at skin temperature to be an oil rather than a solid.

Still other fatty acid molecules, most of them highly unsaturated, are probably esterified to cholesterol to form small amounts of cholesterol esters (Nicolaides and Foster, 1956; Haahti *et al.*, 1963; Wilkinson, 1969).

All of the fatty acids synthesized in sebaceous glands are probably esterified when they leave the glands. No free fatty acids (Kellum, 1967a,b) and no free alcohols have been demonstrated in isolated human sebaceous glands (Nicolaides and Wells, 1957; Haahti, 1961).

The second probable biosynthetic pathway operating in sebaceous glands is the synthesis of squalene and trace amounts of cholesterol, most likely through the metabolic channels of mevalonate and farnesyl pyrophosphate. In human sebaceous glands, the metabolic stops for active conversion of squalene to cholesterol are apparently absent or operating inefficiently. Neither cholesterol nor cholesterol esters can be demonstrated with thin-layer chromatography in isolated human sebaceous glands. Quantitative spectrophotometric measurements of the Liebermann-Burchardt reaction suggested that about 0.5 to 1% of the total sebaceous gland lipid sample consisted of sterol esters (Kellum, 1967a,b). Wilkinson (1969) has confirmed that most of the fatty acids derived from sterol esters of human surface lipid have the unique 6:7 position of unsaturation characteristic of fatty acids in sebaceous glands. Greene et al. (1970) concluded that small amounts of sterol and sterol esters are formed as constituents of sebum, but they did not exclude the alternative possibility that these sterols and sterol esters represent only structural material from sebaceous cells, some of which may have been esterified to fatty acid molecules after leaving the gland.

Lipids in human sebaceous glands are strongly sudanophilic (Fig. 7). The cells of the ducts and the many undifferentiated cells at the periphery of the acini, or between adjacent acini, have discrete perinuclear lipid grandles. Static histological preparations seem to indicate that lipid accumulation in sebaceous acini begins in centrally located cells and progresses centrifugally (Montagna and Noback, 1946a,b; 1947), but we pointed out that the opposite is probably the case. In mature acini, the outermost cells may contain no lipid bodies or they may be inflated with large lipid spherules.

When frozen sections of skin are subjected to secondary osmication, only the sebum within the ducts becomes blackened; the fresh sebum in the center of the acini and the lipid droplets in sebaceous cells are osmiophobic. The osmiophilic sebum probably differs chemically from the osmiophobic new sebum and from the lipid droplets in the sebaceous cells (Montagna, 1949).

Nile blue sulfate colors the mature sebum pink or red, the new sebum purple, and the lipid droplets in sebaceous cells, pink. The lipid droplets in peripheral cells are colored purplish or blue. This is not a specific histochemical test, but a rose color usually indicates neutral lipids or triglycerides. Dark-blue staining favors the presence of fatty acids.

Phospholipids can be demonstrated histochemically in the spongy cytoplasm of mature and degenerating sebaceous cells and in the sebum as well as in cytoplasmic granules of undifferentiated and differentiating sebaceous cells. Analyses of surface lipids show slightly more than 1% phospholipids (Engman and Kooyman, 1934).

An orderly progression of events leads to the maturation of sebaceous cells and the formation of sebum. Histochemical tests demonstrate the accumulation of cholesterol esters, some phospholipids, and possibly tri-glycerides. The stale sebum in comedones and that in sebaceous cysts sometimes contain free cholesterol.

Histochemical demonstrations of cholesterol or cholesterol esters usually give a positive reaction in the sebum. Free cholesterol combines with digitonin to form acetone-insoluble crystals, which are birefringent under polarized light. Normal sebaceous glands contain no histo-chemically demonstrable free cholesterol, but the stagnant sebum of comedones and of early acne cysts abound with it.

Under polarized light, variable amounts of sebaceous lipids are aniso-tropic. The distribution of these lipids corresponds for the most part to the color reaction obtained with tests for cholesterol. Only the fully formed sebum consistently contains birefringent lipids (Suskind, 1951).

In sections of fresh skin viewed under near-ultraviolet light (3600 Å), the sebaceous glands emit a yellow-to-orange fluorescence comparable in distribution to the birefrigence. The sebum in the terminal portions of the ducts fluoresce with a yellow light (Miles, 1958a,b). The new sebum and the degenerating sebaceous cells emit a yellow or white light of low intensity which fades toward the periphery of the glands. The anisotropic, Schulz-positive cholesterol esters may be responsible for the autofluorescence. In some individuals, the sebaceous plugs in comedones emit a reddish fluorescence that derives from porphyrin.

IX. Enzyme Systems

Dissimilarities in enzyme content may reflect differences in the com-position of the sebum. The sebaceous glands of all animals we have studied contain such enzymes as acid phosphatase, nonspecific esterases and lipases, succinic and lactic dehydrogenases, and monomine oxidase; only those of certain animals have alkaline phosphatases and phos-phorylases.

All sebaceous glands show intense reactions for cytochrome oxidase and succinic and lactic dehydrogenase activity, mostly in granular form in the cytoplasm of the undifferentiated cells and no doubt located in the mitochondria (Montagna, 1955). Only some glands have hydroxy-steroid dehydrogenases (Baillie *et al.*, 1965, 1966a,b; Calman *et al.*, 1970).

The application of quantitative histochemical techniques (Lowry, 1953) to skin yields further information on enzyme systems in the sebaceous glands of man and monkeys (Cruickshank *et al.*, 1958;

Hershey *et al.*, 1960; Im and Adachi, 1966a,b; Im *et al.*, 1966; Adachi and Yamasawa, 1966a,b,c,d,e; 1967a,b). When all enzymes participating in major energy-yielding pathways are quantified, their activities in the sebaceous glands are approximately the same as in the epidermis; all data suggest the occurrence of active glycolysis. One difference between the enzyme activities of the epidermis and sebaceous glands is that in the latter, the activities to produce TPNH (such as G6PDH, 6-PGDH, and ICDH) are consistently higher. This coincides with the fact that these glands obviously require much TPNH for reductive synthesis in lipogenesis.

Human sebaceous glands abound in demonstrable phosphorylase and amylo-1, 4-1, 6-transglucosidase activities in the undifferentiated, and transforming cells (Braun-Falco, 1956; Ellis and Montagna, 1958; Takeuchi, 1958). The distribution of these enzymes parallels that of glycogen, and the glands of fetuses and infants, which are full of glycogen, have a much greater enzymic concentration than those of adults. These enzymes are largely peculiar to the sebaceous glands of primates but they are also found in the glands of lions and goats.

With some exceptions (Henrikson, 1965), only the glands of man and of the higher primates consistently contain glycogen (Montagna and Ellis; 1959a, 1960). Yet, the fetal sebaceous glands in all of the animals studied contained it. In man, many of the cells in the excretory ducts and all of the undifferentiated peripheral acinar cells contain glycogen (Figs. 11, 16A). During sebaceous differentiation, glycogen decreases at the same rate that lipid increases (Montagna *et al.*, 1951, 1952). The cells in the glands of older subjects usually have more glycogen than those of younger subjects (Figs. 16A, B).

Demonstrable alkaline phosphatase activity differs notably among the species and even in closely related animals; abundant in the glands of rats, the enzyme is not seen in those of mice. The glands of cats, raccoons, dogs, seal, weasels, lions, South American monkeys, and lemurs have conspicuous alkaline phosphatase reactivity, as do those of cows, pigs, rabbits, guinea pigs, and some Old World monkeys. In man, only the endothelium of the numerous arterioles around them is strongly reactive; in thick frozen sections, the distribution of these vessels can be followed by their reactivity (Fig. 3A). Blood vessels follow the contour of all acini large or small (Fig. 3B). The disproportionately large glands in the face and scalp seem to be more richly vascularized than the smaller ones elsewhere (Ellis, 1968).

Sebaceous glands have pronounced esterase activities, but the presence or absence of the various classes of these enzymes depends on the

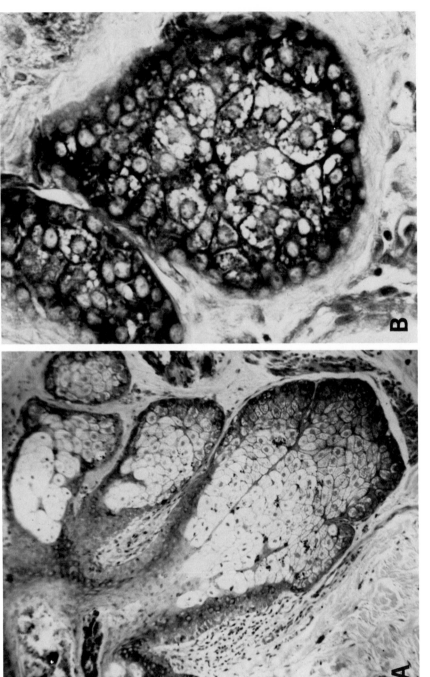

Fig. 16. Glycogen in the glands of the face of (A) a young subject, and (B) an old subject.

techniques and substrates used (Bell and Barnett, 1965) as well as on the species. In man (Montagna and Ellis, 1958), α-esterases, indoxyl esterase, and AS esterases have been demonstrated. In the rat and mouse, all glands, including the preputials, contain large amounts of cholinesterase activity (Montagna and Beckett, 1958; Montagna and Ellis, 1959b). Meibomian glands, whether in man, dog, rat, or goat, generally show nonspecific cholinesterase activity.

Baillie and his colleagues (Baillie et al., 1965, 1966a, 1966b) showed that the glands in areas prone to acne vulgaris and those in other parts of the body differed in their reactivity to steroid dehydrogenases. They found that the glands of the face, nose, forehead, cheeks, neck, back, anterior chest, and epigastric areas are rich in histochemically demonstrable 3β-hydroxysteroid dehydrogenase activity but that the arms (except in the deltoid region), forearm, back of hand, hypogastric and iliac areas, and the entire lower limbs have none. Although there are no sex differences in the reactivity or distribution of hydroxysteroid dehydrogenases, there are marked age changes; maximal reactivity occurred during the first two decades and fell progressively thereafter.

The role of 3α-, Δ^5-3β-, 11β-, 16β-, and 17β-hydroxysteroid dehydrogenases in steroid-secreting organs is not understood, and it is uncertain what these enzymes do in sebaceous glands. Still, it is significant that the enzymes are present only in the glands in areas prone to develop acne vulgaris. Furthermore, the decline in hydroxysteroid dehydrogenase reactivity in these glands after the age of 20 is somewhat parallel to the clinical amelioration in acne after the attainment of maturity.

Androgens, and particularly testosterone, stimulate sebaceous gland activity (Strauss et al., 1962) and the glands of eunuchs respond to adrenal steroids. The presence of Δ^5-3β-hydroxysteroid and 17β-hydroxysteroid dehydrogenases in the glands in areas disposed to acne may indicate that these glands can metabolize DHA and testosterone. At any rate, skin from the shoulders appears to have an appreciably greater conversion rate from DHA to androstenedione than skin from the thighs (Baillie et al., 1966a,b).

Sansone and Reisner (1971) have shown differences in sensitivity to dehydrotestosterone depending on where the glands are located.

X. Innervation

Regardless of techniques, no one has convincingly demonstrated any nerves around sebaceous glands (Hurley et al., 1953; Hellmann, 1955;

Montagna and Ellis, 1957; Thies and Galente, 1957). Although nerves are often seen near the tarsal glands, the palpebrae are so rich in nerves that the relationship may be spatial (Montagna and Ford, 1969). Meibomian glands, however, in man and all other animals studied are completely encircled by complex networks of cholinesterase-rich nerves (Fig. 17) (Montagna and Ellis, 1959b; Montagna and Ford, 1969). Nerves are mentioned here because they can be demonstrated well and easily only with techniques for cholinesterases. What purpose nerves serve around holocrine glands is difficult to say. The only reference to the cholinergic property of these nerves is by Buschke and Frankel (1905), who injected physostigmine into the palpebrae of rabbits and obtained an "emptying" of the glands. Since there are no muscles around the meibomian glands other than those of the palpebrae, how the glands excrete their secretion is not known.

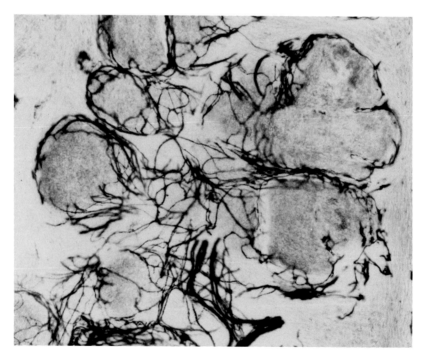

Fig. 17. A few acini from a meibomian gland surrounded by numerous acetylcholinesterase-reactive nerves.

XI. Function of Sebum

Despite attempts to ascribe a function to sebum, no one has been successful, and Kligman (1963) once stated that it is useless and even harmful. The most generally accepted function is that of an emollient of the horny layer. The assumption here is that without sebum the skin would be dry. But there are many valid objections to this theory. Infants, children, and many women, for example, with flawless, supple skin also have small sebaceous glands. Again, sebum may act as a bacteriostatic and fungistatic agent. Yet, the real function of sebum seems to be that of a pheromone, for even fresh sterile sebum has a distinctive odor not to be confused with that of the axillary glands. The scent glands of many mammals are mostly sebaceous, especially modified to produce an odorous secretion that attracts or repels and identifies not only the species but the individual. Aside from axillary odor, the aroma that distinguishes different clean human bodies is sebum.

XII. The Problem of Acne Vulgaris

The most common disease of the skin during the second and third decades of life is acne vulgaris. Most young men and young women suffer in various degrees from this condition. Acne, then, is a predictable and unique disorder of man, and since it is common it may well be considered one of the characteristic features of young human beings. Its onset at puberty coincides with the full maturation of the sebaceous glands, which under the influence of androgens develop and have an increased turnover rate of cells (Sweeney et al., 1969).

The newly formed sebum in the pilosebaceous canal is said to be irritating to tissues and is apparently comedogenic (Kligman et al., 1970). At this stage, sebum contains little or no free fatty acids (Kellum, 1967a,b). When it flows to the surface, however, its triglyceride fraction is hydrolyzed to free fatty acids, partly by the action of bacterial lipases (Freinkel, 1968; Reisner et al., 1968). Since some of these free fatty acids are believed to be potent irritants, perhaps comedogenic fatty acids of different chain lengths elicit these two separate effects. (Kellum believes that the fatty acids of C_{12} and C_{14} series are the most irritating.) Fatty acids with longer chains probably produce hyperkeratosis of the follicular epithelium and an apparent change in the cohesiveness of the epithelial cells of the ducts (Kligman, 1968) and thus cause comedones. The substance inside comedones, the follicular plug, consists of loose cells, sebum, and bacteria. From here, free fatty acids may filter through microscopic breaks in the wall of the follicle and incite inflammatory reactions in the surrounding dermis that result in inflammatory papules. How these papules progress further to form more severe cystic lesions is not known. These altered cutaneous reactivities or this delayed hypersensitivity may play a role in the pathogenesis of the more severe forms of acne. Alterations in the keratinization of the follicle may also be important factors.

Current theories about the pathogenesis of acne have led to some progress in the design of rational therapies. No attempts will be made here to review these methods; only those that appear to attack specific points in the pathogenetic scheme will be briefly sketched.

Because acne depends on many factors other than the size of the sebaceous glands (Strauss and Kligman, 1958, 1960), the extent to which it develops is not an adequate measure of the effect of hormones on these glands. Chemists who analyze surface lipids sometimes forget that mature acne lesions cannot secrete sebum because the sebaceous glands within them are atrophied. Surface lipids in areas afflicted by acne are secreted by the still intact, unaffected glands there.

Since the susceptibility to and severity of acne vulgaris are unrelated to high serum or urine levels of testosterone, an etiologic factor or factors probably resides in the sebaceous glands themselves. Two questions arise: (1) Is testosterone or its metabolites (or both) androgenic for the sebaceous glands (i.e., the formation of tissue-active androgen)? (2) Can sebaceous cells incorporate androgen into their nuclei? (See, for example, the two-step mechanism of hormone action of Jensen *et al.*, 1968.)

A reduction in sebum production, which decreases the amount of substrate available for the action of microbial lipases, is now accomplished with estrogen. Oral contraceptive pills that contain estrogen (0.1 mg Mestranol or ethinyl estradiol) effectively reduce sebum production and control acne; in stubborn cases, higher doses are required. There is considerable interest in the development of antiandrogens or compounds that block the action of testosterone at target organs. Testosterone is converted to the more active metabolite dihydrotestosterone inside sebaceous cells; this conversion occurs at a higher rate in acnegenic areas (Baillie *et al.*, 1966a,b). Thus, the peripheral blockage of testosterone uptake could help to control the disease, but the antiandrogens available today have not been effective. Such compounds as eicosanoic acid derivatives probably interfere with sebum synthesis; given to acne patients, they significantly reduce sebum but have undesirable side effects. Free fatty acids can be controlled with such broad-spectrum antibiotics as the tetracycline group, which reduce the bacterial population inside the pilosebaceous canal and may also interfere with the lipolytic enzymes of these microorganisms. New compounds need to be developed that will either selectively inhibit the microbial flora in the follicles or completely inhibit their lipolytic activity.

Almost all traditional topical therapy of acne causes some irritation to the skin surface and desquamation. The most dramatic of these compounds is vitamin A acid or retinoic acid which stimulates an increase in the turnover of sebaceous and epithelial cells and pronounced vasodilation. Still the compound appears to be superior to other "peeling agents" in reducing follicular plugging.

Other forms of therapy have been designed to control the inflammatory lesions that develop in acne. Andrenocorticosteroids, when administered either systemically or inside the lesions, are usually effective. Cryotherapy, about which too little is known, also reduces inflammatory lesions.

To date, this is the status of our knowledge of the most common and predictable disorder of the skin.

References

Adachi, K., and S. Takayasu. 1972. The mechanism of testosterone action on the sebaceous glands of the Syrian hamster. *In* "Advances in Biology of Skin. Pharmacology and the Skin" (W. Montagna, R. B. Stoughton, and E. J. Van Scott, eds.), Vol. XII, pp. 381–401. Appleton-Century-Crofts, New York.

Adachi, K., and S. Yamasawa. 1966a. Quantitative histochemistry of the primate skin. I. Hexokinase. *J. Invest. Dermatol.* **46**: 473–476.

Adachi, K., and S. Yamasawa. 1966b. Quantitative histochemistry of the primate skin. II. Fructoaldolase. *J. Invest. Dermatol.* **46**: 542–545.

Adachi, K., and S. Yamasawa. 1966c. Quantitative histochemistry of the primate skin. IV. α-Glycerophosphate dehydrogenase. *J. Invest. Dermatol.* **47**: 107–109.

Adachi, K., and S. Yamasawa. 1966d. Quantitative histochemistry of the primate skin. VII. Pyruvate kinase. *J. Invest. Dermatol.* **47**: 289–292.

Adachi, K., and S. Yamasawa. 1966e. Quantitative histochemistry of the primate skin. VIII. Enolase. *J. Invest. Dermatol.* **47**: 293–295.

Adachi, K., and S. Yamasawa. 1967a. Quantitative histochemistry of the primate skin. IX. Phosphoglycerate kinase and phosphoglycerate mutase. *J. Invest. Dermatol.* **49**: 22–30.

Adachi, K., and S. Yamasawa. 1967b. Quantitative histochemistry of the primate skin. X. Phosphoglucoisomerase, phosphofructokinase and triose isomerase. *J. Invest. Dermatol.* **50:** 180–185.

Adachi, K., M. Kano, and S. Takayasu. Androgen receptors in sebaceous glands. (In preparation.)

Anderson, K. M., and S. Liao. 1968. Selective retention of dihydrotestosterone by prostatic nuclei. *Nature (London)* **219:** 277–279.

Andrew, W. 1952. A comparison of age changes in salivary glands of man and the rat. *J. Gerontol.* **7:** 178–190.

Bab, H. 1904. Die Talgdrüsen und ihre Sekretion. *Beitr. Klin. Med. Festschr. Senator,* **70** Geburtstag, pp. 1–37.

Baillie, A. H., K. C. Calman, and J. A. Milne. 1965. Histochemical distribution of hydroxysteroid dehydrogenases in human skin. *Brit. J. Dermatol.* **77:** 610–616.

Baillie, A. H., M. M. Ferguson, and D. McK. Hart. 1966a. "Developments in Steroid Histochemistry." Academic Press, London and New York.

Baillie, A. H., J. Thomson, and J. A. Milne. 1966b. The distribution of hydroxysteroid dehydrogenase in human sebaceous glands. *Brit. J. Dermatol.* **78:** 451–457.

Bell, M. 1970. A comparative study of sebaceous gland ultrastructure in subhuman primates. I. *Galago crassicaudatus, G. senagalensis,* and *G. demidovii. Anat. Rec.* **166:** 213–224.

Bell, M. 1971a. A comparative study of sebaceous gland ultrastructure in subhuman primates. II. Macaques: Crystalline inclusions in the sebaceous cells of *Macaca mulatta, M. nemestrina, M. speciosa, M. fascicularis. J. Cell Biol.* **49:** 932–936.

Bell, M. 1971b. A comparative study of sebaceous gland ultrastructure in subhuman primates. III. Macaques: Ultrastructure of sebaceous glands during fetal development of rhesus monkeys *(Macaca mulatta). Anat. Rec.* **170:** 331–341.

Bell, M., and R. J. Barrnett. 1965. The use of thiol-substituted carboxylic acids as histochemical substrates. *J. Histochem. Cytochem.* **13:** 611–628.

Benfenati, A., and F. Brillanti. 1939. Sulla distribuzione delle ghiandole sebacee nella cute del corpo umano. *Arch. Ital. Dermatol. Sifilogr. Venereol.* **15:** 33–42.

Bizzozero, G., and G. Vassale. 1887. Ueber die Erzeugung und die physiologische Regeneration der Drüsenzellen bei den Säugethieren. *Virchows Arch. A* **110:** 155–214.

Boughton, B., and V. R. Wheatley. 1959. Studies of sebum. 9. Further studies of the composition of the unsaponifiable matter of human-forearm 'sebum.' *Biochem. J.* **73:** 144–149.

Brandes, D., F. Bertini, and E. W. Smith. 1965. The role of lysosomes in cellular lytic processes. II. Cell death during holocrine secretion in sebaceous glands. *Exp. Mol. Pathol.* **4:** 245–265.

Braun-Falco, O. 1956. Über die Fähigkeit der menschlichen Haut zur Polysaccharid-synthese, ein Beitrag zur Histotopochemie der Phosphorylase. *Arch. Klin. Expt. Dermatol.* **202:** 163–170.

Breathnach, A. S. 1971. "An Atlas of the Ultrastructure of Human Skin. Development, Differentiation, and Postnatal Features." Churchill, London.

Brinkmann, A. 1912. Die Hautdrüsen der Säugetiere (Bau and Secretionsverhältnisse). *Ergeb. Anat. Entwicklungsgesch.* **20:** 1173–1231.

Brown, R. A., W. S. Young, and N. Nicolaides. 1954. Analysis of high molecular weight alcohols by the mass spectrometer. The wax alcohols of human hair fat. *Anal. Chem.* **26**: 1653–1654.

Bruchovsky, N., and J. D. Wilson. 1968a. Evidence that dihydrotestosterone is the active form of testosterone. *Clin. Res.* **14**: 74.

Bruchovsky, N., and J. D. Wilson. 1968b. The conversion of testosterone to 5α-androstan-17β-ol-3-one by rat prostate *in vivo* and *in vitro*. *J. Biol. Chem.* **243**: 2012–2021.

Bruchovsky, N., and J. D. Wilson 1968c. The intranuclear binding of testosterone and 5α-androstan-17β-ol-3-one by rat prostate. *J. Biol. Chem.* **243**: 5953–5960

Bullough, W. S., and E. B. Laurence. 1960. Experimental sebaceous gland suppression in the adult male mouse. *J. Invest. Dermatol.* **35**: 37–42.

Bullough, W. S., and E. B. Laurence. 1970. Chalone control of mitotic activity in sebaceous glands. *Cell and Tissue Kinet.* **3**: 291–300.

Buschke, A., and A. Frankel. 1905. Ueber die Funktion der Talgdrüsen und deren Beziehung zum Fettstoffwechel. *Berlin. Klin. Wachdenschr.* **42**: 318–322.

Calman, K. C., A. V. Muir, J. A. Milne, and H. Young. 1970. Survey of the distribution of steroid dehydrogenases in sebaceous glands of human skin. *Brit. J. Dermatol.* **82**: 567–571.

Charles, A. 1960. Electron microscopic observations of the human sebaceous gland. *J. Invest. Dermatol.* **35**: 31–36.

Clara, M. 1929. Morfologia e sviluppo delle ghiandole sebacee nell'uomo. *Ric. Morfol.* **9**:121–182.

Cruickshank, C. N. D., F. B. Hershey, and C. Lewis. 1958. Isocitric dehydrogenase activity of human epidermis. *J. Invest. Dermatol.* **30**: 33–37.

DeSombre, E. R., G. A. Puca, and E. V. Jensen. Purification of an estrophilic protein from calf uterus. *Proc. Nat. Acad. Sci. U.S.* **64**: 148–154.

Donnelly, G. H., and S. Navidi. 1950. Sebaceous glands in the cervix uteri. *J. Pathol. Bacteriol.* **62**: 453–454.

Dorfman, R. I., and A. S. Dorfman. 1962. Assay of subcutaneously administered androgens on the chick's comb. *Acta. Endocrinol.* **41**: 101–106.

Dougherty, C. M. 1948. A sebaceous gland in the cervix uteri. *J. Pathol. Bacteriol.* **60**: 511–512.

Doupe, J., and M. E. Sharp. 1943. Studies in denervation (G)—Sebaceous secretion. *J. Neurol. Psychiat.* **6**: 133–135.

Ebling, F. J. 1957. The action of testosterone and oestradiol on the sebaceous glands and epidermis of the rat. *J. Embryol. Exp. Morphol.* **5**: 74–82.

Ebling, F. J., E. Ebling, and J. Skinner. 1969a. The influence of the pituitary on the response of the sebaceous and preputial glands of the rat to progesterone. *J. Endocrinol.* **45**: 257–263.

Ebling, F. J., E. Ebling, and J. Skinner. 1969b. The influence of pituitary hormones on the response of the sebaceous glands of the rat to testosterone. *J. Endocrinol.* **45**: 245–256.

Ebling, F. J., E. Ebling, and J. Skinner. 1970a. The effects of thyrotrophic hormone and of thyroxine on the response of the sebaceous glands of the rat to testosterone. *J. Endocrinol.* **48**: 83–90.

Ebling, F. J., E. Ebling, J. Skinner, and A. White. 1970b. The response of the sebaceous glands of hypophysectomized-castrated male rats to adrenocorticotrophic hormone and to testosterone. *J. Endocrinol.* **48:** 73–81.

Eisen, A. Z., J. B. Holyoke, and W. C. Lobitz, Jr. 1955. Responses of the superficial portion of the human pilosebaceous apparatus to controlled injury. *J. Invest. Dermatol.* **25:** 145–156.

Ellis, R. A. 1967. Eccrine, sebaceous and apocrine glands. *In* "Ultrastructure of Normal and Abnormal Skin" (A. S. Zelickson, ed.), Chapter 7, pp. 132–162. Lea & Febiger, Philadelphia, Pennsylvania.

Ellis, R. A. 1968. Electron microscopy of sebaceous glands. *In* "Biopathology of Pattern Alopecia," pp. 146–154. Karger, Basel and New York.

Ellis, R. A., and R. C. Henrikson. 1963. The ultrastructure of the sebaceous glands of man. *In* "Advances in Biology of Skin. The Sebaceous Glands" (W. Montagna, R. A. Ellis, and A. F. Silver, eds.), Vol. 4, pp. 94–109. Pergamon, Oxford.

Ellis, R. A., and W. Montagna, 1958. Histology and cytochemistry of human skin. Sites of phosphorylase and amylo-1, 6 glucosidase activity. *J. Histochem. Cytochem.* **6:** 201–207.

Emanuel, S. 1936. Quantitative determinations of the sebaceous glands' function with particular mention of method employed. *Acta Dermato-Venereol.* **17:** 444–456.

Emanuel, S. 1938. Mechanism of the sebum secretion. *Acta Dermato-Venereol.* **19:** 1–15.

Engman, M. F., and D. J. Kooyman. 1934. Lipids of the skin surface. *Arch. Dermatol. Syphilol.* **29:** 12–19.

Epstein, E. H., Jr., and W. L. Epstein. 1966. New cell formation in human sebaceous glands. *J. Invest. Dermatol.* **46:** 453–458.

Fang, S., and S. Liao. 1971. Androgen receptors. Steroid- and tissue-specific retention of a 17β-hydroxy-5α-androstan-3-one protein complex by the cell of nuclei of ventral prostate. *J. Biol. Chem.* **246:** 16–24.

Fang, S., K. M. Anderson, and S. Liao. 1969. Receptor proteins for androgens. On the role of specific proteins in selective retention of 17β-hydroxy-5α-androstan-3-one by rat ventral prostate *in vivo* and *in vitro*. *J. Biol. Chem.* **244:** 6584–6595.

Fawcett, D. W. 1966. "An Atlas of Fine Structure: The Cell, Its Organelles and Inclusions," pp. 378–379. Saunders, Philadelphia, Pennsylvania.

Freinkel, R. K. 1968. Origin of free fatty acids in sebum. I. Role of coagulase negative staphylococci. *J. Invest. Dermatol.* **50:** 186–188.

Geipel, P. 1949. Talgdrüse im Kehlkopf. *Zbl. Allg. Pathol. Anat.* **85:** 69–71.

Greene, R. S., D. T. Downing, P. E. Pochi, and J. S. Strauss. 1970. Anatomical variation in the amount and composition of human skin surface lipid. *J. Invest. Dermatol.* **54:** 240–247.

Haahti, E. 1961. Major lipid constituents of human skin surface with special reference to gas-chromatographic methods. *Scand. J. Clin. Lab. Invest.* **13** (Suppl. 59): 1–108.

Haahti, E., and E. C. Horning. 1961. Separation of human waxes by gas chromatography. *Acta Chem. Scand.* **15:** 930–931.

Haahti, E., and E. C. Horning. 1963. Isolation and characterization of saturated and unsaturated fatty acids and alcohols of human skin surface lipids. *Scand. J. Clin. Lab. Invest.* **15**: 73–78.

Haahti, E., T. Nikkari, and K. Juva. 1963. Fractionation of serum and skin sterol esters and skin waxes with chromatography on silica gel impregnated with silver nitrate. *Acta Chem. Scand.* **17**: 538–540.

Hamilton, J. B. 1941. Male hormone substance: a prime factor in acne. *J. Clin. Endocrinol.* **1**: 570–592.

Hartz, P. H. 1946. Development of sebaceous glands from intralobular ducts of the parotid gland. *Arch. Pathol.* **41**: 651–654.

Haskin, D., N. Lasher, and S. Rothman. 1953. Some effects of ACTH, cortisone, progesterone and testosterone on sebaceous glands in the white rat. *J. Invest. Dermatol.* **20**: 207–212.

Hellmann, K. 1955. Cholinesterase and amine oxidase in the skin: A histochemical investigation. *J. Physiol. (London)* **129**: 454–463.

Henrikson, R. C. 1965. Glycogen and sebaceous transformation. *J. Invest. Dermatol.* **44**: 435.

Hershey, F. B., C. Lewis, Jr., J. Murphy, and T. Schiff. 1960. Quantitative histochemistry of human skin. *J. Histochem. Cytochem.* **8**: 41–49.

Hibbs, R. G. 1962. Electron microscopy of human axillary sebaceous glands. *J. Invest. Dermatol.* **38**: 329–336.

Horner, W. E. 1846. On the odoriferous glands of the Negro. *Amer. J. Med. Sci.* **21**: 13–16.

Hougen, F. W. 1955. The constitution of the aliphatic alcohols in human sebum. *Biochem. J.* **59**: 302–309.

Hurley, H. J., Jr., W. B. Shelley, and G. B. Koelle. 1953. The distribution of cholinesterases in human skin, with special reference to eccrine and apocrine sweat glands. *J. Invest. Dermatol.* **21**: 139–147.

Im, M. J. C., and K. Adachi. 1966a. Quantitative histochemistry of the primate skin. V. Glucose-6-phosphate dehydrogenase and 6-phosphogluconate dehydrogenase. *J. Invest. Dermatol.* **47**: 121–124.

Im, M. J. C., and K. Adachi. 1966b. Quantitative histochemistry of the primate skin. VI. Lactate dehydrogenase. *J. Invest. Dermatol.* **47**: 286–288.

Im, M. J. C., S. Yamasawa, and K. Adachi. 1966. Quantitative histochemistry of the primate skin. III. Glyceraldehyde-3-phosphate dehydrogenase. *J. Invest. Dermatol.* **47**: 35–38.

James, A. T., and V. R. Wheatley. 1956. Studies of sebum. 6. The determination of the component fatty acids of human forearm sebum by gas-lipid chromatography. *Biochem. J.* **63**: 269–273.

Jensen, E. V., and H. T. Jacobson. 1962. Basic guides to the mechanism of estrogen action. *Recent Progr. Horm. Res.* **18**: 387–414.

Jensen, E. V., T. Suzuki, T. Kawashima, W. E. Stumpf, P. W. Jungblut, and E. R. DeSombre. 1968. A two-step mechanism for the interaction of estradiol with rat uterus. *Proc. Nat. Acad. Sci. U.S.* **59**: 632–638.

Johnsen, S. G., and J. E. Kirk. 1952. The number, distribution and size of the sebaceous glands in the dorsal region of the hand. *Anat. Rec.* **112**: 725–735.

Kellum, R. E. 1967a. Short chain fatty acids (below C_{12}) of human skin surface lipids. *J. Invest. Dermatol.* 48: 364–371.

Kellum, R. E. 1967b. Human sebaceous gland lipids. Analysis by thin-layer chromatography. *Arch. Dermatol.* 95: 218–220.

Kellum, R. E., and K. Strangfeld. 1970. Acne vulgaris. Studies in pathogenesis. Fatty acids of *Corynebacterium acnes*. *Arch. Dermatol.* 101: 337–339.

Kligman, A. M. 1963. The uses of sebum? *In* "Advances in Biology of Skin. The Sebaceous Glands" (W. Montagna, R. A. Ellis, and A. F. Silver, eds.), Vol. 4, pp. 110–124. Pergamon, Oxford.

Kligman, A. M. 1968. Pathogenesis of acne vulgaris. II. Histopathology of comedones induced in the rabbit ear by human sebum. *Arch. Dermatol.* 98: 58–66.

Kligman, A. M., and W. B. Shelley. 1958. An investigation of the biology of the human sebaceous gland. *J. Invest. Dermatol.* 30: 99–125.

Kligman, A. M., V. R. Wheatley, and O. Mills. 1970. Comedogenicity of human sebum. *Arch. Dermatol.* 102: 267–275.

Knutson, D. D. 1973. The ultrastructural observations in Acne vulgaris. *In* "Advances in Biology of Skin. Sebaceous Glands and Acne Vulgaris" (W. Montagna, M. Bell, J. Strauss, and A. Shalita, guest eds.), Vol. XIV. *J. Invest. Dermatol.* (Special Issue). (In press.)

Kurosumi, K., T. Kitamura, and K. Kano. 1960. Electron microscopy of the human sebaceous gland. *Arch. Histol.* 20: 235–246.

Kyrle, J. 1925. "Vorlesunger über Histobiologie der menschlichen Haut und ihrer Erkrankungen." Springer-Verlag, Berlin and New York.

La Pava, S. de, and J. W. Pickren. 1962. Ectopic sebaceous glands in the esophagus. *Arch. Pathol.* 73: 397–399.

Lasher, N., A. L. Lorincz, and S. Rothman. 1954. Hormonal effects on sebaceous glands in the white rat. II. The effect of the pituitary adrenal axis. *J. Invest. Dermatol.* 22: 25–31.

Lowry, O. H. 1953. Quantitative histochemistry of brain. Histological sampling. *J. Histochem. Cytochem.* 1: 420–428.

Machado de Sousa, O. 1931. Sur la présence de glandes sebacees au niveau du gland chez l'homme. *C. R. Soc. Biol.* 108: 894–897.

Meza Chavez, L. 1949. Sebaceous glands in normal and neoplastic parotid glands. Possible significance of sebaceous glands in respect to the origin of tumors of the salivary glands. *Amer. J. Pathol.* 25: 627–645.

Miles, A. E. W. 1958a. Sebaceous glands in the lip and cheek mucosa of man. *Brit. Dental J.* 105: 235–248.

Miles, A. E. W. 1958b. The development and atrophy of buccal sebaceous glands in man. 6th Ann. Meeting of Brit. Div. of the Intern. Assoc. for Dental Research. *J. Dental Res.* 37: 757.

Miles, A. E. W. 1963. Sebaceous glands in oral and lip mucosa. *In* "Advances in Biology of Skin. The Sebaceous Glands" (W. Montagna, R. A. Ellis, and A. F. Silver, eds.), Vol. 4, pp. 46–77. Pergamon, Oxford.

Montagna, W. 1949. Anisotropic lipids in the sebaceous glands of the rabbit. *Anat. Rec.* 104: 243–254.

Montagna, W. 1955. Histology and cytochemistry of human skin. VIII. Mitochondria in the sebaceous glands. *J. Invest. Dermatol.* 25: 117–121.

Montagna, W. 1962. "The Structure and Function of Skin," 2nd Edition. Academic Press, New York.

Montagna, W. 1962. The skin of lemurs. *Ann. N. Y. Acad. Sci.* **102**: 190–209.

Montagna, W. 1970. Histology and cytochemistry of human skin. XXXV. The nipple and areola. *Brit. J. Dermatol.* **83** (Special Issue): 2–13.

Montagna, W., and E. B. Beckett. 1958. Cholinesterases and alpha esterases in the lip of the rat. *Acta Anat.* **32**: 256–261.

Montagna, W., and R. A. Ellis. 1957. Histology and cytochemistry of human skin. XII. Cholinesterase in the hair follicles of the scalp. *J. Invest. Dermatol.* **29**: 151–157.

Montagna, W., and R. A. Ellis. 1958. L'histologie et la cytologie de la peau humaine. XVI. Repartition et concentration des esterases carboxyliques. *Ann. Histochim.* **3**: 1–17.

Montagna, W., and R. A. Ellis. 1959a. The skin of primates. I. The skin of the potto *(Perodicticus potto). Amer. J. Phys. Anthropol.* **17**: 137–162.

Montagna, W., and R. A. Ellis. 1959b. Cholinergic innervation of the Meibomian gland. *Anat. Rec.* **135**: 121–128.

Montagna, W., and R. A. Ellis, 1960. The skin of primates. III. The skin of the slow loris *(Nycticebus coucang). Amer. J. Phys. Anthropol.* **19**: 1–22.

Montagna, W., and D. M. Ford. 1969. Histology and cytochemistry of human skin. XXXIII. The eyelid. *Arch. Dermatol.* **100**: 328–335.

Montagna, W., and R. J. Harrison. 1957. Specializations in the skin of the seal *(Phoca vitulina). Amer. J. Anat.* **100**: 81–113.

Montagna, W., and C. R. Noback. 1946a. The histology of the preputial gland of the rat. *Anat. Rec.* **96**: 41–54.

Montagna, W., and C. R. Noback. 1946b. The histochemistry of the preputial gland of the rat. *Anat. Rec.* **96**: 111–128.

Montagna, W., and C. R. Noback. 1947. Histochemical observations on the sebaceous glands of the rat. *Amer. J. Anat.* **81**: 39–62.

Montagna, W., and J. S. Yun. 1962. The skin of primates. X. The skin of the ring-tailed lemur *(Lemur catta). Amer. J. Phys. Anthropol.* **20**: 95–118.

Montagna, W., and J. S. Yun. 1972. The glands of Montgomery. *Brit. J. Dermatol.* **86**: 126–133.

Montagna, W., H. B. Chase, and J. B. Hamilton. 1951. The distribution of glycogen lipids in human skin. *J. Invest. Dermatol.* **17**: 147–157.

Montagna, W., H. B. Chase, and W. C. Lobitz, Jr. 1952. Histology and cytochemistry of human skin. II. The distribution of glycogen in the epidermis, hair follicles, sebaceous glands and eccrine sweat glands. *Anat. Rec.* **114**: 231–248.

Montagna, W., K. Yasuda, and R. A. Ellis. 1961. The skin of primates. V. The skin of the black lemur *(Lemur macaco). Am. J. Phys. Anthrop.* **19**: 115–130.

Montagna, W., R. A. Ellis, and A. F. Silver. 1963. "Advances in Biology of Skin. The Sebaceous Glands," Vol. 4. Pergamon, Oxford.

Moss, C. W., V. R. Dowell, Jr., V. J. Lewis, and M. A. Schekter. 1967. Cultural characteristics and fatty acid composition of *Corynebacterium acnes. J. Bacteriol.* **94**: 1300–1305.

Nicholson, G. W. 1918–1919. Sebaceous glands in the cervix uteri. *J. Pathol. Bacteriol.* **22**: 252–254.

Nicolaides, N. 1961. Gas chromatographic analysis of the waxes of human scalp skin surface fat. *J. Invest. Dermatol.* **37**: 507–510.

Nicolaides, N. 1963. Human skin surface lipids—origin, composition and possible function. In "Advances in Biology of Skin. The Sebaceous Glands" (W. Montagna, R. A. Ellis, and A. F. Silver, eds.), Vol. 4, pp. 167–187. Pergamon, Oxford.

Nicolaides, N. 1965. Skin lipids. IV. Biochemistry and function. J. Amer. Oil Chem. Soc. 33: 404–409.

Nicolaides, N., and R. C. Foster, Jr. 1956. Esters in human hair fat. J. Amer. Oil Chem. Soc. 33: 404–409.

Nicolaides, N., and G. C. Wells. 1957. On the biogenesis of the free fatty acids in human skin surface fat. J. Invest. Dermatol. 29: 423–433.

Nicolaides, N., R. E. Kellum, and P. V. Woolley, III. 1964. The structure of the free unsaturated fatty acids of human skin surface fat. Arch. Biochem. Biophys. 105: 634–639.

Nikkari, T., and M. Valavaara. 1969. The production of sebum in young rats: effects of age, sex, hypophysectomy and treatment with somatotrophic hormones and sex hormones. J. Endocrinol. 43: 113–118.

Palay, S. L. 1958. The morphology of secretion. In "Frontiers in Cytology" (S. L. Palay, ed.), Chapter 11, pp. 305–342. Yale Univ. Press, New Haven, Connecticut.

Pelfini, C., S. Sacchi, and F. Serri. 1970. Is Wolff's unicellular sebaceous gland a melanocyte? Histochemical and electron microscopical observations. J. Invest. Dermatol. 54: 94 (Abstr.).

Perkins, O. C., and A. M. Miller. 1926. Sebaceous glands in the human nipple. Amer. J. Obstet. Gynecol. 11: 789–794.

Pontén, B. 1960. Grafted skin. Observations on innervation and other qualities. Acta Chir. Scand. Suppl. 257: 1–78.

Reisner, R. M., D. Z. Silver, M. Puhvel, and T. H. Sternberg. 1968. Lipolytic activity of Corynebacterium acnes. J. Invest. Dermatol. 51: 190–196.

Rogers, G. E. 1957. Electron microscope observations of the structure of sebaceous glands. Exp. Cell Res. 13: 517–520.

Rony, H. R., and S. J. Zakon. 1943. Effect of androgen on the sebaceous glands of human skin. Arch. Dermatol. Syphilol. 52: 323–327.

Rothman, S. 1954. "Physiology and Biochemistry of the Skin." Univ. Chicago Press, Chicago, Illinois.

Saito, K., and Y. Asada. 1967. Fatty acid composition of skin surface lipids and resident bacterial lipids. Biochem. Biophys. Acta, 137: 581–583.

Sansone, G., and R. M. Reisner, 1971. Differential rates of conversion of testosterone to dihydrotestosterone in acne and in normal human skin—a possible pathogenic factor in acne. J. Invest. Dermatol. 56: 366–372.

Schaffer, J. 1927. Die Drusen. I. Teil. In "Handbuch der mikroskopischen Anatomie des Menschen" (W. von Möllendorff, ed.), Vol. 2, Part 1, pp. 132–148. Springer-Verlag, Berlin and New York.

Serri, F., and M. W. Huber. 1963. The development of the sebaceous glands in man. In "Advances in Biology of Skin. The Sebaceous Glands" (W. Montagna, R. A. Ellis, and A. F. Silver, eds.), Vol. 4, pp. 1–18. Pergamon, Oxford.

Stamm, R. H. 1914. Über den Bau und die Entwicklung der Seitendrüse del Waldspitzmaus (Sorex vulgaris L.) Mindeschrift for Japetus Steenstrup. Kobenhavnn, pp. 1–24 (Cited from Schaffer, 1927).

Staricco, R. G. 1960. The melanocytes and the hair follicle. J. Invest. Dermatol. 35: 185–194.

Strauss, J. S., and A. M. Kligman. 1958. Pathologic patterns of the sebaceous glands. J. Invest. Dermatol. 30: 51–61.

Strauss, J. S., and A. M. Kligman. 1960. The pathologic dynamics of acne vulgaris. *Arch. Dermatol.* **82**: 779–790.

Strauss, J. S., and P. E. Pochi. 1969. Recent advances in androgen metabolism and their relation to the skin. *Arch. Dermatol.* **100**: 621–636.

Strauss, J. S., A. M. Kligman, and P. E. Pochi. 1962. The effect of androgens and estrogens on human sebaceous glands. *J. Invest. Dermatol.* **39**: 139–155.

Suskind, R. R. 1951. The chemistry of the human sebaceous gland. I. Histochemical observations. *Invest. Dermatol.* **17**: 37–54.

Sweeney, T. M., R. J. Szarnicki, J. S. Strauss, and P. E. Pochi. 1969. The effect of estrogen and androgen on the sebaceous gland turnover time. *J. Invest. Dermatol.* **53**: 8–10.

Takayasu, S., and K. Adachi. 1970. Hormonal control of metabolism in hamster costovertebral glands. *J. Invest. Rermatol.* **55**: 13–19.

Takayasu, S., and K. Adachi. 1972. The *in vivo* and *in vitro* conversion of testosterone to 17β-hydroxy-5α-androstan-3-one (dihydrotestosterone) by the sebaceous gland of hamsters. *Endocrinology* **90**: 73–80.

Takeuchi, T. 1958. Histochemical demonstration of branching enzyme (amylo-1, 4-1, 6-transglucosidase) in animal tissues. *J. Histochem. Cytochem.* **6**: 208–216.

Thies, W., and L. F. Galente. 1957. Zur histochemischen Darstellung der cholinesterasen in vegetativen Nervensystem der Haut. *Hautarzt* **8**: 69–75.

Thody, A. J., and S. Shuster. 1971a. Sebotrophic activity of β-lipotrophin. *J. Endocrinol.* **50**: 533–534.

Thody, A. J., and S. Shuster. 1971b. The effect of hypophysectomy on the response of the sebaceous gland to testosterone propionate. *J. Endocrinol.* **49**: 329–333.

Thody, A. J., and S. Shuster. 1971c. Effect of adrenalectomy and adrenocorticotrophic hormone on sebum secretion in the rat. *J. Endocrinol.* **49**: 325–328.

Toft, D., and J. Gorski. 1966. A receptor molecule for estrogens: isolation from the rat uterus and preliminary characterization. *Proc. Nat. Acad. Sci. U.S.* **55**: 1574–1581.

Wakil, S. J., E. L. Pugh, and F. Sauer. 1964. The mechanism of fatty acid synthesis. *Proc. Nat. Acad. Sci. U.S.* **52**: 106–114.

Weitkamp, A. W., A. M. Smiljanic, and S. Rothman. 1947. The free fatty acids of human hair fat. *J. Amer. Chem. Soc.* **69**: 1936–1939.

Wheatley, V. R. 1952. The chemical composition of sebum. "Livre Jubilaire 1901–1951" de la Société Belge de Dermatologie et de Syphiligraphie. Imprimerie Medicale et Scientifique, Bruxelles, pp. 90–102.

Wheatley, V. R. 1953. Some aspects of the nature and function of sebum. *St. Bartholomew's Hosp. J.* **57**: 5–9.

Wheatley, V. R., D. C. Chow, and F. D. Keenan, Jr. 1961. Studies of the lipids of dog skin. II. Observations of the lipid metabolism of perfused surviving dog skin. (Preliminary and Short Report). *J. Invest. Dermatol.* **36**: 237–239.

Wilkinson, D. I. 1969. Esterified cholesterol in surface lipids: Methods of isolation and fatty acid content. *J. Invest. Dermatol.* **53**: 34–38.

Wolff, E. 1951a. Unicellular sebaceous glands in basal layer of normal human epidermis. *Lancet* **i**: 888–889.

Wolff, E. 1951b. Xanthelasma palpebarum: tumour of sebaceous glands. *Brit. J. Dermatol.* **63**: 296–302.

Yun, J. S., and W. Montagna. 1964. The skin of primates. XX. Development of the appendages in *Lemur catta* and *Lemur fulvus*. *Amer. J. Physiol. Anthropol.* **22**: 399–406.

11

Apocrine Glands

I. Introduction

A varied assortment of sometimes totally unrelated structures is gathered under the category of apocrine gland, a somewhat meaningless designation since as a rule apocrine glands do not, as the term implies, slough off the apical cytoplasm into the lumen. Moreover, because these glands secrete slowly and sparsely and rarely respond to thermal stimulation, their product should not be called sweat.

Krause (1844) was the first to observe that the apocrine glands in man are more numerous and active in the axilla and in the external auditory meatus (ceruminous glands), the circumanal area, and the eyelids (glands of Moll). Horner (1846) and Rolin (1846) recognized and described more clearly than Krause the large axillary glands, and finally Kölliker (1853) reported their histology with perspicacious clarity. Years later Schiefferdecker (1917) called them epicrine or apocrine glands, and Pinkus (1958) described them as occurring singly or in clusters in the mons, perineum, genitalia, face, scalp, abdomen, and elsewhere.

Among mammals the apocrine sweat glands are found throughout the hairy skin of Bovidae, Ovidae, Equidae, Suidae, and most carnivores.

Although profusely distributed and active, the glands of horses have not been carefully described, perhaps because they differ so much from those of man. However, under stress and heat horses do "sweat," and the glands of dogs, cattle, pigs, and hippopotamuses sometimes respond to thermal stimuli.

The ratio of eccrine to apocrine glands in nonhuman primates suggests that the former are gradually replacing the latter in these animals. Schiefferdecker (1922) believed that Europeans, alleged to have the fewest apocrine glands, are the most advanced members of the human race, whereas the Australian Negritoes, with the most apocrine glands, are the most primitive. However, Schiefferdecker's analyses were not exact; moreover, whereas some ethnic differences no doubt exist, so do individual differences within the same racial group, even in Europeans (Woollard, 1930). Limited studies suggest that apocrine glands are better developed and more numerous in Negroes than in Caucasians, and, regardless of race, women are said to have more of them than men (Homma, 1926). For whatever reason, studies of these glands have not been popular; hence little progress in understanding them has been made during the last ten years.

As the opening statement indicated, the apocrine glands of man, like others in the human body, are different enough to warrant special classification. However, a thorough consideration of each of the various glands called "apocrine" is outside the scope of this chapter. We will limit our discussion to certain basic similarities in the properties of the apocrine glands in the cavum axillae.

The largest and most numerous of the human glands are found in the axilla, where huge apocrine glands and eccrine sweat glands in a one-to-one ratio form the axillary organ. Besides man, only the chimpanzee and gorilla have an axillary organ, which in this sense can be regarded as primarily a hominoid characteristic.

Human axillary apocrine glands have peculiarities not found in those of any other mammals. Models reconstructed from serial sections and teased whole glands from the axilla show them to be completely coiled (Horn, 1935; Sperling, 1935; Hurley and Shelley, 1954, 1960). The secretory segment in each is so large and compact that adjacent loops fuse together and are often joined by shunts. Furthermore, the tubule is so variably distended and so many diverticula extend from it that the viscus looks more like a sponge than a tubule (Figs. 1 and 2). The secretory segments of the smaller glands rest entirely within the dermis, but those of the larger ones often extend deep into the subcutaneous fat, sometimes as much as 5 mm below the surface. As Horner (1846)

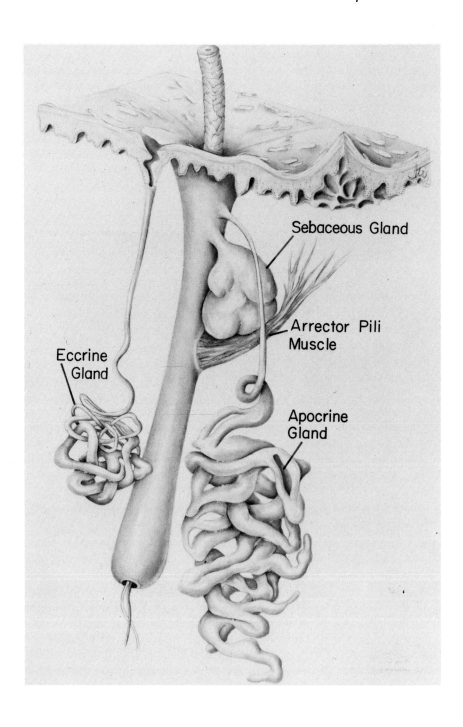

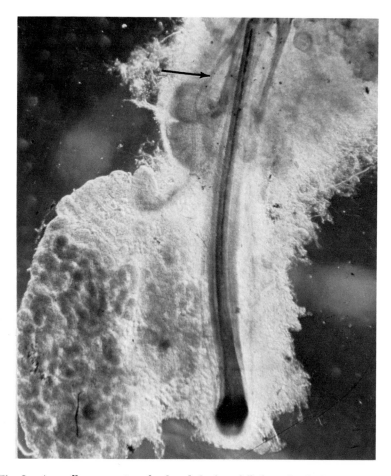

Fig. 2. An axillary apocrine gland and the hair follicle with which it is associated, macerated and dissected, to show its coil on the left of the picture, its duct (arrow) opening into a pilary canal, and its relative size in comparison with a hair follicle. (Courtesy of Dr. W. B. Shelley.)

showed years ago, some glands measure as much as 2 mm in diameter at the widest part and are, therefore, acroscopic. The duct of each gland rests close and parallel to a hair follicle and usually opens inside the pilary canal some distance above the entrance of the duct of a sebaceous gland (Figs. 3 and 4). Sometimes two glands open into one pilary canal

Fig. 1. Diagram of an apocrine gland in relationship to other cutaneous append-ages.

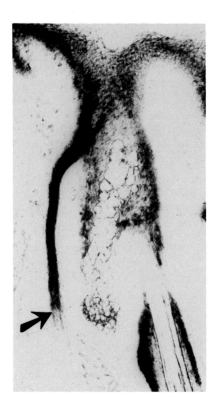

Fig. 3. The duct of the ceruminous gland runs parallel to the hair follicle and opens above the sebaceous gland. There is an abrupt transition (arrow) between the duct and secretory segment. Amylophosphorylase preparation.

(Hurley and Shelley, 1960), and some open directly onto the surface. All apocrine glands have a straight narrow duct that runs parallel to the hair follicle.

Since in man and many animals apocrine glands develop as appendages of hair follicles and open into pilosebaceous canals, they are defined, not always correctly, by some authors according to their association with hair follicles. Even in man, the ducts of some glands open directly onto the surface of the epidermis, either because they have become separated from their points of origin or because the glands have developed from the epidermis. Among the nonhuman primates, the apocrine glands of all Lemuridae open directly onto the surface of the

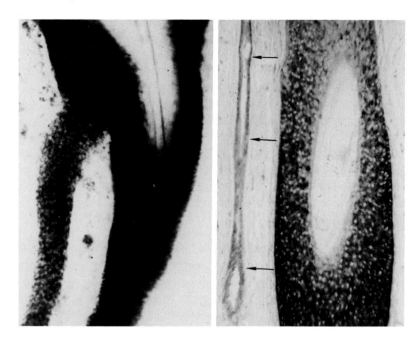

Fig. 4. The figure on the left shows the duct of an axillary sweat gland opening into a pilary canal. The figure to the right shows a duct (arrows) parallel to a hair follicle.

epidermis. In many monkeys and even in the anthropoid apes, they often open near the orifices of a pilary canal, seldom inside it.

II. Development

The development of sweat glands has been systematically studied by a number of authors, and the account that follows is a brief summary of their findings supplemented with our own observations (Carossini, 1912; Steiner, 1926; Borsetto, 1951; Pinkus, 1958). In 5- to 6-month-old fetuses, the anlagen of apocrine glands develop as solid epithelial buds or bulges above the sebaceous gland from the side of hair follicles. At this stage, fully formed hair follicles have already produced a hair that emerges at the surface of the skin, and the sebaceous glands are at least partially differentiated. Authors generally agree that in the scalp and face most hair follicles develop glandular anlagen that usually regress later. The primordial buds elongate into solid cords with somewhat flask-shaped terminations. By the sixth fetal month, the base of the cord begins to coil, and except in the axilla, external auditory meatus, eyelids, and anogenital areas, most of the glandular vestiges are gradually resorbed. In the remaining glands, a lumen appears in the presumptive duct, produced by a partial cornification and shriveling up of the center cells. Isolated clefts then appear in the center of the coil and later become extensive and confluent, forming a continuous lumen. By the seventh and eighth months, the glands are larger and convoluted, have a bigger lumen, and resemble mature apocrine sweat glands. Finally at birth, they are recognizably apocrine; the secretory coil has penetrated deep into the hypodermal fat and is more or less encapsulated in connective tissue. However, myoepithelial cells are difficult to recognize.

Because apocrine glands develop their structural and functional properties slowly and are small at birth, some authors have concluded that they are formed postnatally. Furthermore, since they are functionless in the neonate, they can easily be missed altogether unless the skin specimens are removed from the cavum axillae, an extremely small area in the newborn. Hence there has been a great deal of unnecessary confusion about the identity of the two types of glands. However, anyone familiar with their histology cannot possibly mistake one for the other. In children up to one year of age, the coils of the apocrine glands lie deep in the dermis, whereas those of the eccrine sweat gland are closer to the surface (Montagna, 1959). Although the diameter of the two types is about the same during infancy, apocrine glands have a large lumen lined with identical low cuboidal cells, and eccrine glands have a small lumen lined with pyramidal dark and clear cells (Fig. 5). Even in children apocrine glands, unlike eccrine glands, have a very thick basement membrane. Axillary glands generally remain small and relatively undifferentiated up to 7 years of age, but in some children they are precociously large and apparently functional even at 5 years. In preadolescents, an incipient accumulation of fluid in the lumen causes some of the glands to become distended and even cystic (Montagna, 1959). From the seventh year through adolescence, the glands become progressively larger and attain the general structural and functional properties of adults. Although morphologically the glands of 10-year-old children are large and well-differentiated, limited pharmacological observations indicate that they are not functional (Hurley and Shelley, 1960). Still, most girls at that age, begin to emit the characteristic axillary odor even before the axillary hairs have become coarse. Despite

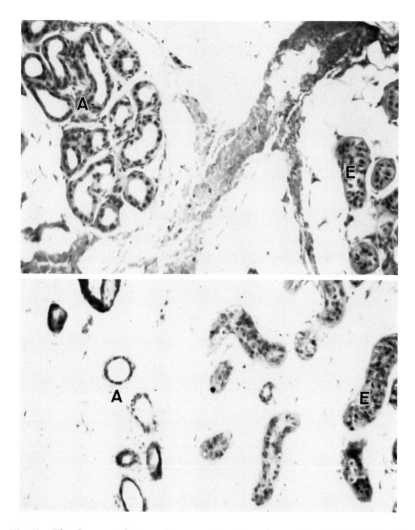

Fig. 5. The figure at the top shows eccrine (E) and apocrine glands (A) from the axilla of a boy 4 years old; at the bottom is a sample of axilla from a 5-year-old boy. In both tissues the apocrine glands are to the left, the eccrine glands to the right.

their small size, the glands of infants and children are similar to those of adults and have similar histochemical properties. In old age they, like all organs, gradually degenerate, but this involution does not seem to be related to the deterioration of the sex hormones (Montagna, 1959;

Hurley and Shelley, 1960). The details of aging changes are discussed later in this chapter.

III. Structure of the Gland

The apocrine glands are tubular glands consisting of a secretory coil embedded in the dermis and a duct that conveys secretory product to the pilosebaceous canal.

IV. The Secretory Coil

The secretory coil is lined with either cuboidal or columnar cells. The shape of the cell is variable and may depend upon the secretory activity (compare Figs. 6 and 7). Segments of some tubules or sometimes entire tubules are so distended that their epithelium is cuboidal or even simple squamous (Fig. 6). All epithelial cells rest upon a bed of large, loosely dovetailed myoepithelial cells (Figs. 6, 7, and 8), outside of which is a thick, hyalin basement membrane. This, in turn, as in the eccrine sweat glands, is invested with delicate elastic fibers.

Secretory cells have one, sometimes two large spherical nuclei near the base. When well-fixed, these nuclei are round and turgid and stain deeply with most basic dyes and with the Feulgen reaction; the one or two large nucleoli are flanked by two strongly basophilic, Feulgen-positive satellite bodies. The basophilic property of the nucleolus is destroyed by digestion with ribonuclease, whereas that of the rest of the nucleus is only slightly diminished.

Ultrastructurally, the nucleus of the secretory cell is surrounded by an envelope of thick inner and thin outer membranes, both crammed with fine granules (Yasuda and Ellis, 1961). The granules attached to the outer membrane are about 150 Å in diameter; those on the inner membrane are so small that they are difficult to distinguish from particulate chromatin (Kurosumi *et al.*, 1959). The narrow space between the two varies in width and is occasionally vesiculated. Nuclear pores are occasionally covered by indefinite single-membrane diaphragms (Watson, 1955).

Adult apocrine cells do not usually divide, but when one mitotic figure is found, others are generally present in nearby cells. During mitosis, cells become roughly spherical and relatively free from cyto-

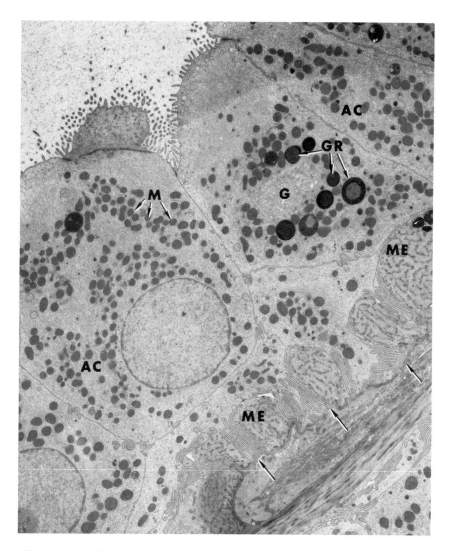

Fig. 6. Overall view of the cells comprising the secretory segment of apocrine sweat gland. At regular intervals, myoepithelial cells (ME) rest on a thick basal lamina (arrows). Apocrine cells (AC) have a basal nucleus above which is the prominent Golgi complex (G). The rest of the cytoplasm contains many mitochondria (M) and an assortment of granules (GR). The luminal border has numerous microvilli. × 4200. (Courtesy of Dr. M. Bell.)

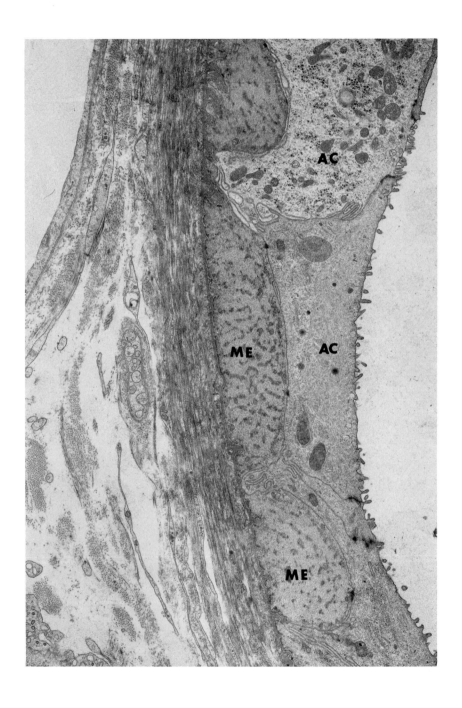

plasmic granules. In view of the high cytoplasmic activity and the great amount of cellular sloughing, the paucity of mitotic figures is puzzling. Some segments of tubules are sometimes filled with castoff, still viable whole cells. With all basic dyes used, some basophilia appear mostly at the base of the cells. Toluidine blue, methylene blue, and azure B, used in solutions of 1/5000 to 1/2000 and buffered to pH 4.0 and 5.0, stain little more than nucleic acid. At their base, the cells have a cyto- plasm normally stippled with fine basophilic granules. When copious, these granules are aligned longitudinally and give the effect of striations or fine reticula. The rest of the cytoplasm is weakly basophilic, and above the nucleus the negative image of the Golgi zone does not stain at all (Fig. 8); the luminal border of each cell is relatively free of basophilic granules. Ribonuclease destroys all basophilia.

The basophilic material in the cytoplasm of the secretory cells is inversely proportional to the number of granules and apparently directly proportional to the number of mitochondria. Morphologically, nucleic acid, together with mitochondria, is probably responsible for the syn- thesis of pigmented and chromophobic secretion granules (Montes et al., 1960).

Mitochondria are easily demonstrated in axillary glands, even in tissues fixed in Zenker-formol without postchromatin. Heidenhain's hematoxylin, Regaud's hematoxylin, aniline acid fuchsin methyl green, and various other methods clearly show the mitochondria to be large, numerous, and pleomorphic.

In the tall columnar cells with few cytoplasmic granules, mitochondria are in the form of rodlets, filaments, granules, and spherules. More often they look like long filaments aligned parallel to the axis of the cell. Large pleomorphic mitochondria are found even in the cuboidal or flat cells that line the dilated tubules.

On the other hand, in the tall columnar cells replete with granules above the nucleus, the mitochondria look like large granules (Fig. 6) or moniliform filaments. In stained preparations, it is difficult to deter- mine whether particulate elements are large mitochondria or secretion granules; many granules, like mitochondria, are fuchsinophilic. The largest are always around the Golgi region. In the glands of some aged individuals, the number of mitochondria and cytoplasmic granules is conspicuously reduced. The mitochondria at the base of the cells are

Fig. 7. A portion of an apocrine sweat gland with flat, almost squamous secretory cells (AC); the size and shape reflects secretory activity. The myoepithelial cells (ME) punctuate the base of the gland. × 6000. (Courtesy of Dr. M. Bell.)

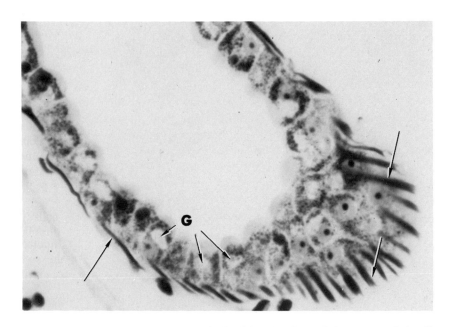

Fig. 8. Secretory cells rest on a bed of loosely dovetailed myoepithelial cells (arrows), cut here at different planes. The clear area above the nucleus is the negative image of the Golgi apparatus (G).

small and somewhat typical; above the nucleus, however, they become large, attaining 2 to 3 μm in diameter, and have electron opaque matrices and poorly delineated cristae. Transitional stages lie between these two. Large, atypical mitochondria also stain with the periodic-Schiff reaction (Yasuda *et al.*, 1962; Munger, 1965; Ellis, 1967). These bodies, then, must be the acidophil, nonpigmented, PAS-positive granules seen with the light microscope (Montagna *et al.*, 1953; Montes *et al.*, 1960).

In the glands of other animals, Brinkmann (1908) thought mitochondria were replaced by or transformed into secretion granules, and Nicolas *et al.* (1914) stated that in man the granules arise directly from mitochondria. However, in stained preparations, mitochondria were found to be most numerous in cells with few or no granules (Fig. 6) and scarce in cells replete with granules (Minamitani, 1941a,b; Ota, 1950). Mitochondrial transformation into different types of secretory granules has been claimed in a wide variety of cells, but none of these claims has been substantiated.

Bergen (1904) first described a "canalicular system" (Golgi apparatus) between the nucleus and the free border of apocrine cells. Osmium tetroxide techniques revealed a system of twisted rodlets and granules above the nucleus and about the same size as the nucleus (Melczer, 1935; Minamitani, 1941b; Mitchell and Hamilton, 1949; Ota, 1950), which corresponds to the area of the negative image of the Golgi apparatus (Fig. 4). The distribution of the Golgi element between the secretion granules distally and the nucleus proximally suggested to Minamitani (1941b) that the secretion granules are formed below the Golgi apparatus, at the expense of the mitochondria, and that the Golgi element is concerned with the maturation of the granules. Moreover, during the ripening process, some granules acquire pigment, lipid, and iron by way of the Golgi apparatus. However, many granules contain none of these substances in large enough quantities to be seen microscopically; though present in the small yellow pigmented granules, ionic iron is only rarely found in the large brown ones. Furthermore, the Golgi apparatus is as well developed in cells that contain neither iron nor pigment as in cells rich in these substances. The Golgi apparatus also contains lipid granules and PAS-positive substances that could be phospholipids, glycoproteins, or mucopolysaccharides (Fig. 9).

The cytoplasm of the secretory cells has a somewhat opaque matrix that contains granules of different sizes and shapes. The large, naturally brown pigmented granules stain blue-green with basic dyes; the smaller yellow ones remain unstained. Acidophilic secretion granules are numerous in the supranuclear cytoplasm.

Numerous dense granules above the nucleus have been described as "rough" (Charles, 1959) or "dark granules" (Kurosumi et al., 1959). Most of them are found around the Golgi zone and vary greatly in size and shape (Fig. 6). Because of an apparent ultrastructural similarity between them and keratin and because of the presence of SH– and S–S groups in them, Munger (1965) believed them to be "presumptive keratin granules" or "keratin crystalloids." However, this seems to be unlikely. Biempca and Montes (1965) believed that they are lysosomes, and Yasuda et al. (1962) and Ellis (1967) regarded them as large basophil granules containing lipid, pigment, and iron.

Another kind of granule with light or opaque inclusions is smaller, somewhat spherical, has a single limiting membrane, and contains eccentrically displaced (artifact?) masses embedded in a pale matrix (Ellis, 1967).

Throughout the cytoplasm but most numerous near the atypical

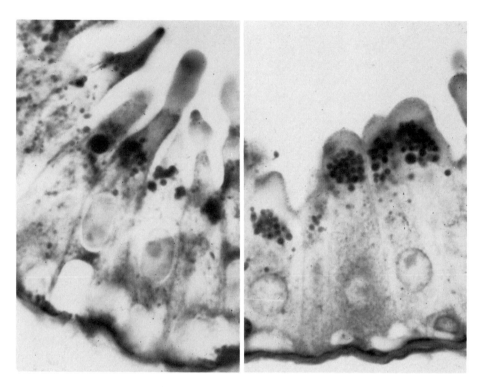

Fig. 9. PAS-positive granules in the cells of axillary glands. Both figures show some reaction in the apical granule-free cytoplasm.

mitochondria are profiles of RER. Since they vary from cell to cell, Ellis (1967) believed that they reflect the state of cellular activity. Annulate lamellae, rarely found in somatic cells, were seen near the nucleus by Gross (1966). No one knows the significance of these structures. Whorls of tonofilaments, as well as profiles of smooth endoplasmic reticulum (SER), ribosomes, and a few microtubules, are found in the supranuclear cytoplasm.

The luminal cytoplasm of the cells is nearly chromophobic. Ito and Iwashige (1953) reported that apocrine cells secreted into the lumen a basophilic substance regarded by them as ribonucleic acid. The lumen of some tubules contained secretions that stained weakly with basic dyes buffered to pH 6.0 or above. Since ribonuclease does not destroy their stainability, the secretion is probably free of ribonucleic acid (Montes *et al.,* 1960).

V. Myoepithelial Cells

The large, spindle-shaped myoepithelial cells, sandwiched between the secretory epithelium and the basement membrane (Figs. 6, 7, and 8), are oriented somewhat parallel to the length of the tubule. These cells, which are believed to arise from the ectoderm, resemble smooth muscle fibers and are 5 to 10 μm in the widest diameter and 50 to 100 μm in length. In transverse sections of tubules, the cell body insinuated between the bases of the secretory cells appears to be triangular. Myoepithelial cells are so loosely dovetailed that parts of the secretory cells rest against the basement membrane through the interstices.

The cells are best developed in tubules lined with the tallest epithelium; in dilated ones lined with flat, elongated epithelial cells, they are sometimes impossible to demonstrate. Montes et al. (1960) differed with this statement, but their evidence was not conclusive. The delicate longitudinal fibrils in these cells are more numerous and coarse than those in the smooth muscle fibers of the cutaneous arterioles and of the arrectores pilorum (Bunting et al., 1948). Rows of phospholipid granules and glycogen particles are aligned parallel to the axis of the cell (Montagna et al., 1951). Under polarized light, the cells show longitudinally oriented anisotropic striae similar to those in smooth muscle. In old people, the myoepithelial cells are said to become swollen, vacuolated, and laden with lipid and pigment (Ito et al., 1951). More often than not, however, they reflect the general state of the glands and sometimes are in excellent condition even in aged subjects (Winkelmann and Hultin, 1958; Montagna, 1960).

Under the electron microscope, "myofilaments" are aligned parallel to the long axis of the cells as in smooth muscle fibers (Fig. 10). Hurley and Shelley (1954, 1960) reported seeing myoepithelial cells contract like muscle fibers and in peristaltic waves when a droplet of secretion appeared on the surface. Many have reported that pharmacological, mechanical, and electrical stimuli, which normally initiate contraction in smooth muscle, induced contraction in the myoepithelium to expel apocrine secretion. Although these cells may contract, their major function, like that of the delicate elastic fibers outside the basement membrane, is probably supportive and protective. There are no myoepithelial cells around the duct.

As seen in the electron microscope, the Golgi apparatus consists of an aggregate of vacuoles and vesicles outlined by single smooth-surfaced membranes with no typical "Golgi lamellae" (Fig. 6). The spherical,

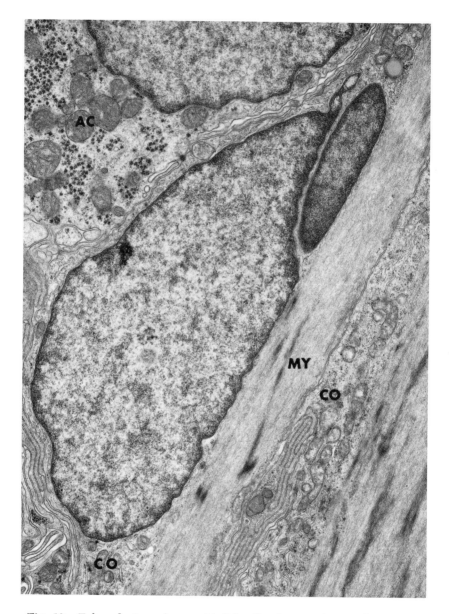

Fig. 10. Enlarged view of myoepithelial cells showing the myofilaments (MY) aligned parallel to the long axis of the cells. The cytoplasmic organelles (CO) are often segregated around the nucleus. At the top is a portion of secretory cell (AC). × 13,000. (Courtesy of Dr. M. Bell.)

or frequently pleomorphic, vesicles and vacuoles contain material whose density is similar to that in the many small vesicles in the "clear" apical cytoplasm (Yasuda and Ellis, 1961; Ellis, 1967). Sparse, fine granules are scattered through the Golgi area, which in turn is surrounded by granules and mitochondria.

VI. The Structure of the Duct

The upper part of the secretory segment, whose diameter is less than one-third that of the lower coils, narrows abruptly and emerges into the extremely slender duct (Fig. 3). There is no transition between the epithelium of the duct and that of the secretory segment. The duct runs a relatively straight course parallel to the hair follicle and is composed of two layers of cells, the flattened luminal ones having a slightly cornified border or cuticle. The multilayered terminal part is funnel-shaped (Fig. 3). Despite some reports to the contrary, there are myoepithelial cells around the duct.

VII. Lipids

With Sudan dyes, the large, darkly pigmented masses color lightly, some of the smaller ones color intensely, and some not at all. None of the nonpigmented granules contain sudanophilic lipids. Immersion overnight in acetone and in pyridine at 60°C for as long as 18 hours removes only a small part of the sudanophil lipid. Acid hematein, a quasispecific test for demonstrating phospholipids, shows parallel rows of reactive granules in the myoepithelial cells. In unstained frozen sections under near ultraviolet light, the large pigment droplets emit an orange-yellow light. None of the glands show birefringent lipids.

The large, darkly pigmented droplets are sudanophilic even in paraffin sections. Similar granules can be seen in the lumen only when whole epithelial cells are sloughed. Carbol-fuchsin shows acid-fastness mostly in the large, darkly pigmented droplets.

VIII. Iron

Intraepithelial ionic iron is a distinctive feature of the axillary apocrine glands of man (Fig. 11). Sparse in the glands of the mons, iron is

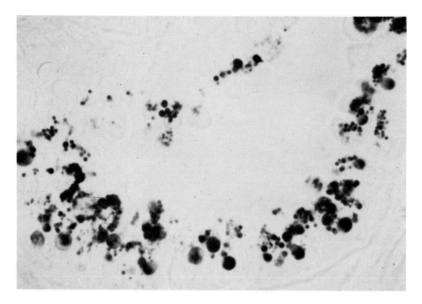

Fig. 11.   Iron-containing granules in the columnar cells of an axillary apocrine gland.

more abundant in the external auditory meatus and copious in the active mammary glands of some mammals.

Although Homma (1925) and Manca (1934) reported iron in most of the glands they studied, actually some specimens abound in it, and others have little or none. Moreover, different coils of the same gland may contain different amounts of iron. When abundant, it is usually found in all tubules and frequently in the connective tissue macrophages surrounding them. Some have contended that there is more iron in the glands of middle-aged people (this has never been confirmed) and that it increases appreciably in subjects with pulmonary tuberculosis (Manca, 1934).

Iron-containing granules mostly occur alongside and above the nucleus, clustered around the clear supranuclear negative image of the Golgi apparatus. Whether in the tall or in the low cuboidal cells, iron nearly always occurs in the form of small granules. Reports to the contrary notwithstanding, it is rarely present in large granules and is found in the lumen only when whole epithelial cells are discarded. When the luminal content is clear and free of all cell particulates, there is no iron. Rough chemical analyses of apocrine secretion, showing only small quantities of iron, corroborate this finding.

There is some relation between the abundance of iron and that of pigmented granules. Iron is mostly associated with small yellow pigment granules; large dark brown granules are either free of it or surrounded by a thin reactive film through which the brown pigment is often discernible. Despite this tenuous relationship, the amount of yellow or brown granules in the cells is not a true index of the presence of iron. However, when iron is present, it is found only in the pigmented cells, regardless of size. Manca (1934) believed that the pigment that contains iron is hemosiderin. Zorzoli (1950), who saw no association between pigment and iron, concluded that iron is associated with some substance other than pigment.

Iron is not always present, even when the small yellow pigmented granules are numerous and the pigment is not hemosiderin. There is no clue about the fate of iron. Under the electron microscope, it has not been seen in the apical cytoplasm of the cells; however, small, dense particles arranged in patterns characteristic of ferritin have been seen in what may correspond to the yellow granules.

IX. Miscellaneous Histochemical Properties

Unlike eccrine glands, human apocrine glands have only a moderate amount of succinic dehydrogenase activity (Montagna and Formisano, 1955), despite a relatively strong reaction in the ducts. Cytochrome oxidase activity is moderate in the duct and copious in the secretory coil, and monoamine oxidase abounds in the duct and secretory epithelium (Yasuda and Montagna, 1960). The duct is also rich in amylophosphorylase activity (Fig. 3), but the secretory coil is unreactive except for the myoepithelial cells (Yasuda et al., 1958; Montagna, 1959; Ellis and Montagna, 1959). There are intense concentrations of nonspecific esterase (Montagna and Ellis, 1958, 1959) and ATPase, and the apical border of the tall epithelial cells and the myoepithelial cells show some alkaline phosphatase. The endothelium of the vascular bed around the glands is phosphatase-reactive (Ellis et al., 1958; Montagna, 1959). Acid phosphatase reactivity can be seen in the duct and in the apical portion of the secretory cell. The glands are strongly reactive for β-glucuronidase and aminopeptidase activity (Adachi and Montagna, 1961).

Previously the nerves around the apocrine glands were thought to contain no cholinesterase (Shelley and Hurley, 1953; Thies and Galente, 1957; Montagna and Ellis, 1958; Hurley and Shelley, 1960). However, improved methods have shown acetylcholinesterase nerves around the axillary gland (Montagna, 1962), the glands of Moll (Montagna and Ford, 1969), and those in other places (Figs. 12, 13, and 14). In each of these areas, the number of nerves that can be demonstrated with these techniques differs.

Many of the apocrine glands on the general body surface of non-human primates have no cholinesterase-containing nerves, whereas those in specialized skin areas are rich in them. Such nerves are also found around the axillary apocrine glands in chimpanzees and gorillas, in the glands of the antebrachial organs of ringtail lemurs, around those of the brachial organs of the lorises, and the body glands of horses (Fig. 14).

Whereas the axillary glands of children sometimes contain glycogen (Montagna, 1959), those of adults rarely do except for the myoepithelial cells (Montagna *et al.*, 1951, 1953). Such wide variations make the numerous assertions and denials about the presence or absence of glycogen pointless.

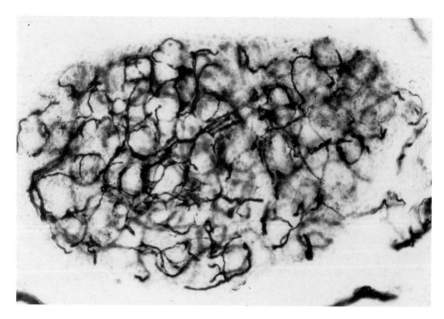

Fig. 12.   Nerves around glands of Moll. Acetylcholinesterase preparation.

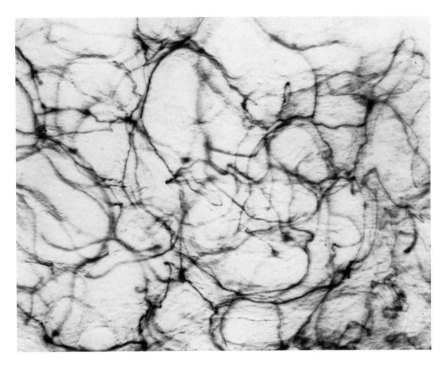

Fig. 13. Nerves around axillary glands. Acetylcholinesterase preparation.

X. Secretory Process

Apocrine secretion is a viscid substance ranging from a milky pale gray or clear white to a reddish, yellowish, or even black exudate that dries in shiny, gluelike droplets. With some variation, it fluoresces with a white or yellow light of moderate to low intensity. The secretion, considerably less than 1 ml, is slow to respond to a given stimulus. Again, with wide variations among individuals, secretion is not continuous and is characterized by long latent periods between active cycles (Shelley, 1951; Hurley and Shelley, 1960).

The name *apocrine* was given to these glands because it was believed on morphological grounds that the free end of each cell is pinched off into the lumen when the myoepithelial cells contract. The remaining part of the cell was thought to repair its luminal border, synthesize

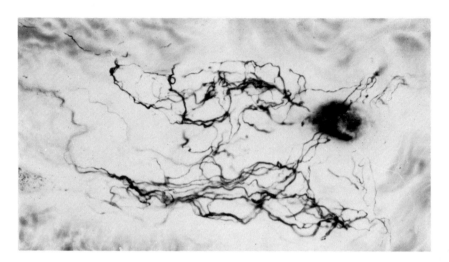

Fig. 14. Acetylcholinesterase-reactive nerves around two apocrine glands of a horse.

more secretion products, and repeat the process (Schiefferdecker, 1922; Hoepke, 1927; Schwenkenbecker, 1933). Some authors believed that axillary secretion can also be eccrine and holocrine (Holmgren, 1922; Ota, 1950; Rothman, 1954). The whole matter of the mechanism of apocrine secretion has been obfuscated by too many senseless arguments. When all the evidence is in, all the mechanisms of secretion from merocrine to holocrine seem to be present in apocrine glands.

The dilation of some segments of apocrine glands is a common phenomenon, particularly in old age, when parts of a gland become cystic and the character of their secretion is so changed that Winkelmann and Hultin (1958) called it mucoid metaplasia. These changes are not necessarily characteristic of old age since they occur even in children and young adults. "Mucoid metaplasia" and cystic dilatation are found predictably in women in their late 20's and increasingly so in both men and women later in life; they are common in people over 50. Much of the axillary organ, however, even in the oldest subjects studied remains morphologically intact (Montagna, 1959). Although gonadal hormones seem to play an initial role in the development and maintenance of these organs, once developed the glands are relatively independent. In women ovariectomized 10 or more years, either with or without replacement estrogenic therapy, these glands were morphologically normal (Klaar,

1926; Montagna, 1962). Topical applications of estrogens or androgens to the axilla or implantation of pellets of these hormones had no effect on the glands (Shelley and Cahn, 1955; Shelley and Hurley, 1953). Yet the axillary organ of adult idiot women remains infantile, and axillary hairs do not grow (Shelley and Butterworth, 1955) even in some female subjects with relatively normal menstrual cycles. Thus, axillary hairs and apocrine glands differentiate at the same pace, and both are probably controlled by the same agents. Furthermore, gonadal hormones in amounts sufficient to maintain the menstrual cycle seem to be incapable of promoting the development of the axillary apocrine glands.

Glands undergo the aging changes just described but not always as units. Some segments of a gland change while others apparently remain normal. In old subjects, whole glands undergo degenerative changes while others remain intact. These changes are not specific nor of greater magnitude than any of the other aberrations in aging skin (Montagna, 1959). Some authors have stated, largely on the basis of histological evidence, that the axillary apocrine glands of women undergo periods of greater or lesser activity that parallel the menstrual cycle and pregnancy (Talke, 1903; Waelsch, 1912; Loeschcke, 1925; Schaffer, 1926; Richter, 1932; Cavazzana, 1947; Cornbleet, 1952; Montes et al., 1960), but again these pronouncements lacked supportive data. When biopsy specimens were removed from the same women at weekly intervals during the menstrual cycle, no changes were seen in the axillary organ (Klaar, 1926; Montagna, 1956, 1959; Montes et al., 1960). Even the specimens removed at monthly intervals from pregnant and puerperal women showed a normal range of individual variations (Montagna, 1956; Montes et al., 1960). Thus if cyclic differences do occur in the axillary apocrine glands, they are not reflected morphologically.

The characteristic axillary odor emanates largely from apocrine secretion and is probably due to bacterial degradation (Shelley and Hurley, 1953; Hurley and Shelley, 1960). In Japanese subjects, axillary odor is said to diminish and disappear in old age (Ito et al., 1951). The data on Caucasians are incomplete, but Hurley and Shelley (1960) found a strong, normal axillary odor in 4 out of 5 healthy men between 65 and 70 years of age.

XI. Composition of Apocrine Secretion

Perhaps the reason we know so little about the composition of apocrine secretion is that the gland secretes only small amounts and only at long intervals. Hurley and Shelley (1960) calculated that each gland secretes about 0.01 cm^3 every 24 to 48 hours and between periods of secretion is refractory to stimulation. They suggested that the cells slowly elaborate and accumulate their product until the secretion is expelled by the rhythmic contraction of the myoepithelial cells. This last conclusion is probably wrong.

Shelley and Hurley (1953), using spot tests, demonstrated protein, reducing sugars, and ammonia in the milky droplets of apocrine secretion. However, Richter's (1932) report of cholesterol in the secretion is debatable. In some cases (e.g., chromidrosis), the glands may secrete chromogens, such as indoxyl, a malodorous compound which, when exposed to the air, is oxidized to blue indigo perhaps as a result of bacterial decomposition. Such a process could be the source of blue apocrine secretion (Rothman, 1954). The presence of fatty acids and aromatic substances can only be deduced from various data. In 1954, Rothman, who concluded that at that time information on the composition of apocrine secretion had not advanced much beyond that of the twenties, suggested that skin odors belong to the category of "caprylic odors" and originate from free, volatile fatty acids. The purest caprylic odor, of course, emanates from the pubic areas; that of the axilla is more pungent, and that of the scalp milder.

In any discussion on the function of apocrine secretion, the adaptive or survival function arises. Despite the risk of broad generalizations, observations from other animals can supply some clues. Glands from various parts of the body probably secrete different substances at different intervals. Axillary glands, for example, secrete substances that either are malodorous or become so. The ceruminous glands in the ears add their secretion to the copious sebum in the external ear canal to form cerumen. What the other glands do is still a matter of conjecture. Kligman and Shehadeh (1964) even stated that the numerous apocrine pubic glands do not function at all: they were unable to stimulate them pharmacologically and could not detect odors by sniffing. This conclusion was surprising enough to have warranted a reinvestigation of the entire problem, which as yet remains unsolved.

Rothman (1954) called attention to the strong axillary odor in adolescents, which becomes less intense in maturity. Parenthetically, juvenile stumptailed macaques have a rank stench reminiscent of human axillary odor, which they lose after adolescence. Shelley and Hurley (1953) detected no odor in sterile apocrine secretions, which are said to become malodorous as a result of bacterial action. Although Rothman

(1954) denied this result, it is significant that most effective deodorants contain antibiotics.

Considered as an odor-producing surface, the human axilla is a perfectly tailored organ. Small amounts of viscid material are secreted by the apocrine glands and dissolve in the watery eccrine sweat which spreads them over a wide surface. Axillary hairs harbor microorganisms that attack the proper substances and the whole area is kept almost constantly moist.

This is not the place to discuss social customs, but we cannot leave the subject of odor without emphasizing that biologically it is as much a human attribute as any other. For too long and in the name of overfastidiousness, we have overlooked (and sometimes decried) the role of cutaneous chemical communication. Human beings have characteristic, as well as specifically individual and topographic odors of pheromones, which play a significant and subtle role in communication between microsmatic animals and could once more assume the significance they must once have had in human communications.

XII. Pharmacological Responses of Apocrine Sweat Glands

Pharmacological studies of the axillary apocrine glands have been few and for the most part difficult to interpret. Aoki (1962) made careful observations in a limited number of subjects, and Goodall (1970) analyzed apocrine secretion by using a large number of adrenergic blocking agents. In general, intradermal injections of adrenaline or noradrenaline in concentrations of 10^{-4} and 10^{-5} induce apocrine secretion (Shelley and Hurley, 1953; Aoki, 1962; Goodall, 1970). Injections of acetyl-β-methylcholine chloride, acetylcholine chloride, and carbamylcholine chloride cause such profuse local eccrine sweating as to mask the identification of apocrine sweat spots in the pilary orifices; in some subjects, however, apocrine secretion can be seen in response to Mecholyl in concentrations of 10^{-4} or 10^{-5} (Aoki, 1962) (Fig. 15). In others, acetylcholine and Carcholin in concentrations of 10^{-4} and 10^{-5} also produce apocrine secretion within the orifices of the hair follicles at the site of injection about 1 minute after injection. It is surprising that Shelley and Hurley (1953) and Hurley and Shelley (1960) found cholinergic agents completely ineffective in producing apocrine secretion in the axilla.

After intradermal injections of Mecholyl, apocrine secretion collected with capillary tubes from the orifices of hair follicles is turbid and

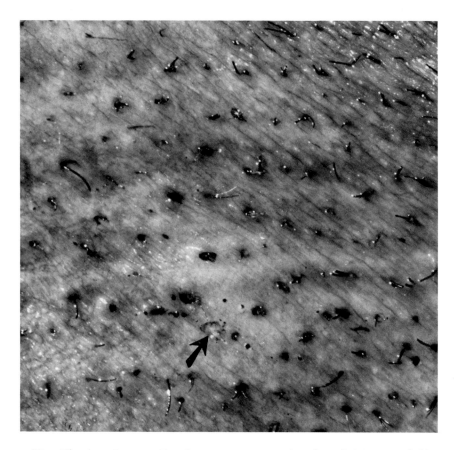

Fig. 15. Apocrine secretion in response to an intradermal injection of 10⁻⁴ Mecholyl in the axilla of a Negro. The large black spots at the base of each hair represent apocrine secretion. The arrow indicates the site and the direction of the injection. (Courtesy of Dr. T. Aoki.)

slightly yellowish. Since apocrine secretion occurs after both adrenergic and cholinergic stimulation, the glands probably produce only one type of secretion (Hurley and Shelley, 1954, 1960) although others have contended that they produce both apocrine and eccrine types of secretion (Rothman, 1954; Sulzberger and Herrmann, 1954; Kuno, 1956).

When intradermal injections of atropine or dihydroergotamine (DHE) in concentrations of 10^{-4}, 10^{-6}, were followed 5 minutes later by Mecholyl injected in the same site, both apocrine secretion and eccrine sweating were inhibited at the site. DHE alone did not inhibit the effect of Mecholyl on either apocrine or eccrine sweating. However, the high doses used here were unphysiological, and the apocrine response caused by Mecholyl, acetylcholine, and Carcholin was not necessarily due to cholinergic stimulation. Furthermore, these responses may have been the result of a simple squeezing out of secretion already pooled in the sweat gland.

Goodall (1970) found no response of apocrine glands to cholinergic agents. After having injected 0.05 ml 1:10,000 adrenaline or noradrenaline, he reported "an immediate expulsion of apocrine sweat" and attributed it to the contraction of myoepithelial cells in the "duct." The problem here is that there are *no* myoepithelial cells in the duct. Of the numerous adrenergic blocking agents he used topically, he found trimethaphan camphorsulfate phenoxylinganine HCl and N-(2-bromethyl)-N-ethyl-1-naphthaline-methylanine HBr the most effective in inhibiting apocrine secretion. Apparently none of the adrenolytic and adrenergic agents had any effect on eccrine sweat glands. Aoki (1962) found that a combination of nicotine and acetylcholine caused a repeatable axon reflex secretion in the glands of the slow loris (*Nycticebus coucang*).

XIII. The Phylogeny of Sweat Glands

The phylogenetic history of any organ system cannot be reconstructed by a study of its extant forms. Because eccrine glands in nonhuman primates appear to have evolved gradually over the general body surface and to have "replaced" the apocrine glands, one cannot thereby assume that apocrine glands are primitive. Furthermore, it is inherently fallacious to assume that the glands in the skin of "primitive" mammals are necessarily primitive. For example, the skin of Ornithorhynchus is highly specialized and uniquely adapted for an aquatic existence; significantly, Ornithorhynchus has well-differentiated eccrine and apocrine glands (Montagna and Ellis, 1960). In view of this, one cannot say which of these two glands is the more primitive. In addition, if we state that apocrine glands are more primitive than eccrine glands, we assume that the latter have become progressively modified and specialized and that the former have stood still. In reality, the two types of glands are separate organs with different functions; therefore to set up areas of comparison and contrast between them is a futile exercise in illogic. From systematic studies of the skin of primates, we can deduce that both eccrine sweat glands and apocrine glands are equally highly specialized in man.

All prosimian primates have generalized types of "apocrine" glands in the hairy skin; their "eccrine" glands, which differ from those of other primates, are restricted to the friction surfaces.

Furthermore, in the Lorisidae eccrine and apocrine glands have many morphological and histochemical characteristics in common. At first, the ratio of glands to hair follicles is usually 1:4 to 20. Gradually in the evolutionary process the number of glands in the hairy skin increases until there is practically a one-to-one ratio, and eventually, in some parts of the human skin, apocrine glands may be more numerous than hair follicles.

The development of apocrine glands from hair follicles and the opening of their ducts inside the pilary canal are of some interest. In many prosimians, the glands open not inside the pilary canals but directly onto the surface, often by a slightly coiled terminal segment. In ring-tailed lemurs, large, active apocrine glands open directly onto the surface of antebrachial organs, which are proximal extensions of the palmar epidermis. Even in apes apocrine glands open near, but not inside, the pilary canals. If the glands in the hairy skin of other mammals are also considered, the only pattern that emerges is the association of the glands with hair follicles.

In man, eccrine glands have histochemical features, such as abundant glycogen, succinic dehydrogenase, amylophosphorylase, and numerous cholinesterase-reactive nerves (Fig. 16), that are not shared with apocrine glands. In the lorises, however, eccrine and apocrine glands have practically identical characteristics.

In both man and the higher nonhuman primates, the diameter of the duct of the eccrine sweat gland is usually only slightly smaller than that of the secretory coil, whereas the diameter of the duct of apocrine glands is only a fraction of that of the secretory coil. In the friction surfaces of prosimians and marmosets, the diameter of the duct of eccrine sweat glands (and for the most part, of apocrine glands) is much narrower than that of the secretory coil. The duct of the eccrine glands of man and of other higher primates is very long and forms a sizable part of the glomerate segment, but in prosimians it is straight and narrow and directly attached to the coiled, dilated secretory segment.

Like the prosimians, the Cebidae from Central and South America have eccrine glands *only* on the friction surfaces and apocrine glands

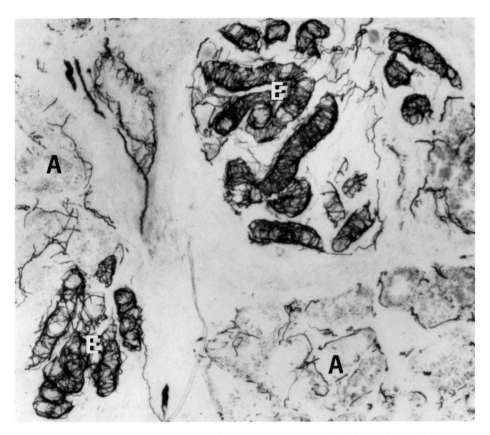

Fig. 16. Section from the axilla showing eccrine sweat glands (E) surrounded by numerous acetylcholinesterase-reactive nerves and apocrine glands (A) with fewer nerves.

only on the hairy skin. Yet the tree shrew (*Tupaia*), the most primitive of all primates, has eccrine glands in the hairy skin. The distribution of apocrine glands in nonhuman primates is peculiar to each species. For example, whereas the great apes and most other primates have only a few of them on the scalp, the macaques and mangabeys have many of them. It is difficult to interpret this phylogenetically. The fact that in these animals the scalp, like the human axilla, produces a characteristic odor suggests that part of it emanates from apocrine secretion.

In man, eccrine and apocrine glands are equally highly specialized and neither is more primitive than the other. In fact, both glands have probably attained a degree of development and specialization not reached by any other mammal.

References

Adachi, K., and W. Montagna. 1961. Histology and cytochemistry of human skin. XXII. Sites of leucine aminopeptidase (LAP). *J. Invest. Dermatol.* **37**: 145–152.

Aoki, T. 1962. Stimulation of human axillary apocrine sweat glands by cholinergic agents. *J. Invest. Dermatol.* **38**: 41–44.

Bergen, F. von. 1904. Zur Kenntnis gewisser Strukturbilder ("Netzapparate," "Saftkanälchen," "Trophospongien") in Protoplasma verschiedener Zellenarten. *Arch. Mikroskop. Anat. Entwicklungsmech.* **64**: 498–574.

Biempca, L., and L. F. Montes. 1965. Secretory epithelium of the large axillary sweat glands. A cytochemical and electron microscope study. *Amer. J. Anat.* **117**: 47–72.

Borsetto, P. L. 1951. Osservazioni sullo sviluppo delle ghiandole sudoripare nelle diverse regioni della cute umana. *Arch. Ital. Anat. Embriol.* **56**: 332–348.

Brinkmann, A. 1908. Die "Rückendrüsen von Dicotyles." *Z. Anat. Entwicklungsgesch.* **36**: 281–307.

Bunting, H., G. B. Wislocki, and E. W. Dempsey. 1948. The chemical histology of human eccrine and apocrine sweat glands. *Anat. Rec.* **100**: 61–77.

Carossini, G. 1912. Lo sviluppo delle ghiandole sudoripare, particolarmente ne'suoi rapporti collo sviluppo dell'apparato pilifero, nelle diverse regioni della pelle dell'uomo. *Arch. Ital. Anat. Embriol.* **11**: 545–603.

Cavazzana, P. 1947. Indagini sul comportamento morfo-funzionale delle ghiandole glomerulari apocrine della pelle e dell'ascella durante le fasi del ciclo mestruale e nella gravidanza. *Riv. Ital Ginecol.* **30**: 114–134.

Charles, A. 1959. An electron microscopic study of the human axillary apocrine gland. *J. Anat.* **93**: 226–232.

Cornbleet, T. 1952. Pregnancy and apocrine gland diseases: hidradenitis, Fox-Fordyce disease. *Arch. Dermatol. Syphilol.* **65**: 12–19.

Ellis, R. A. 1967. Eccrine, sebaceous and apocrine glands. *In* "A. S. Zelickson's Ultrastructure of Normal and Abnormal Skin," pp. 132–162. Lea & Febiger, Philadelphia, Pennsylvania.

Ellis, R. A., and W. Montagna. 1959. Comparative histochemistry of the sweat glands of primates. *J. Histochem. Cytochem.* **7**: 308.

Ellis, R. A., W. Montagna, and H. Fanger. 1958. Histology and cytochemistry of human skin. XIV. The blood supply of the cutaneous glands. *J. Invest. Dermatol.* **30**: 137–145.

Goodall, McC. 1970. Innervation and inhibition of eccrine and apocrine sweating. *J. Clin. Pharmacol.* **10**: 235–246.

Gross, B. G. 1966. Annulate lamellae in the axillary apocrine glands of adult man. *J. Ultrastruct. Res.* **14**: 64–73.

Hoepke, H. 1927. Die Drüsen der Haut. *In* "Handbuch der mikroskopischen Anatomie des Menschen" (W. von Möllendorff, ed.), Vol. 3, pp. 55–66. Springer-Verlag, Berlin and New York.

Holmgren, E. 1922. Die Achseldrüsen des Menschen. *Anat. Anz.* **55**: 553–565.

Homma, H. 1925. Über positive Eisenfunde in den Epithelien der apokrinen Schweissdrüsen menschlicher Axillarhaut. *Arch. Dermatol. Syphilis* **148**: 463–469.

Homma, H. 1926. On apocrine sweat glands in white and negro men and women. *Bull. Johns Hopkins Hosp.* **38**: 365–371.

Horn, G. 1935. Formentwicklung und Gestalt der Schweissdrüsen der Fussohle des Menschen. *Z. Mikrosk. Anat. Forsch.* **38**: 318–329.

Horner, W. E. 1846. On the odoriferous glands of the negro. *Amer. J. Med. Sci.* **21**: 13–16.

Hurley, H. J. and W. B. Shelley. 1954. The human apocrine sweat gland: two secretions? *Brit. J. Dermatol.* **66**: 43–48.

Hurley, H. J., and W. B. Shelley. 1960. "The Human Apocrine Sweat Gland in Health and Disease." Thomas, Springfield, Illinois.

Ito, T., and K. Iwashige. 1953. Zytologische und histologische Untersuchungen über die apokrinen Achselschweissdrüsen von gesunden Menschen höheren Alters. *Arch. Histol.* **5**: 455–475. (In Japanese.)

Ito, T., K. Tsuchiya, and K. Iwashige. 1951. Studien über die basophile Substanz (Ribonukleinsäure) in den Zellen der menschlichen Schweiss drüsen. *Arch. Anat. Jap.* **2**: 279–287. (In Japanese.)

Klaar, J. 1926. Zur Kenntnis des weiblichen Axillarorgans beim Menschen. *Wein. Klin. Wochenschr.* **39**: 127–131.

Kligman, A. M., and N. Shehadeh. 1964. Pubic apocrine glands and odor. *Arch. Dermatol.* **89**: 461–463.

Kölliker, A. 1853. *In* "Manual of Human Histology" (G. Busk and T. Huxley, eds. and translators), Vol. 1. Sydenham Society, London.

Krause, C. 1844. Haut. *Wagner's Handbook Physiol.* **2**: 108.

Kuno, Y. 1956. "Human Perspiration." Thomas, Springfield, Illinois.

Kurosumi, K., T. Kitamura, and T. Iijima. 1959. Electron microscope studies on human axillary apocrine sweat glands. *Arch. Histol.* **16**: 523–566.

Loeschcke, H. 1925. Über zyklische Vorgänge in den Drüsen des Achselhöhlenorgans und ihre Abhängigkeit vom Sexualzyklus des Weibes. *Virchows Arch. A* **255**: 283–294.

Manca, P. V. 1934. Ricerche sulla struttura delle ghiandole apocrine. *G. Ital. Dermatol. Sifilol.* **75**: 283–294.

Melczer, N. 1935. Über das Golgi-Kopsche Binnennetz der menschlichen apokrinen Schweissdrüsenzellen. *Dermatol. Wochenschr.* **100**: 337–342.

Minamitani, K. 1941a. Zytologische und histologische Untersuchungen der Schweissdrüsen in menschlicher Achselhaut. Über das Vorkommen der besonderen Formen der apokrinen und ekkrinen Schweissdrüsen in Achselhaut von Japanern. *Okajimas Folia Anat. Jap.* **20**: 563–590.

Minamitani, K. 1941b. Zytologische und histologische Untersuchungen der Schweissdrüsen in der menschlicher Achselhaut. Zur Zytologie der apokrinen Schweissdrüsen in der menschlichen Achselhaut. *Okajimas Folia Anat. Jap.* **21**: 61–94.

Mitchell, H. H., and T. S. Hamilton. 1949. The dermal excretion under controlled environmental conditions of nitrogen and minerals in human subjects, with particular reference to calcium and iron. *J. Biol. Chem.* **178**: 345–361.

Montagna, W. 1956. Ageing of the axillary apocrine sweat glands in the human female. *In* "Ciba Foundation Colloquium on Ageing. Ageing in Transient Tissues," Vol. 2, pp. 188–197. Churchill, London.

Montagna, W. 1959. Histology and cytochemistry of human skin. XIX. The development and fate of the axillary organ. *J. Invest. Dermatol.* **33**: 151–161.

Montagna, W. 1960. Cholinesterases in cutaneous nerves of man. *In* "Advances in Biology of Skin. Cutaneous Innervation" (W. Montagna, ed.), Vol. 1, pp. 74–87. Pergamon, Oxford.

Montagna, W. 1962. Histological, histochemical, and pharmacological properties. *In* "Advances in Biology of Skin. Eccrine Sweat Glands and Eccrine Sweating" (W. Montagna, R. A. Ellis, and A. F. Silver, eds.), Vol. 3, pp. 6–29. Pergamon, Oxford.

Montagna, W., and R. A. Ellis. 1958. L'histologie et la cytologie de la peau humaine. XVI. Repartition et concentration des esterases carboxyliques. *Ann. Histochim.* **3**: 1–17.

Montagna, W., and R. A. Ellis. 1959. L'istochimica degli annessi cutanei. *Minerva Dermatol.* **34**: 475–494.

Montagna, W., and R. A. Ellis. 1960. Sweat glands in the skin of *Ornithorhynchus paradoxus. Anat. Rec.* **137**: 271–278.

Montagna, W., and D. M. Ford. 1969. Histology and cytochemistry of human skin. XXXIII. The eyelid. *Arch. Dermatol.* **100**: 328–335.

Montagna, W., and V. R. Formisano. 1955. Histology and cytochemistry of human skin. VII. The distribution of succinic dehydrogenase activity. *Anat. Rec.* **122**: 65–78.

Montagna, W., H. B. Chase, and J. B. Hamilton. 1951. The distribution of glycogen and lipids in human skin. *J. Invest. Dermatol.* **17**: 147–157.

Montagna, W., H. B. Chase, and W. C. Lobitz, Jr. 1953. Histology and cytochemistry of human skin. V. Axillary apocrine sweat glands. *Amer. J. Anat.* **92**: 451–470.

Montes, L. F., B. L. Baker, and A. C. Curtis. 1960. The cytology of the large axillary sweat glands in man. *J. Invest. Dermatol.* **35**: 273–291.

Munger, L. B. 1965. The cytology of apocrine sweat gland. II. Human. *Z. Zellforsch. Mikrosk. Anat.* **68**: 837–841.

Nicolas, J., C. Regaud, and M. Favre. 1914. Sur la fine structure des glandes sudoripares de l'homme, particuliérement en cas que concerne les mitochondries et les phénomènes de sécrétion. *17th Intern. Congr. Med., Sect. 13, Dermatol. Syphilis, London,* pp. 105–109.

Ota, R. 1950. Zytologische und histologische Untersuchungen der apokrinen Schweissdrüsen in den normalen, keinen Achselgeruch (Osmidrosis axillae) gebenden Achselhaüten von Japanern. *Arch. Anat. Jap.* **1**: 285–308.

Pinkus, H. 1958. Embryology of hair. *In* "The Biology of Hair Growth" (W. Montagna and R. A. Ellis, eds.), pp. 1–32. Academic Press, New York.

Richter, W. 1932. Beiträge zur normalen und pathologischen Anatomie der apokrinen Hautdrüsen des Menschen mit besonderer Berücksichtigung des Achselhöhlenorgans. *Virchows Arch. A* **287**: 277–296.

Rolin, M. C. 1846. Sudoriparous glands of the axilla. *Amer. J. Med. Sci.* **11**: 439.

Rothman, S. 1954. "Physiology and Biochemistry of the Skin." Univ. of Chicago Press, Chicago, Illinois.

Schaffer, J. 1926. Über die Hautdrüsen. *Wien. Klin. Wochenschr.* **38**: 1–5.

Schiefferdecker, P. 1917. Die Hautdrüsen des Menschen und des Säugetieres, ihre biologische und rassenanatomische Bedeutung, sowie die Muscularis sexualis. *Biol. Zentr.* **37**: 534–562.

Schiefferdecker, P. 1922. Die Hautdrüsen des Menschen und des Säugetieres, ihre Bedeutung, sowie die Muscularis sexualis. *Zoologica* **72**: 1–154.

Schwenkenbecker, A. 1933. Cited from Comel (1933).

Shelley, W. B. 1951. Apocrine sweat. *J. Invest. Dermatol.* **17**: 255.

Shelley, W. B., and T. Butterworth. 1955. The absence of the apocrine glands and hair in the axilla in mongolism and idiocy. *J. Invest. Dermatol.* **25**: 165–167.

Shelley, W. B., and M. M. Cahn. 1955. Experimental studies on the effect of hormones on the human skin with reference to the axillary apocrine sweat gland. *J. Invest. Dermatol.* **25**: 127–131.

Shelley, W. B., and H. J. Hurley. 1953. The physiology of the human axillary apocrine sweat gland. *J. Invest. Dermatol.* **20**: 285–297.

Sperling, G. 1935. Die Form der apokrinen Haardrusen des Menschen. *Z. Mikrosk. Anat. Forsch.* **38**: 241–252.

Steiner, K. 1926. Über die Entwicklung der grossen Schweissdrüsen beim Menschen. *Z. Anat. Entwicklungsgesch.* **78**: 83–97.

Sulzberger, M. B., and F. Herrmann. 1954. "The Clinical Significance of Disturbance in the Delivery of Sweat." Thomas, Springfield, Illinois.

Talke, L. 1903. Über die grossen Drüsen der Achselhöhlenhaut des Menschen. *Arch. Mikrosk. Anat. Entwicklungsmech.* **61**: 537–555.

Thies, W., and F. Galente. 1957. Zur histochemischen Darstellung der Cholinesterasen im vegetativen Nervensystem der Haut. *Hautarzt* **8**: 69–75.

Waelsch, L. 1912. Über Veränderungen der Achselschweissdrüsen während der Gravidität. *Arch. Dermatol. Syphilis* **114**: 139–160.

Watson, M. L. 1955. The nuclear envelope, its structure and relation to cytoplasmic membranes. *J. Biophys. Biochem. Cytol.* **1**: 257–270.

Winkelmann, R. K., and J. V. Hultin. 1958. Mucinous metaplasia in normal apocrine glands. *A.M.A. Arch. Dermatol.* **78**: 309–313.

Woollard, H. H. 1930. The cutaneous glands of man. *J. Anal.* **64**: 415–421.

Yasuda, K., and R. A. Ellis. 1961. Electron microscopy of human apocrine sweat glands. (Personal communication.)

Yasuda, K., and W. Montagna. 1960. Histology and cytochemistry of human skin. XX. The distribution of monoamine oxidase. *J. Histochem. Cytochem.* **8**: 356–366.

Yasuda, K., H. Furusawa, and N. Ogata. 1958. Histochemical investigation on the phosphorylase in the sweat glands of axilla. *Okajimas Folia Anat. Jap.* **31**: 161–169.

Yasuda, K., R. A. Ellis, and W. Montagna. 1962. The fine structural relationship between mitochondria and light granules in the human apocrine sweat glands. *Okajimas Folia Anat. Jap.* **38**: 455.

Zorzoli, G. 1950. Ricerche istochimiche sul pigmento intracellulare delle ghiandole axcellari dell'uomo. Nota I. *Boll. Soc. Ital. Biol. Sper.* **26**: 1–3.

12

Eccrine Sweat Glands[*]

I. Introduction

Observing the watery exudate that issues from skin pores, Malpighi (1628–1694) deduced that they are the orifices of sweat glands. More than a century later, Purkinje and his pupil Wendt (1833) and Breschet and Roussel de Vouzzeme (1834), working independently, discovered and described the sweat glands. However, it was almost a hundred years later that Schiefferdecker (1922), recognizing that eccrine glands are distinct from apocrine glands, gave them distinct names.

Eccrine sweat glands are found on the hairy skin of relatively few mammals. In primates the distribution is unpredictable. They are found in tree shrews but not in prosimians; there are none in the Cebidoidea (New World monkeys) except the Atelinae (South American monkeys), and they appear in variable numbers in the cercopithecoids (Old World monkeys), being more abundant in the great apes and most abundant in man. They are found on the friction surfaces (palms and soles) of all primates, on the digital pads of many animals, and on the snouts of others. Their orifices pierce the ridges of epidermal dermatoglyphics, characteristic of the palms and soles of all primates and marsupials, the

---

[*] Written mostly by Richard L. Dobson.

underside of the prehensile tail of some South American monkeys (Atelinae), and the knuckle pads of gorillas and chimpanzees. In horses and cows and other species, the glands that aid in evaporative heat loss are apocrine in nature (Jenkinson, 1969).

Studies on the actual numbers of sweat glands per square centimeter of surface show a wide range of individual variations. About three million glands are distributed throughout the human body (Szabó, 1967) on an average of 143 to 330 cm² (Kuno, 1956). Only the glans penis and clitoris, labia minora, and the inner surface of the prepuce have no eccrine glands. In infants, the glands are distributed somewhat uniformly over the body; in adults, they are most numerous on the palms and soles, then in decreasing order on the head, the trunk, and the extremities. They are more numerous on the flexor and ventral surfaces than on the extensor and dorsal surfaces. No effort has been made to determine whether any relationship exists between the size of a human being and the number of glands in his skin.

No new sweat glands appear to be formed after birth; thus, they have their highest density at birth and gradually decrease as the body grows. The skin of the thigh of a full-term human neonate has about 12 times more eccrine glands than that of the average adult (Szabó, 1967) and the skin of a 1-year-old infant contains 8 times more per unit area that that of an adult (Thomson, 1954). These data correlate with those on the surface area of the adult body, which is about 7 times that of a newborn and 5.5 times that of a 1-year-old infant. Thus, individual differences in the population density of glands in the adult body are probably due to differences in total body size, the larger surface areas being associated with a smaller count and vice versa. However, a small person does not necessarily have a relatively higher output of sweat than a large one, since there are marked individual variations in glandular activity, and the number of glands does not correspond with the amount of sweat secretion. Table I summarizes the number of sweat glands in some ethnic groups. Reports that Negroes have more sweat glands than Europeans are not substantiated by information on American Negroes. African Negroes, however, have more active sweat glands than other racial groups (Table I). Although this could be due to their chronic exposure to a hot, humid environment, Caingang Indians (Brazil), who inhabit similar tropical environments, have about the same number of active sweat glands as Europeans (Roberts et al., 1970). Counts of sweat glands after thermal stimulation give lower results than those obtained from actual anatomical counts (Table I).

TABLE I

Regional Counts of Sweat Glands [a,b]

Region	Thermal stimulation						Anatomical
	European	African	Korean	Syrian-Iranian	Caingang Indians (Brazil)	American Negroes	American Negroes and Caucasians
Forehead	177.8 ± 5.8	352.7 ± 10.8	290	137.7 ± 5.3	—	158.9 ± 4.0	360 ± 50
Cheek	—	—	—	—	—	—	320 ± 60
Chest	79.6 ± 2.0	151.5 ± 6.0	115	88.2 ± 2.7	94 ± 8	80.5 ± 8.0	175 ± 35
Abdomen	87.1 ± 2.6	149.7 ± 5.7	125	85.2 ± 2.0	84 ± 10	71.2 ± 8.2	190 ± 5
Scapula	80.4 ± 2.7	174.7 ± 5.9	130	94.3 ± 4.2	70 ± 8	73.0 ± 7.6	—
Lumbar	85.8 ± 3.4	173.7 ± 5.9	175	79.8 ± 2.6	61	67.3 ± 6.8	160 ± 30
Upper arm	81.5 ± 2.5	114.8 ± 4.2	115	112.8 ± 2.7	125 ± 16	73.3 ± 1.2	150 ± 20
Forearm	105.8 ± 5.0	201.4 ± 6.9	140	100.3 ± 1.8	116 ± 18	79.5 ± 1.2	225 ± 25
Hand (dorsum)	—		215	—	—	—	
Palm	—		270	—	258	—	—
Fingertip	—		—	—	359 ± 28	—	—
Thigh	61.6 ± 6.8	98.5 ± 4.0	145	72.7 ± 6.4	—	61.6 ± 6.8	120 ± 10
Calf	—		110		—		150 ± 15
Foot (dorsum)	—		175		—		250 ± 5
Sole	—		—		—		620 ± 120

[a] Glands/cm^2 ± S.E. mean.
[b] Modified from Roberts et al. (1970).

II. Development

The anlagen of eccrine sweat glands appear in the following order: during the fourth fetal month, in the palms and soles; early in the fifth fetal month in the axilla; and later that month, in other parts of the body. Except for the friction surfaces, they are arranged around hair follicles, but not necessarily in the characteristic patterns described by Horstmann (1952). The chronological sequence of development suggests a threefold classification: glands on the palms and soles, the axilla, and the general body surface. In addition to this sequence, they develop first on the forehead and scalp and gradually on the trunk and limbs. Kuno (1956) suggested that phylogenetically the sweat glands that appear first antedate the late comers. However, glands develop at different times even in the same body region.

In fetuses 16 to 19 weeks of age, proliferative cell buds grow down from the crests of the ridges on the underside of the volar surface of the hands and feet, nail folds and eponychium, and the entire dorsal surface of the distal phalanges. At this time there are no other gland primordia although hair follicles and their sebaceous glands are well differentiated. Regardless of chronology or location, anlagen develop in the basal layer of the epidermis as small condensations of cells that resemble those that form the hair follicles (Pinkus, 1927) except that they are smaller and narrower and have no aggregates of dermal cells at their bases. These buds grow into columns of cells that plow their way to the hypodermis; the cells of the cord are rich in glycogen but those of their clavate tip are not. At the dermis-hypodermis junction, the cell column becomes tortuous, twisted, and then glomerate.

The cells in the center of the straight or undulating part of the cord become cornified, and as they shrink, establish a lumen. In the seventh and eighth months, "vacuoles" appear in the center of the twisted cords; adjacent vacuoles coalesce to form elongated clefts, which in turn coalesce and establish a lumen. Eventually, by the seventh and eighth month, the lumen becomes continuous with that in the duct, establishing a hollow gland. By the end of the eighth month, the lumen broadens and the presumptive secretory cells assume the characteristic appearance of the adult glands (Tsuchiya, 1954).

Myoepithelial cells cannot be discerned during fetal life. Even the characteristic loose connective tissue stroma around the entire glands is not fully differentiated until after birth. In 9-month-old fetuses, the glands are morphologically similar to those of an adult (Borsetto, 1951; Tsuchiya, 1954), and pharmacological studies (Foster et al., 1969; Green and Behrendt, 1969) report that they can be stimulated with sudorific drugs.

III. Blood Supply

None of the many studies (Eichner, 1954) of the blood vessels to the eccrine sweat glands are satisfactory. The injection of vessels with colored or opaque substances is limited to the flow of these materials, but when successful they do show at least some of the vasular bed. Unfortunately, such treatment is not feasible in most specimens. The various techniques for alkaline phosphatase activity when applied to thick frozen sections are convenient and show most capillaries and arterioles, whose endothelium is probably reactive for the enzyme.

In thick preparations, the network of capillaries and arterioles that surround the gland can be clearly seen (Fig. 1). The entire glomerate

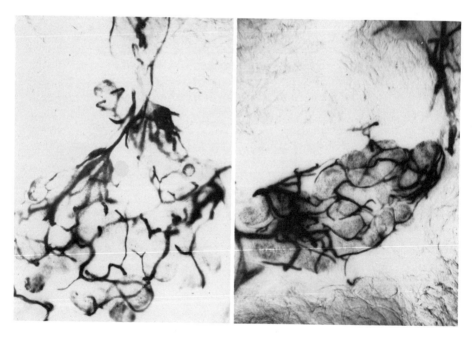

Fig. 1. Coils of two sweat glands surrounded by alkaline phosphatase-reactive vessels. In these thick frozen sections the coiled segment of the ducts are as well vascularized as the secretory coil.

part of the gland, i.e., the secretory segment and the coiled duct, is richly supplied with blood vessels, which closely follow the contours of the coils (Ellis *et al.*, 1958). Frequently, all of the capillaries of one gland arise from one small artery or arteriole (Fig. 2), but sometimes they come from several arterioles. When the coil of the glands is loose, nearby vessels form capillary shunts. Tightly coiled tubules are surrounded by a denser plexus of vessels.

Capillaries and roughly parallel arterioles wind loosely around the straight (undulating) part of the duct as it ascends to the surface, with crossshunts connecting them. At the base of the papillary body, the vessels branch to form baskets of capillaries or loops around the epidermal cone that contains the terminal coiled part of the duct and join laterally with the loops from the arcades of the superficial capillary network under the epidermis. The capillary networks around the sweat glands apparently remain unaltered even in the skin of aged individuals.

IV. Structural Features

Eccrine sweat glands are simple tubes that extend from the epidermis to the dermis; some reach only the lower part of the dermis and others extend into the hypodermis. The glands, then, can be separated into shallow and deep. Occasionally, the terminal part of the duct branches and at other times the ducts of two adjacent glands fuse beneath the epidermis and deliver their contents to the surface through a common duct (Giacometti and Machida, 1965; Spearman, 1968). The course of the duct within the dermis is somewhat helical or undulating (Wells and Landing, 1968), not straight, as sometimes depicted.

Only in carefully dissected and isolated, stained, and untreated (Fig. 3) specimens can the gross morphology of eccrine glands be seen. Each tubule consists of a duct and secretory segment; the latter, roughly half of the tubular length, is irregularly and often tightly coiled. The duct consists of a distal convoluted segment within the glomerate portion and a proximal undulating or helical segment that traverses the dermis. Inside the epidermis, the duct pursues a spiral, inverted conical course. In preparations of split epidermis, when the undersurface is highly sculptured, the ducts enter through a cone of epidermis; when the rete ridges are shallow, a shallow moat or slight crinkling surrounds the ducts. The clock-dial or cockade patterns of epidermal folds described by Horstmann (1952) around the penetration of the duct into the

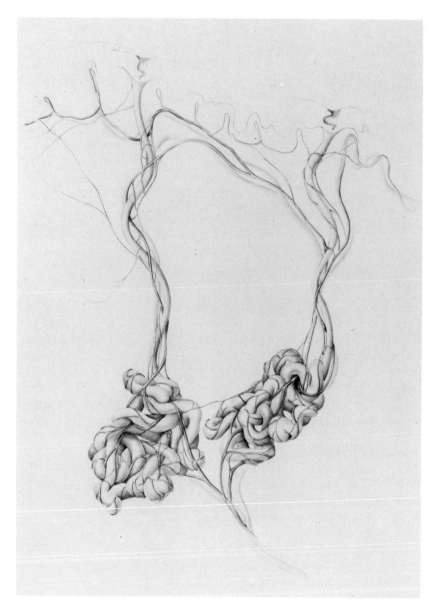

Fig. 2. Diagram of two adjacent sweat glands and the blood vessels around them; reconstructed from thick frozen sections.

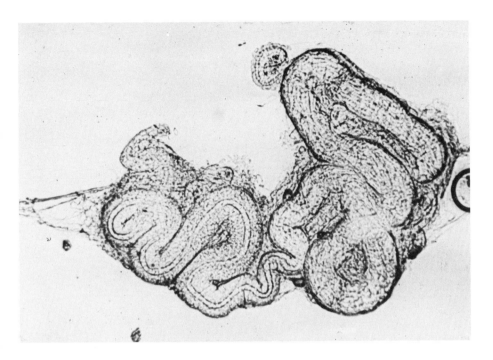

Fig. 3. A sweat gland dissected from abdominal skin. Although unstained, the ductal part on the left is easily recognized by the prominent cuticle that lines its lumen. The secretory part is on the right.

epidermis are rare (Fig. 4). A duct, then, has three segments: a spiral section within the epidermis called the "epidermal sweat duct unit" (Lobitz *et al.*, 1954a), an undulating or helical midportion, and a coiled segment continuous with and equal in length to the secretory coil (Sato and Dobson, 1970).

In different parts of the body, the diameter of the duct is variably smaller than that of the secretory coil. In the palms and soles, however, and to a lesser extent in the axilla, the duct is less than one-third the diameter of the secretory coil; thus the transition between duct and gland, somewhat like that in apocrine glands, is rather abrupt. The compactness of the basal coil varies according to the level of the dermis in which the glands are located; for example, those within the dermis are more compact than those deeper in the fatty tissue.

In histological sections, glandular nests are composed of different proportions of duct and secretory tubules, depending on the angle of the

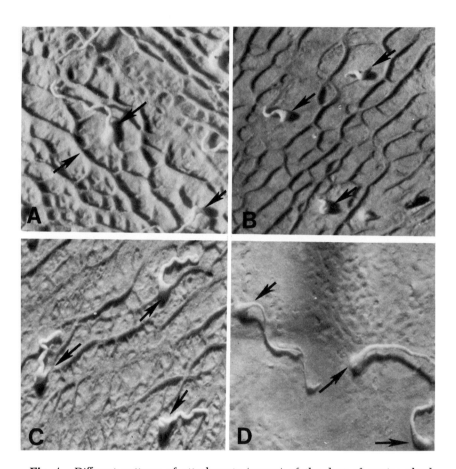

Fig. 4. Different patterns of attachments (arrows) of the duct of eccrine glands to the epidermis in split skin preparations. (A) In the skin from the scrotum the duct is surrounded by a moat which outlines a rough clock-dial pattern. (B) Three glands, also from the scrotum, with no rete ridges attached to them. (C) Ducts from the breast traversed by a single ridge. (D) Three ducts from the face, with no ridge patterns around them.

cut. Sections through the higher levels of the coil contain more duct than secretory tubules, whereas those through the lower levels sometimes contain mostly secretory tubules (Fig. 5). The epithelium of the duct consists of two compact layers of almost equally small cuboidal cells. The luminal cells, which stain less intensely with basic dyes than

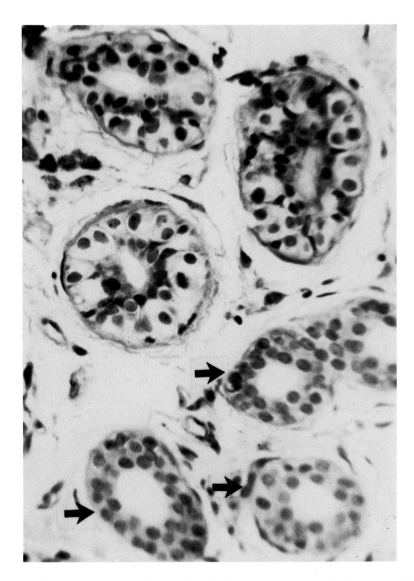

Fig. 5. Section through the basal coil of an eccrine sweat gland in the axilla. There are as many sections of the coiled duct (arrows) as there are of secretory segment. Stained with toluidine blue.

the basal ones, have a variably hyalinized border. The secretory tubule is lined with one layer of truncated pyramidal cells, misshapen and heaped in such a way as to suggest two or more layers. The apparent stratification is accentuated by the differences in size. The nucleus of the larger cells is at the base, that of the small cells is toward the distal end of the cytoplasm. This disposition has led some observers to distinguish between large or "basal" cells and small or "superficial" cells (Ito and Iwashige, 1951; Kuno, 1956), but all cells, "basal" and "superficial," rest upon a layer of myoepithelial cells and a basement membrane. Both types have distinct histological properties. The smaller cells, whose nucleus is misplaced toward the lumen, contain cytoplasmic granules with a strong affinity for basic dyes (Fig. 6); hence, they are called "dark" cells. Because the cytoplasm of the larger cells stains faintly or not at all with basic dyes, they are called "clear" or "pale" cells (Montagna et al., 1953). These cells also contain sparse cytoplasmic granules, stain less intensely with iron hematoxylin, and normally have acidophilic properties. Treatment with ribonuclease before staining neutralizes the cytoplasmic basophilia of the granules in the dark cells and thus indicates that the basophilic substance contains ribonucleic acid (Ito et al., 1951; Montagna et al., 1953). When stained with Giemsa or eosinmethylene blue, some of the granules at the apices of the dark cells take up basic dye, others the acid one. The cytoplasm of the clear cells contains evenly and diffusely distributed small basophilic and eosinophilic granules. The proportion of dark cells to clear varies but is relatively consistent in glands from the same specimen. In most specimens, dark and clear cells are distinct, but most of those that are not are clear. A hyalin or cuticular border often lines the luminal border of the clear cells. Occasionally, the secretory cells have a clear cytoplasm riddled with evenly spaced vacuoles that contain neither glycogen nor lipids. Such examples occur in individuals of all ages, in apparently normal or abnormal skin (Holyoke and Lobitz, 1952), and in all glands of a given individual (Montagna, 1962). Without correlative functional studies, nothing is known about the significance of this phenomenon.

Between the secretory cells and the basement membrane is a layer of myoepithelial cells aligned parallel to the axis of the tubule. These do not form a solid sheet but are loosely dovetailed, with interstices between them.

Fig. 6. Above is an eccrine secretory coil with dark (D), shoved toward the lumen and clear cells (C). Stained with toluidine blue. Below, enlarged secretory segment showing clear (C) and dark cells (D), each touching the basal lamina.

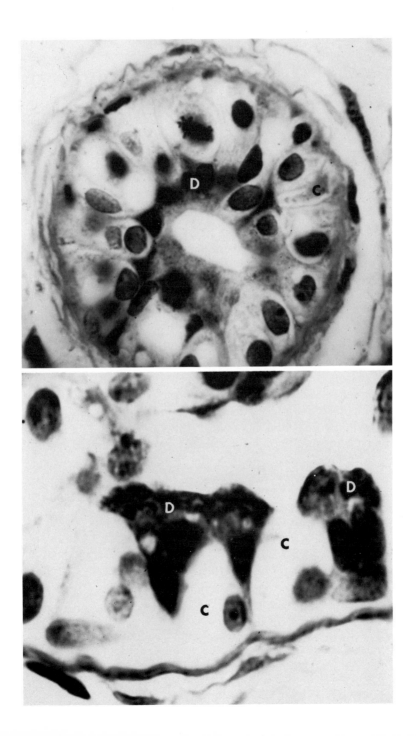

A fairly thick basement membrane lies on the outside of the secretory segment which, in unstained specimens, is yellow, hyalin, and refractile. When the tubules are sectioned transversely and viewed under polarized light, the basement membrane is birefringent. It resembles collagen and stains neither with basic dyes nor with the van Gieson or Mallory triacid stains. It blackens with ammoniacal silver nitrate especially where it is in contact with the bases of the secretory and myoepithelial cells and is reactive to the periodic acid-Schiff (PAS) routine. In the latter preparations, the membrane has a distinct inner PAS-reactive layer and an outer, more extensive unreactive part. In aging persons only the outer, hyalin, PAS-negative basement membrane thickens. The duct has a thin and inconspicuous basement membrane. Outside the basement membrane, delicate elastic fibers are wound tightly around the secretory coil (Montagna and Giacometti, 1969). These fibers, seen before only by Terebinsky (1908) but overlooked by others, are finer than the nerves that surround them, and their orientation is the same as that of the nerve fibers (Fig. 7). Only some of the fibers are aligned along the axis of the tubules. Fewer and less well-oriented elastic fibers surround the coiled segment of the duct; there are practically none around the undulating segment.

The transition between the secretory coil, which is single-layered, and the duct, which is double-layered, is abrupt except that the luminal cells of the duct near the transition contain granules that resemble those in the secretory cells. The cells of the duct are cuboidal and the luminal cells sometimes slightly flattened with a faintly yellow and hyalin cuticular border facing the lumen. This border varies in thickness in different glands and becomes particularly distinct immediately after the coiled duct straightens.

The segment between the glomerate and the epidermal parts of the duct has the narrowest lumen in the entire duct. The luminal cells here are cuboidal or columnar, and the cuticular border is always well developed. In transverse section, the luminal cells have a pentagonal shape and are joined together with a thick cement substance (Fig. 8).

On friction surfaces, eccrine ducts open at the apex of ridges, only rarely in the furrows (Fig. 9); in either case, the opening is regularly spaced in the center of epidermal ridges (Hambrick and Blank, 1954). The orifices of the ducts and the surrounding elevation of the stratum corneum form cup-shaped depressions. The details of this structure become evident when plastic impressions of the skin surface are studied with the scanning electron microscope (Johnson *et al.*, 1970). On the

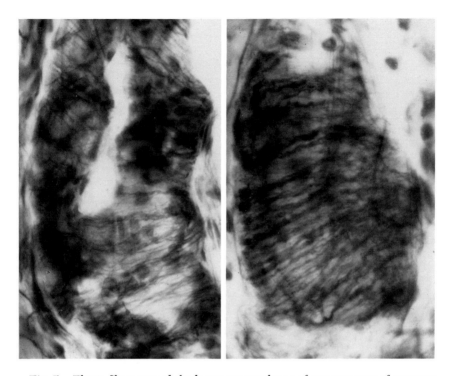

Fig. 7. Elastic fibers around the basement membrane of two segments of secretory coils.

palms and soles, a central nipplelike projection of the stratum corneum is seen within the terminal spiral part of the duct (Fig. 10). On the forearms and trunk, the terminal ducts open in elevations bounded by the fine, flat scales of the horny layer (Fig. 11A). When cut obliquely, the loops of the terminal spirals of the duct look almost horizontal.

When the undersurface of the epidermis is heavily sculptured, sweat ducts usually join at the side of the epidermal ridges, which in the palms and soles are mostly flattened (Blaschko's ridges). Elsewhere there are many different patterns of entry. Upon entering the epidermis, the ducts form spirals that become progressively larger as they approach the surface (Hambrick and Blank, 1954) (Fig. 11B). In the thick palmar and plantar epidermis, the duct forms a steep spiral (Fig. 12), whereas in thinner epidermis it is more conical. It has been suggested that the intraepidermal part of the sweat duct has a more or less fixed length which must adjust its shape to conform to variations in epidermal thickness (Pinkus, 1939). Whether the cells that line the intraepidermal

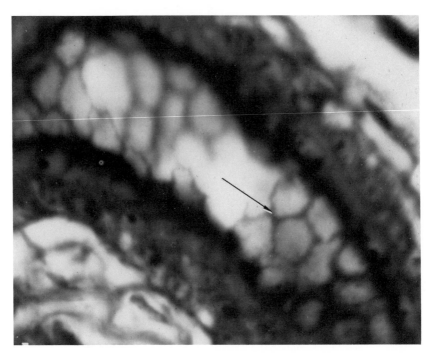

Fig. 8. The coiled segment of the duct. The outlines of the cells are pentagonal (arrow). From the skin of the axilla, stained with Regaud's hematoxylin.

duct are derived from the epidermis or from the duct itself remains to be answered; Pinkus (1939) believed that the epidermis resembles a "cast which is pierced by biologically separate structures [the sweat ducts] around which it is moulded." Because of some cytological differences between these cells and those of the surrounding epidermis, the terminal part of the duct and the immediately adjacent cells have been named "the epidermal sweat duct unit" (Lobitz *et al.*, 1954a,b). There is absolutely no evidence that this structure is a separate biological entity, but it may have different properties from those of the surrounding epidermis: it often remains unaltered when it traverses such damaged epidermis as in solar keratosis (Pinkus, 1939). In psoriatic lesions, the intraepidermal part of the duct is often but not always straightened.

As a rule, the cells of the sweat duct contain no melanin so that in heavily pigmented skin the intraepidermal sweat ducts contrast sharply with the epidermis.

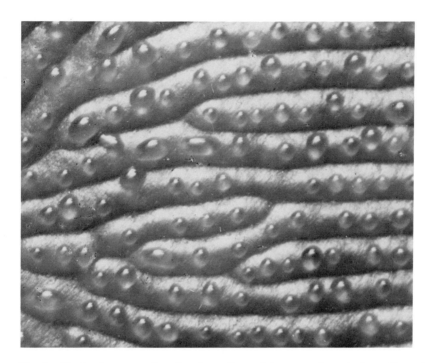

Fig. 9. Beads of sweat gathered at the orifices of the sweat ducts in the epidermal ridges of the palm. (Courtesy of Dr. W. C. Lobitz, Jr.)

The cells in contact with the keratinized cells around the lumen of the intraepidermal duct contain large keratohyalin granules. These keratinized cells probably originate from the base of the spiral, where the luminal cells regularly have a hyalin border but no keratohyalin.

If the epidermal duct and its surrounding epidermis are removed experimentally, the exposed dermis and duct form a thin necrotic crust (Lobitz *et al.*, 1954b). Three days later the basal cells near the end of the ductal stump become swollen and crowded around the lumen and cause it to become tortuous. The cells that stream out from the lumen outline a single layered tortuous tube in the shape of a cone beneath the necrotic scab of the wound. If cut experimentally, the midportion of the duct behaves in the same way (Lobitz *et al.*, 1956). Since little or no mitotic activity occurs in the basal cells at this time, the growth of the apical cone must be attributed principally to an increase in and reshifting of the volume of the cells already there. After the third day,

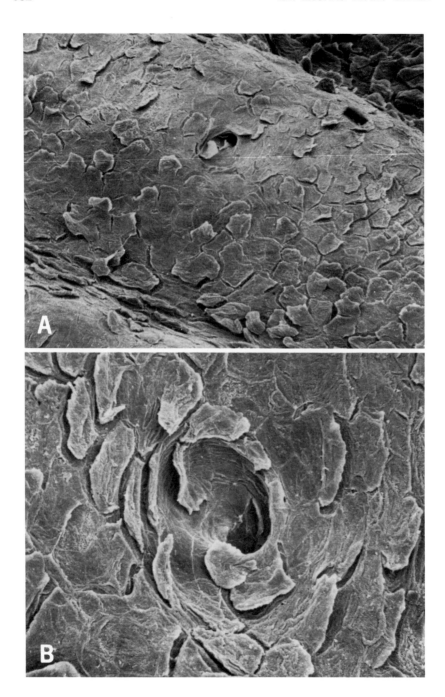

there is a burst of mitotic activity in the basal cells of the duct at the base of the cone. Many cells also radiate laterally from the cone to form an epithelial layer under the necrotic crust. When the crust is sloughed, the regenerated epidermis and sweat duct units appear normal. This may confuse the issue of cellular identity but it must be remembered that all epidermal derivatives retain the potentiality to form keratinizing epidermis.

Unlike hair follicles and sebaceous glands, whose cells proliferate rapidly, those of eccrine glands are replaced infrequently. The clear and dark cells of the secretory coil and the basal cell of the duct divide only occasionally, except after injury (Lobitz et al., 1956; Lobitz and Dobson, 1957). Repeated episodes of thermal sweating, particularly in experimentally salt-depleted subjects, also stimulate mitotic activity in the duct (Dobson et al., 1961).

V. Ultrastructure

The three types of cells in the secretory coil—myoepithelial, clear or serous cells, and dark or mucous cells—have definite histological and ultrastructural characteristics (Fig. 13). The elongated myoepithelial cells, which are thickened only in the center around the ovate nucleus, form an incomplete layer interposed between the basement membrane and the secretory epithelium. The cytoplasm is filled with myofilaments about 50 Å in diameter embedded in a low-density matrix (Fig. 14). Cytoplasmic organelles are found mostly in the small amount of peri-nuclear cytoplasm not filled with myofilaments. There are variable numbers of separate or aggregated glycogen granules. All cells contain smooth endoplasmic reticulum and some have rough endoplasmic reticulum. The Golgi apparatus is above the nucleus, facing the luminal surface. Mitochondria are mostly rod-shaped, and irregularly shaped particles of lipofuchsin can be found anywhere within the cells. The occasional small vesicles with dense, granular material are probably lysosomes.

The inner surface of the basement membrane follows the somewhat serrated peripheral edge of the cells but its outer part is smooth. The

Fig. 10. (A) Scanning electron micrograph of orifice of sweat glands on the sole. (B) A much enlarged orifice, showing the terminal part of the helix. (Courtesy of Dr. W. H. Fahrenbach.)

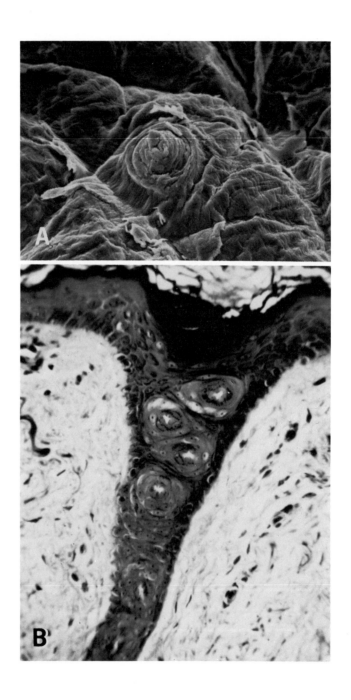

almost smooth apical surface of the myoepithelial cell seldom interdigitates with the folded surfaces of the secretory cells that rest on them.

The clear cells rest mostly on the basement membrane; they are
characterized by secretory granules, an elaborate folding of the basal
membrane, and an intricate system of intercellular canaliculi (ICC)
between adjacent cells (Figs. 13, 14, 15, 16). A basal process connects
them with myoepithelial cells and also with the basement membrane.
The apical end of the clear cells is narrower than the basal part and its
free border has few, short microvilli. The ovoid or round nucleus lies
near the center of the cell.

The basal part of the plasma membrane of clear cells forms complex,
long villous folds that interdigitate with those of adjacent clear cells.
Continuous junctional complexes interrupt the interdigitating villous
processes between clear cells and delineate the intercellular canaliculi
(ICC) (Fig. 16). The channels between clear cells are best demonstrated by treatment with lanthanum before sectioning (Hashimoto,
1971); they open into the lumen of the secretory coil and are closed by
a tight junction (zona occludens) at the basal end of the cell.

The cytoplasm usually contains numerous rosettes of glycogen particles. In addition, many spherical or ovoid mitochondria are scattered at
random throughout the cytoplasm along with much smooth endoplasmic
reticulum, but only a little rough endoplasmic reticulum. The Golgi
apparatus consists of closely packed, parallel smooth membranes and
small peripheral vesicles close to the ICC. The cytoplasm usually contains lipid droplets and pigmented granules of unknown origin. In aged
persons the sweat glands may have polyvesicular lipid inclusions
(Kurosumi *et al.*, 1960).

Most dark cells have an inverted pyramidal shape with a broad apical
surface. As mucin-producing cells, they vary in appearance probably
as a result of the state of their secretory cycle; the luminal border of
the plasma membrane forms short microvilli. Adjacent dark cells make
contact by villous folds. When a dark cell abuts a clear cell, its surface
is mostly smooth, but sometimes a few slender processes interdigitate
with those of the clear cell. Predominantly rough endoplasmic reticulum
packs the cytoplasm of dark cells. The large Golgi apparatus contains
numerous membrane-limited vacuoles (Fig. 15), whose size and content

Fig. 11. (A) Scanning electron micrograph of the terminal spiral of the duct in
the scapular area. (Courtesy of Professor Sam Shuster.) (B) "Epidermal sweat duct
unit" from the skin of the scalp. The spiral duct has been cut many times.

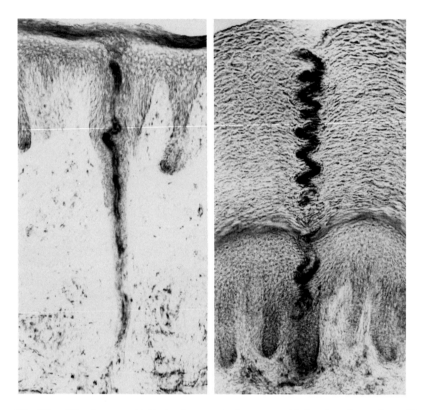

Fig. 12. AS esterase activity in the terminal segment of the duct from the axilla on the left, and in that of the palm on the right.

seem to correlate with cycles of synthesis and secretion. The largest vacuoles contain a flocculent electron-lucent substance, the intermediate ones a somewhat more opaque material, and the smallest and most numerous the most opaque substance. These last probably contain the final secretory product since small, dense spheres occasionally occur within the lumen of the secretory coil.

Mitochondria in the dark cells are few, short, rodlike, or filamentous. Lipid droplets similar to those in the clear cells lie between the secretory granules. Microtubules parallel to the long axis of the cells are most numerous near the plasma membrane. Tonofilaments are usually present in the supranuclear cytoplasm, and lysosomes are found occasionally. Paired centrioles consistently occur just beneath the plasma membrane at the secretory surface of the cells.

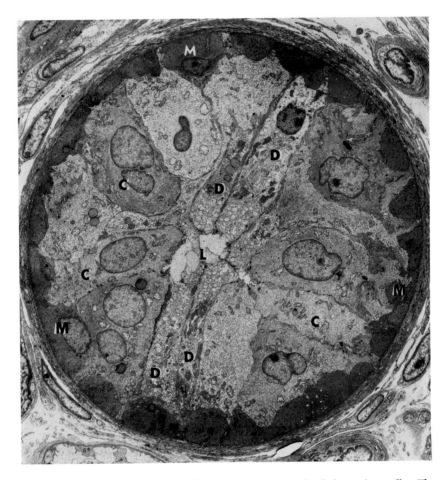

Fig. 13. A cross section through an eccrine sweat gland from the axilla. The dark cells (D) are more slender than the clear cells (C), but cells of both types extend from the basal portion of the gland to the lumen (L). The myofilaments of the myoepithelial cells (M) lie parallel to the longitudinal axis of the sweat gland and therefore appear in cross section here. × 1775. (Courtesy of Dr. M. Bell.)

VI. The Duct

The coiled and straight (undulating) parts of the duct are similar morphologically, although the epithelial cells gradually change in appearance as they proceed toward the epidermis. Except at their outer periphery, the basal cells of the duct have microvilli that interdigitate

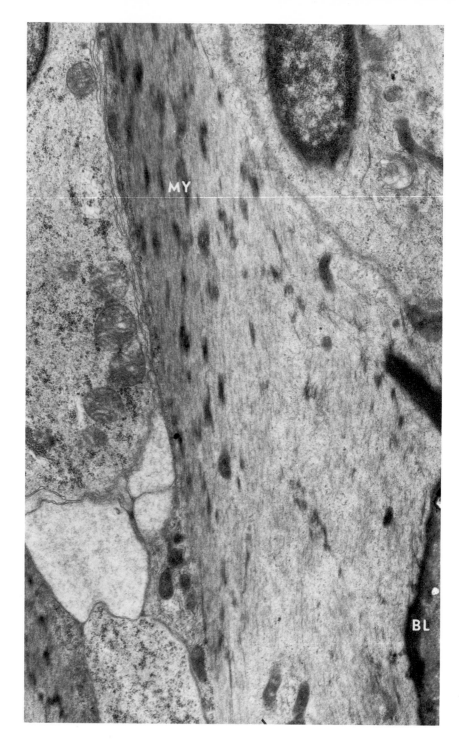

irregularly with those of adjacent basal cells to form a complex system of uneven channels. Villous processes are more numerous in the coiled part of the duct and have wide intercellular channels. This may reflect some functional significance; Sato and Dobson (1971a) found that only the coiled part of the duct has high metabolic activity. The numerous mitochondria in the basal cells have dilated cristae and an opaque matrix. There are also glycogen granules, some rough and smooth endoplasmic reticulum, a few microtubules, a small Golgi apparatus, and a few membrane-limited vesicles containing opaque granular material in the basal cells. Toward the junction of the coiled with the "straight" part of the duct, the basal cells have fewer mitochondria and more tonofilaments. As the duct rises toward the epidermis, the number of mitochondrial gradually decreases and the tonofilaments increase.

The luminal cells make contact by interdigitating plasma membranes that contain desmosomes (Fig. 17). The apical cytoplasm has a prominent cuticle with low microvilli. The cuticular border becomes gradually more conspicuous from the coiled duct outwards; in the straight duct, a layer of dense tonofilaments lies under a zone of finer filaments. The cytoplasm of luminal cells contains pleomorphic mitochondria, granular and smooth endoplasmic reticulum, and glycogen granules, but no distinct Golgi apparatus. The bases of these cells have villous folds that interdigitate with those of the basal cells. The distal ends of the luminal cell are fused by tight junctions.

The part of the sweat duct that traverses the epidermis resembles the intradermal part; it has an inner ring of luminal cells and two to three layers of peripheral cells that separate it from the surrounding epidermal cells (Zelickson, 1961). In the lower parts of the epidermis, short, bent, and twisted microvilli cover the luminal border of the surface cells. A cuticular border is less prominent than in the duct below the epidermis. Under the luminal surface are round vesicles and many lysosomes. The cells around the luminal ones keratinize earlier than the neighboring epidermal cells (Lee, 1960).

Fig. 14.   Electron micrograph of a myoepithelial cell in the secretory segment of an eccrine sweat gland. The cell rests upon the basement membrane (BL): the basal cytoplasm is filled with myofilaments (MY) oriented longitudinally. The nucleus is in the apical portion of the cell. Small mitochondria in the fibrous and vesiculated cytoplasm of the basal portions of several clear cells are also shown. × 18,000.

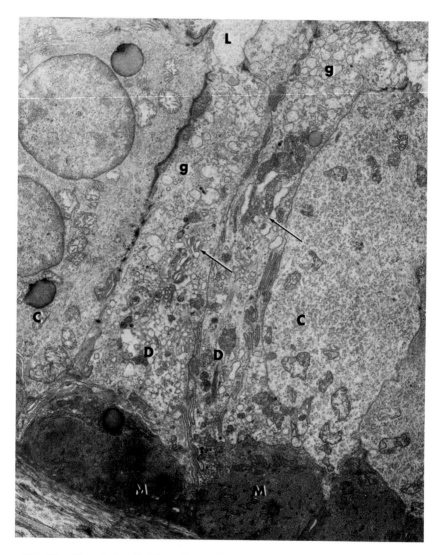

Fig. 15. Two dark cells (D) with interdigitations at the surfaces abutting adjacent clear cells (C) and myoepithelial cells (M). Tight junctions occur at the apical portion of lateral membranes approaching the lumen (L). The dark cells contain many granules (g), the contents of which were eluted during fixation. Golgi zones (arrows) are prominent in the dark cells. × 5250. (Courtesy of Dr. M. Bell.)

VII. Cytochemistry

A. The Secretory Coil

Myoepithelial cells, which contain few mitochondria, have low reactivities for cytochrome oxidase (Braun-Falco, 1961) and succinic dehydrogenase (Montagna and Formisano, 1955) and are only weakly reactive for monoamine oxidase (Yasuda and Montagna, 1960). They contain rows of small glycogen granules and high concentrations of amylophosphorylase (Ellis and Montagna, 1958). Esterase activity appears to be associated with lipid and lipofuchsin granules as well as with lysosomes (Montagna and Ellis, 1958).

Viewed under the light microscope, the clear cells have an accumulation of glycogen around the intercellular canaliculi, but this is no doubt an artifact of fixation since the electron microscope does not confirm such localization (Ellis, 1962). Lipid droplets, usually in the supranuclear cytoplasm, contain phospholipids, an acetone-extractable sudanophilic material, and a fluorescent component (Montagna et al., 1953). The electron microscope confirms the presence of triglyceride droplets and lipofuchsin granules (Ellis, 1962). The clear cells contain abundant oxidative mitochondrial enzymes: succinic dehydrogenase, cytochrome oxidase, and NADH- and NADPH-tetrazolium reductase. Much carbonic anhydrase activity has been demonstrated, but the technique used is unreliable (Braun-Falco and Rathjevs, 1955). However, this enzyme must be present since acetazoleamide profoundly affects the concentrations of sweat bicarbonate (Slegers and Moon, 1968). The monoamine oxidase in the clear cells may be responsible for inactivating the adrenergic amines that stimulate secretory activity (Yasuda and Montagna, 1960). When sweating is stimulated pharmacologically, the already abundant phosphorylase activity (Braun-Falco, 1956) increases, but the total phosphorylase activity does not change; this suggests that phosphorylase b serves as a reserve from which phosphorylase a can be rapidly formed (Smith and Dobson, 1966). Glycogen synthetase exists in two forms, one of which depends on glucose 6-phosphate. The clear cells of glands stimulated with methacholine contain less glycogen synthetase activity independent of glucose 6-phosphate than inactive glands (Smith, 1970).

The high β-glucuronidase activity and lesser concentrations of esterase are paradoxical since these enzymes are usually associated with

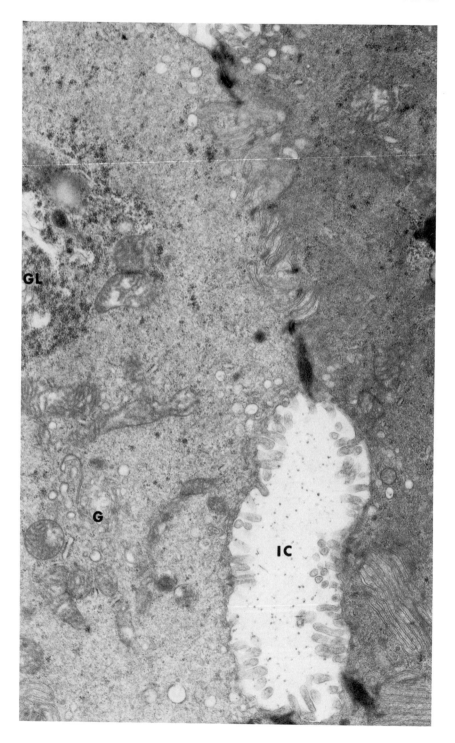

lysosomes, which are notably lacking in clear cells. Acid phosphatase, a characteristic lysosomal enzyme, has also been detected. Although the technique used to demonstrate β-glucuronidase has been challenged, its presence in the sweat glands with biochemical methods has been established (Gibbs and Griffin, 1968).

Alkaline phosphatase activity and sialomucin are localized along the intracellular canaliculi (Constantine and Mowry, 1966).

The dark cells contain variable amounts of lipid, some of which are not extractable with acetone. Pigment granules are autofluorescent; they contain no iron, are PAS-negative, and increase in size and number with age (Ellis, 1962). With the electron microscope, these pigment granules appear to be the small opaque components of the liquid droplets seen with the light microscope.

Much of the apical part of the cells is filled with PAS-positive, diastase-resistant granules (Formisano and Lobitz, 1957), which stain well with basic dyes, are often metachromatic, and whose basophilia is unaffected by digestion with ribonuclease. These granules also stain with Hale's colloidal iron and with alcian blue, an indication that an acid mucopolysaccharide component, at least some of which is sialomucin, is present (Constantine and Mowry, 1966).

Dark cells, with fewer mitochondria than clear cells, are less reactive for cytochrome oxidase and succinic dehydrogenase. Some phosphorylase activity is probably present; but because the reaction in the clear cells is intense, localization is difficult. α-Naphthol and AS esterase activity, as well as nonspecific esterases, occurs entirely within the dark cells. Small cytoplasmic granules contain acid phosphatase but no alkaline phosphatase activity.

B. The Duct

The basal cells of the duct, which contain large numbers of mitochondria, react strongly for succinic dehydrogenase and cytochrome oxidase. The concentrations of monoamine oxidase, β-glucuronidase, and phosphorylase are high. Phosphorylase a activity does not appear to change after sweating (Smith and Dobson, 1966). Traces of acid phosphatase, nonspecific esterase, and aminopeptidase activity have been demonstrated. The basal cells do not contain lipid droplets or pigment granules; under resting conditions, they have moderate amounts of glycogen.

Fig. 16. Clear cells showing slender processes of plasma membranes frequently interdigitate at the lateral cell margins and form intercellular canaliculi (IC). Golgi zones (G) and glycogen (GL) are prominent in the cells.

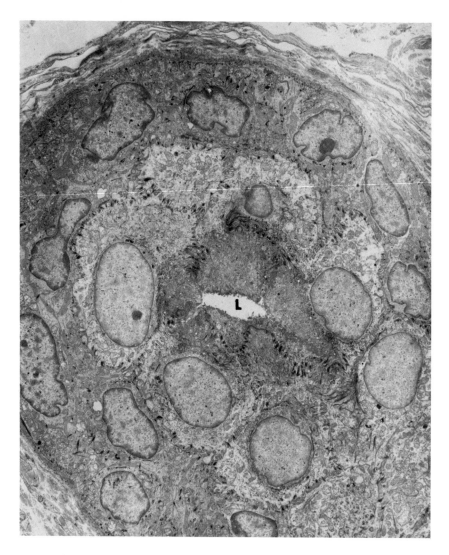

Fig. 17. A cross section through the straight portion of an eccrine sweat duct. The cells of the superficial and basal layers interdigitate by numerous slender processes. A relatively wide cuticular border immediately surrounds the narrow lumen (L). × 3100. (Courtesy of Dr. M. Bell.)

Except for their hyalinized border, the luminal cells are histochemically similar to the basal cells. The luminal borders contain much nonspecific esterase activity. A thin band of glycogen is present just beneath the cuticle, a region that reacts for sulfhydryl and disulfide linkages, and a PAS-positive diastase-resistant material, which may represent a glycocalyx, lines the luminal border.

VIII. Pharmacology

The nerves around the eccrine sweat glands are postganglionic, sympathetic fibers of the paraganglia, although physiologically and pharmacologically the glands behave as if they were surrounded by parasympathetic or cholinergic nerves (Dale, 1933). When the abdominal sympathetic trunk is stimulated, acetylcholine (ACh) is released from the nerves that supply the eccrine glands (Dale and Feldberg, 1934). The nerves, which are wound like a delicate skein around the secretory coil, become sparser around the coiled duct and form a loose lattice around the ascending part of the duct (Figs. 18, 19).

Sweat glands respond to ACh in dilutions as low as 1×10^{-9} (Foster, 1969). This concentration corresponds with findings from in vitro studies of isolated palmar sweat glands, which increase their metabolic activity after exposure to acetyl-β-methylcholine (Sato and Dobson, 1970).

IX. Factors That Influence the Responsiveness of the Sweat Gland to Cholinergic Drugs

Iontophoretic stimulation with pilocarpine in infants 1 to 6 days produces a low sweating rate in both sexes. By 1 to 2 months of age, sweat rates almost double and in girls remain at this level until they are mature women. In boys, the rate further increases during puberty until the age of forty. Thus, after puberty, men produce about one-third more sweat than women. Similar differences in sweating are obtained after thermal stimulation (Herrmann et al., 1952; Brown and Dobson, 1967).

By 2 weeks of age, all infants respond to intracutaneous injections of ACh, whereas premature infants with a postconceptual age of less than 225 days do not (Foster et al., 1969).

The rate of sweating is lower in old than in young men but in old women is more like that in men (Silver et al., 1964). Since the number of active sweat glands does not differ significantly, these differences in sweat rate are due to variations in the output of sweat by individual glands. A higher concentration of ACh is necessary to activate sweating in women than in men. For example, 80% of men respond to ACh, 1×10^{-6}, whereas only 50% of the women sweat visibly after the injec-

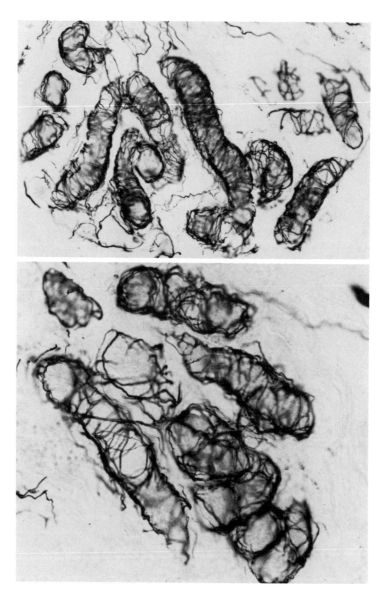

Fig. 18. The specimens above and below show the numerous nerves that surround the glomerate portion of eccrine sweat glands.

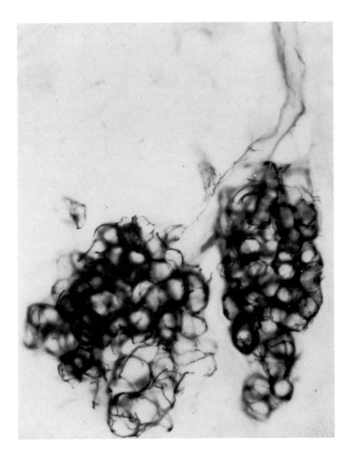

Fig. 19. Acetylcholinesterase-reactive nerves around a coil of an eccrine sweat gland. The coiled duct, above, has fewer nerves around it than the secretory segment.

tion of ACh, 1×10^{-3} (Janowitz and Grossman, 1950). Negro men and women seem to be more responsive to ACh than Caucasians (Gibson and Shelley, 1948).

"Conditioning," i.e., eliciting in sweat glands a different response to the same stimulus but under different conditions, greatly influences the responsiveness of eccrine glands to cholinergic agents (Rothman, 1954). The major conditioning factors are skin temperature, season, and acclimatization. When the skin is heated, iontophoretic stimulation with subthreshold concentrations of pilocarpine causes active sweating (Mac-

Intyre *et al.*, 1968). After local heating, physostigmine, which inhibits cholinesterase, produces augmented sweating which can be blocked by atropine. The possibility that local heating increases the amount of ACh released by each nerve impulse arriving at the preglandular nerve terminals correlates well with the fact that ACh release and resynthesis in other tissues depend on temperature (Salt *et al.*, 1965). An objection to this interpretation is that the effect of local heating remains after local nerve function has been impaired by procainization; perhaps local heating acts at the neuroglandular junction or on the glandular cells themselves (Ogawa, 1970).

The common phenomenon of increased sweating during the summer has been investigated by many authors (Kuno, 1956). Such seasonal variation is also seen in the responsiveness of sweat glands to ACh. After intracutaneous injections of ACh, the rate of sweating is less during the winter and spring than during the summer (Warndorff, 1970).

Repeated daily exposures to heat induce a marked increase in the rate of sweating and a decrease in the sodium content of the sweat, both features of a process known as acclimatization. To attribute these alterations to an increased secretion of aldosterone (Conn, 1949) cannot be defended since aldosterone regularly causes a decrease in the rate of sweating (Sato and Dobson, 1970). More likely, the increased rate of sweating with acclimatization is due to an augmented secretory response to the same neural stimulus. If sweating is locally inhibited with atropine during repeated exposures to heat, subsequent exposure without atropine does not cause an increase in the sweat rate in these previously inhibited glands (Fox *et al.*, 1964). Neither the sensitivity of the sweat glands to atropine (Craig *et al.*, 1969) nor the concentration of ACh necessary to produce sweating after intracutaneous injection (Collins *et al.*, 1966) is altered by acclimatization. Thus, the increased output of the secretory cells during acclimatization is probably elicited by changes, as yet unknown, within the cells themselves. Glycogen depletion of the clear cells generally occurs after a single exposure to heat or after intracutaneous injection of a drug but not after subsequent daily exposures to the same stimulus (Dobson, 1960; Collins *et al.*, 1966).

X. Repeated Injections of Cholinergic Drugs

Repeated intracutaneous injections of ACh or methacholine produce either an increase or a depression in the responsiveness of sweat glands, depending on concentration and frequency of application. Methacholine (2×10^{-4}) injected into the forearm of unacclimatized men twice a day at 50-minute intervals for 10 days increases the maximum rate of sweating from 33 to 98%. Although less sensitive to the drug, women have similar results (Johnson et al., 1969). This phenomenon has been called "training," that is, the increase in sweat rate is due to increased performance per gland rather than to a recruitment of previously inactive glands.

Repeated intracutaneous injections of large doses of ACh or methacholine (400 µg) produce a progressive decrease in the number of active sweat glands and in the sweat rate (Collins et al., 1959). Although this decreased sweating has been attributed to sweat gland "fatigue" (Thaysen and Schwartz, 1955), it may be due to a neuroglandular block caused by the desensitization of the receptor sites on the secretory cells (Collins et al., 1959).

XI. Arterial Occlusion

Both the production of sweat by individual sweat glands and the number of active glands decline after occlusion of the arterial blood supply (Randall et al., 1948; Van Heyningen and Weiner, 1952). During ischemic periods, sweat glands are less responsive to intracutaneous injections of cholinergic drugs (Collins et al., 1958) or to direct heating (MacIntyre et al., 1968).

XII. Handedness

Asymmetry in the quantity of thermal sweat occurs in both men and women (Kral et al., 1969). All right-handed and 97% of the left-handed subjects produce more sweat on one forearm than on the other. In 78% of the left-handed and in 60% of the right-handed subjects, the greater rate of sweating was on the side opposite the one with locomotor dominance. Furthermore, significantly more sweat is produced by right-handed persons than by left-handed ones.

XIII. Effects of Cholinergic Drugs *in Vitro*

Pilocarpine increases the oxygen consumption of human eccrine sweat glands *in vitro*, which produce large quantities of lactate (Schulz et al.,

1965) that is increased by Mecholyl (Wolfe *et al.*, 1970). In eccrine glands isolated from monkeys and man, the production of lactate is increased by the addition of glucose to the incubation medium; it is further stimulated by Mecholyl and inhibited by atropine. Ouabain (1×10^{-5} moles/liter), a specific inhibitor of Na^+-K^+-ATPase, markedly inhibits both CO_2 and lactate production even with Mecholyl (Figs. 19, 20) (Sato and Dobson, 1971b).

XIV. Secretory Activity Stimulated by Cholinergic Drugs

When the nerve supply to sweat glands is stimulated or when the glands are exposed to exogenously administered cholinergic drugs such as ACh, the increased permeability of the basement membrane causes an outward leakage of potassium and an inward leakage of sodium (Slegers, 1968). Whereas direct measurements have not been made on clear cells, in isolated salivary glands a few seconds after stimulation the intracellular potassium concentrations dropped from about 150 mEq/liter to about 110 mEq/liter, whereas intracellular sodium concentrations rose from 40 to 80 mEq/liter (Schneyer and Schneyer,

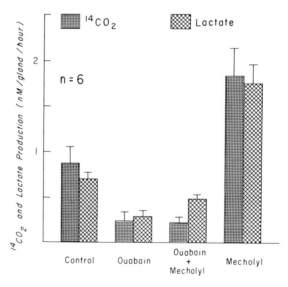

Fig. 20. Calculated ATP production of eccrine sweat glands *in vitro* after the addition of variable concentrations of Mecholyl or epinephrine.

1963). When the regression line of sodium vs. potassium concentrations was calculated for the sweat of cats, an intercept of 80 mEq/liter was obtained on the ordinate (Slegers, 1968) and can be regarded as the hypothetical intracellular sodium concentration at the time of stimulation.

The increase in intracellular sodium concentration is sufficient to activate Na^+-K^+-ATPase (Sato et al., 1971). The hydrolysis of ATP supplies energy to the sodium pumps presumably located along the intracellular canaliculi. The transport of Na^+ into these channels creates an osmotic gradient resulting in the movement of water across clear cells into the canaliculi. The amount of ATP produced after the eccrine glands have been stimulated by Mecholyl has been calculated from the results obtained in vitro (Figs. 20 and 21) (Sato and Dobson, 1971b). If the contributions of intermediates of the glycolytic pathway and the TCA cycle are assumed to be negligible, the total production of ATP can be estimated from the amount of lactate and CO_2 produced: one mole of lactate is equivalent to one mole of ATP and 6 moles of CO_2 are equivalent to 38 moles of ATP. Aerobic glycolysis is by far the more efficient and more important source of energy for the sweat gland.

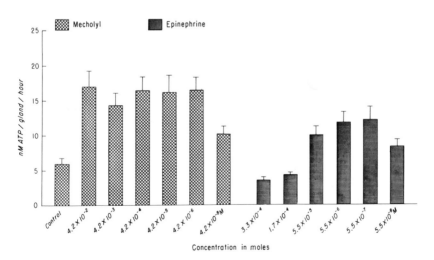

Fig. 21. The effect of various concentrations of epinephrine on CO_2 and lactate production by eccrine sweat glands in vitro.

XV. Adrenergic Stimulation of Sweating

Most human beings sweat after intracutaneous injections of epinephrine. The fact that these effects are blocked by α-adrenergic blocking agents (dibenamine, tolazoline) but not by atropine indicates that the sweating after epinephrine is not caused by a release of ACh (Chalmers and Keele, 1951).

Measurements of the response of sweat glands on the flexor surface of the forearm after intracutaneous injections of epinephrine indicate that men and women sweat equally (Silver *et al.*, 1964). Since the number of active glands does not change, the age-related decrease in sweat rate is due to diminished output per gland.

Unlike cholinergic drugs, increased temperature of the skin does not alter the response of the sweat glands to epinephrine (Ogawa, 1970).

After the iontophoresis of 0.2% pilocarpine combined with different concentrations of epinephrine, the individual response of palmar sweat glands differs markedly from that in the forearm. Epinephrine, 1×10^{-4} or 10^{-5}, plus pilocarpine, 0.2%, produces a significant increase in the sweat rate of the palmar glands despite mild to moderate vasoconstriction. In the forearm this combination of drugs depresses the sweat rate.

After complete denervation, human sweat glands are unresponsive to epinephrine (Janowitz and Grossman, 1951; Randall and Kimura, 1955).

XVI. Effects of Epinephrine *in Vitro*

In isolated human and monkey sweat glands, epinephrine stimulates lactate production under both aerobic and anaerobic conditions, but the percentage increase is much less in the latter case (Sato and Dobson, 1971b). High concentrations of epinephrine (3.3×10^{-4} M) markedly stimulate lactate production but depress CO_2 production. The increase in lactate production declines gradually with decreasing concentrations of epinephrine, whereas CO_2 production, following an inverted U pattern, reaches maximum levels at an epinephrine concentration of 5.5×10^{-7} M (Fig. 22). The stimulating effects of epinephrine are blocked by tolazoline but not by atropine. Methacholine stimulates more ATP production than epinephrine. The inhibition of ATP production by higher concentrations of epinephrine (3.3×10^{-4} to 1.7×10^{-4} M) correlates well with the observation that only lower concentrations of epinephrine

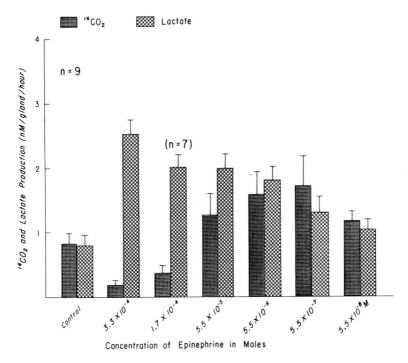

Fig. 22. Both CO_2 and lactate production are inhibited by ouabain even in the presence of Mecholyl. This suggests a linkage between sodium transport and glucose metabolism.

induce sweating *in vivo* (Haimovici, 1950). The relatively small amount of ATP production stimulated by epinephrine compared with a cholinergic drug may also explain the lower rate of sweating produced by epinephrine *in vivo*.

XVII. Possible Mechanisms of Adrenergic Sweating

A. *Contraction of Myoepithelial Cells*

At a moderate rate, sweat is secreted as single bursts of 6 to 7/minute; at higher rates, the frequency and amplitude of the pulsations are damped. Adrenergic drugs may help to expel sweat by stimulating the myoepithelial cells to contact rhythmically or may even express pre-formed sweat droplets (Rothman, 1954). Whether the myoepithelial cells around the eccrine secretory coil contract is not known, but similar cells in other glands do contract (Travill and Hill, 1963). After intracutaneous injections of methylene blue, the dye appears in the sweat when the glands are stimulated with epinephrine; since the dye had to be drawn into the cells from the extracellular fluid, epinephrine probably induces true secretory activity (Hurley and Witkowski, 1961). Furthermore, the prolonged sweat response to epinephrine (1 hour or more) virtually precludes the theory that only preformed sweat is delivered to the surface. The secretory pressure exerted by an actively sweating eccrine gland averages about 400 mm Hg (Schulz, 1969). Pressures of this magnitude can easily be generated by the osmotic gradients in the secretory cell during the secretory process. Thus, myoepithelial contraction is an unnecessary and therefore an unlikely explanation for the expulsion of sweat. Rather, the myoepithelium probably serves as a supportive structure to resist large internal osmotic forces and to insure the structural integrity of the intercellular canaliculi upon whose patency the entire secretory process may depend. The myoepithelium, then, may best be regarded as the cytoskeleton of the secretory coil.

B. Ultimate Liberation of Acetylcholine

Since eccrine sweat glands are innervated entirely by sympathetic fibers, epinephrine probably exerts its effects by stimulating the release of ACh (Fig. 19). However, evidence for this is only indirect and evidence against it is strong. For example, sweating induced by epinephrine is unaffected by atropine, whether *in vivo* (Chalmers and Keele, 1951) or *in vitro* (Sato and Dobson, 1971b). α-Adrenergic blocking agents curtail epinephrine-induced secretory activity both *in vivo* (Chalmers and Keele, 1951) and *in vitro* (Sato and Dobson, 1971b). Furthermore, epinephrine inhibits the response of thermally responsive sweat glands to cholinergic drugs (Sato and Dobson, 1971a).

Epinephrine probably stimulates cells by either activating phosphorylase, enhancing membrane permeability, or stimulating glucose oxidation. Cyclic AMP (cyclic adenosine 3′,5′ monophosphate), the intracellular mediator of many hormonal reactions, stimulates the activity of phosphorylase b kinase, which then converts the inactive phosphorylase b into the active phosphorylase a (Morel, 1969). Phosphorylase a, in turn, splits off glucose units from glycogen by cleaving 1, 4 glycosidic linkages. Eccrine clear cells contain this enzyme, whose activity increases when secretory activity is stimulated by methacholine (Smith and Dobson, 1966).

The cyclic AMP system is supposedly stimulated only by the activation of the β-receptor; in fact, adenyl cyclase itself may be the β-receptor (Robison *et al.*, 1967). As originally defined, the α-receptor is most responsive to norepinephrine or epinephrine and is blocked by phenoxybenzamine or phentolamine. Except in the heart, β-receptors are most responsive to isoproterenol and least receptive to norepinephrine. However, this varies from species to species; and in smooth muscle, cyclic AMP levels increase after exposure to epinephrine, the classic α-receptor. According to these observations, both α- and β-receptors may be related to adenyl cyclase (Robison *et al.*, 1967). In this context, the activation of sweat glands by α-adrenergic agents (epinephrine and norepinephrine) and its inhibition by α-blocking agents do not thereby exclude the role of cyclic AMP in its metabolism. *In vitro* studies of eccrine sweat glands showing that epinephrine without glucose does not stimulate their metabolic activity (Sato and Dobson, 1971b) suggest that sweat glands depend on an exogenous source of energy rather than on the phosphorylase-mediated release of glucose from endogenous glycogen. In some glandular systems, epinephrine also increases cellular permeability to glucose (Burgen, 1965), a phenomenon purported to be mediated by cyclic AMP; hence it is at least possible that similar events occur in sweat glands.

Norepinephrine markedly increases active sodium transport and osmotic water flow across frog skin (Bastide and Jard, 1968), but these effects can also be produced by cyclic AMP without norepinephrine. How this is effected is unknown, but probably the permeability of the membrane to sodium is increased during sweating. Under these circumstances, the mechanism of action of adrenergic drugs is analagous to that of cholinergic agents.

Epinephrine stimulates both the aerobic and the anaerobic glycolysis of glucose. *In vitro*, high concentrations of epinephrine increase lactate production (anaerobic glycolysis) but significantly decrease CO_2 production (aerobic glycolysis) (Sato and Dobson, 1971b). At lower concentrations (5.5×10^{-5} or 10^{-6} M), both pathways are stimulated but only to half the levels stimulated by methacholine. At optimum concentrations, epinephrine produces only about half as much ATP as methacholine (Sato and Dobson, 1971b). Since this correlates well with the maximum sweat rate induced *in vivo* by epinephrine, the major effect of adrenergic drugs on the secretory coil may be to stimulate glycolysis.

XVIII. Comments

Despite its almost universal role as an indispensable thermoregulator in man, sweating occurs in such variable degrees that it is difficult if not impossible to state anything definite about it. This statement does not apply to insensitive sweating about which data are almost completely lacking.

But elsewhere this variability is fairly obvious. For example, in hot dry environments some persons sweat profusely, others practically not at all as devotees of the sauna will attest. Like palmar sweating, axillary sweating, stimulated either by heat or by physical stress, is characterized by extreme individual differences that cannot be explained by factors of sex or age. Though long assumed to be similar to palmar and digital sweating, plantar sweating and that between the toes has not been adequately documented. Moreover, many of the phenomena that control sweating are still only vaguely understood. For example, we do not know whether certain eccrine sweat glands on both hairy skin and friction surfaces respond best or solely to heat, stress, or still unknown substances in food and spices. We do know, however, that eccrine glands respond to both cholinergic and adrenergic drugs, some being more sensitive to one than to the other.

In other mammals, the sweat glands are nearly torpid or represent a whole array of different mechanisms for dissipating body heat. Horses, cows, and some dogs rely partially on their apocrine glands for this service. In general only primates possess large numbers of true eccrine glands in their hairy skin. Despite this lack of adequate information, we do know that primate sweat glands secrete poorly or not at all and do not respond appreciably to either cholinergic or adrenergic stimulation. Studies on the palmar and digital glands of these animals are, however, more rewarding. Thus, except for their palms and soles, primates have provided no really good experimental models for studies of sweating. Man is the exception, but because of all the variables, even he is not an ideal experimental subject.

References

Bastide, F., and S. Jard. 1968. Actions de la noradrénline et de l'ocytocine sur le transport actif de sodium et la perméabilité àl'eau de la peau do grenouille. Rôle du 3′,5′-AMP cyclique. Biochim. Biophys. Acta 150: 113–123.

Borsetto, P. L. 1951. Osservazioni sullo sviluppo della ghiandole sudoripare nelle diverse regioni della cute umana. Arch. Ital. Anat. Embriol. 56: 332–348.

Braun-Falco, O. 1956. Über die Fähigkeit der menschlichen Haut zur Polysaccharid-synthese, ein Beitrag zur Histotopochemie der Phosphorylase. *Arch. Klin. Exp. Dermatol.* **202**: 163–170.

Braun-Falco, O. 1961. Zur Histotopographie der Cytochromoxydase in normaler und pathologisch verändertea Haut sowie in Hauttumoren. *Arch. Klin. Exp. Dermatol.* **214**: 172–224.

Braun-Falco, O., and B. Rathjevs. 1955. Über die histochemische Darstellung der Kohlen-säureanhydratase in normaler Haut. *Arch. Klin. Exp. Dermatol.* **201**: 73–82.

Brown, G., and R. L. Dobson. 1967. Sweat sodium excretion in normal women. *J. Appl. Physiol.* **23**: 97–99.

Burgen, A. S. V. 1965. The secretion of non-electrolytes by the salivary glands. *In* "Research on Pathogenesis of Cystic Fibrosis" (P. A. di Sant 'Agnese, ed.), pp. 112–118. Wickersham, Washington, D.C.

Chalmers, T. M., and C. A. Keele. 1951. Physiological significance of the sweat response to adrenaline in man. *J. Physiol. (London)* **114**: 510–514.

Collins, K. J., F. Sargent, and J. S. Weiner. 1958. The effect of ischaemia on the response of sweat glands to acetyl-β-methylcholine and acetylcholine. *J. Physiol. (London)* **142**: 32–33.

Collins, K. J., F. Sargent, and J. S. Weiner. 1959. Excitation and depression of eccrine sweat glands by acetylcholine, acetyl-β-methylcholine and adrenaline. *J. Physiol. (London)* **148**: 592–614.

Collins, K. J., G. W. Crockford, and J. S. Weiner. 1966. The local training effect of secretory activity on the response of eccrine sweat glands. *J. Physiol. (London)* **184**: 203–214.

Conn, J. W. 1949. Electrolyte composition of sweat. Clinical implications as an index of adrenal cortical function. *Arch. Intern. Med.* **83**: 416–428.

Constantine, V. A., and R. W. Mowry. 1966. Histochemical demonstration of sialomucin in human eccrine sweat glands. *J. Invest. Dermatol.* **46**: 536–541.

Craig, F. N., E. G. Cummings, H. L. Froehlich, and P. F. Robinson. 1969. Inhibition of sweating by atropine before and after acclimatization to heat. *J. Appl. Physiol.* **27**: 498–502.

Dale, H. H. 1933. Nomenclature of fibers in the autonomic nervous system and their effects. *J. Physiol. (London)* **80**: 10–11P.

Dale, H. H., and W. Feldberg. 1934. The chemical transmission of secretory impulses to the sweat glands of the cat. *J. Physiol. (London)* **82**: 121–128.

Dobson, R. L. 1960. The effects of repeated episodes of profuse sweating on the human eccrine glands. *J. Invest. Dermatol.* **35**: 195–198.

Dobson, R. L., D. C. Abele, and D. M. Hale. 1961. The effect of high and low salt intake and repeated episodes of sweating in the human eccrine sweat gland. *J. Invest. Dermatol.* **36**: 327–335.

Eichner, F. 1954. Zur Frage der Motivfildung in der menschlichen Haut. *Anat. Anz.* **100**: 303–310.

Ellis, R. A. 1962. The fine structure of eccrine sweat glands. *In* "Advances in Biology of Skin. Eccrine Sweat Glands and Eccrine Sweating" (W. Montagna, R. A. Ellis, and A. F. Silver, eds.), Vol. 3, pp. 30–53. Pergamon, Oxford.

Ellis, R. A., and W. Montagna. 1958. Histology and cytochemistry of human skin. XV. Sites of phosphorylase and amylo-1, 6-glucosidase activity. *J. Histochem. Cytochem.* **6**: 201–207.

Ellis, R. A., W. Montagna, and H. Fanger. 1958. Histology and cytochemistry of human skin. XIV. The blood supply of the cutaneous glands. *J. Invest. Dermatol.* **30**: 137–145.

Formisano, V., and W. C. Lobitz, Jr. 1957. "The Schiff-positive, nonglycogen material" in the human eccrine sweat glands. I. Histochemistry. *A.M.A. Arch. Dermatol.* **75**: 202–209.

Foster, K. G. 1969. Analysis of sweat gland responses to intradermal injections of acetylcholine. *J. Physiol. (London)* **205**: 11–12P.

Foster, K. G., E. N. Hey, and G. Katz. 1969. The response of the sweat glands of the newborn baby to thermal stimuli and to intradermal acetylcholine. *J. Physiol. (London)* **203**: 13–29.

Fox, R. H., R. Goldsmith, I. F. G. Hampton, and H. E. Lewis. 1964. The nature of the increase in sweating capacity produced by heat acclimatization. *J. Physiol. (London)* **171**: 368–376.

Giacometti, L., and H. Machida. 1965. Histology and cytochemistry of human skin. XXV. Common abnormalities in the eccrine sweat glands of man. *Arch. Dermatol.* **91**: 73–74.

Gibbs, G. E., and G. D. Griffin. 1968. Quantitative determination of beta-glucuronidase in sweat gland and other skin components in cystic fibrosis. *J. Invest. Dermatol.* **51**: 200–203.

Gibson, T. E., and W. B. Shelley. 1948. Sexual and racial difference in the response of sweat glands to acetylcholine and pilocarpine. *J. Invest. Dermatol.* **11**: 137–142.

Green, M., and H. Behrendt. 1969. Sweating capacity of neonates. Nicotine-induced axon reflex sweating and the histamine flare. *Amer. J. Dis. Child.* **118**: 725–732.

Haimovici, H. 1950. Evidence for adrenergic sweating in man. *J. Appl. Physiol.* **2**: 512–521.

Hambrick, G. W., Jr., and H. Blank. 1954. Whole mounts for the study of skin and its appendages. *J. Invest. Dermatol.* **23**: 437–453.

Hashimoto, K. 1971. Demonstration of the intercellular spaces of the human eccrine sweat gland by lanthanum. I. The secretory coil. *J. Ultrastruct. Res.* **36**: 249–262.

Herrmann, F., P. H. Prose, and M. B. Sulzberger. 1952. Studies on sweating. V. Studies of quantity and distribution of thermogenic sweat delivery to the skin. *J. Invest. Dermatol.* **18**: 71–86.

Holyoke, J. B., and W. C. Lobitz, Jr. 1952. Histologic variations in the structure of human eccrine sweat glands. *J. Invest. Dermatol.* **18**: 147–167.

Horstmann, E. 1952. Über den papillarkörper der menschlichen Haut und seine regionalen Unterschiede. *Acta Anat.* (Basel) **14**: 23–42.

Hurley, H. J., and J. A. Witkowski. 1961. Mechanism of epinephrine-induced eccrine sweating in human skin. *J. Appl. Physiol.* **16**: 652–654.

Ito, T., and K. Iwashige. 1951. Zytologische Untersuchung über die ekkrinen Schweissdrüsen in menschlicher Achselhaut mit besonderer Berücksichtigung der apokrinen Sekretion derselben. *Okajimas Folia Anat. Jap.* **23**: 147–165.

Ito, T., K. Tsuchiya, and K. Iwashige. 1951. Studien über die basophile Substanz (Ribonukleinsäure) in den Zellen der menschlichen Schweissdrüsen. *Arch. Histol.* **2**: 279–287. (In Japanese.)

Janowitz, H. D., and M. I. Grossman. 1950. The response of the sweat glands to some locally acting agents in human subjects. *J. Invest. Dermatol.* **4**: 453–458.

Janowitz, H. D., and M. I. Grossman. 1951. An exception to Cannon's law. *Experientia* **7**: 275.

Jenkinson, D. M. 1969. Sweat gland function in domestic animals. *In* "Exocrine glands: Proceedings Internat. Cong. of Phys. Sci. 14th Satellite Symp." (S. Y. Botelho, F. B. Brooks, and W. B. Shelley, eds.), pp. 201–216. Univ. of Pennsylvania Press, Philadelphia, Pennsylvania.

Johnson, B. B., R. E. Johnson, and F. Sargent. 1969. Sodium and chloride in eccrine sweat of men and women during training with acetyl-beta-methylcholine. *J. Invest. Dermatol.* **53**: 116–121.

Johnson, C., R. Dawber, and S. Shuster. 1970. Surface appearance of the eccrine sweat duct by scanning electron microscopy. *Brit. J. Dermat.* **83**: 655–660.

Kral, J. A., A. Zenisek, and J. Kopechka. 1969. Crossed locomotor and sudomotor innervation. *Int. Z. Angew. Physiol.* **27**: 165–170.

Kuno, Y. 1956. "Human Perspiration." Thomas, Springfield, Illinois.

Kurosumi, K., T. Kitamura, and K. Kano. 1960. Electron microscopy of the human eccrine sweat gland from an aged individual. *Arch. Histol.* **20**: 253–269.

Lee, M. M. C. 1960. Histology and histochemistry of human eccrine sweat glands with special reference to their defense mechanisms. *Anat. Rec.* **136**: 97–105.

Lobitz, W. C., Jr., and R. L. Dobson. 1957. Responses of the secretory coil of the human eccrine sweat gland to controlled injury. *J. Invest. Dermatol.* **28**: 105–120.

Lobitz, W. C., Jr., J. B. Holyoke, and W. Montagna. 1954a. The epidermal eccrine sweat duct unit. A morphologic and biologic entity. *J. Invest. Dermatol.* **22**: 157–158.

Lobitz, W. C., Jr., J. B. Holyoke, and W. Montagna. 1954b. Responses of the human eccrine sweat duct to controlled injury. Growth center of the "epidermal sweat duct unit." *J. Invest. Dermatol.* **23**: 329–344.

Lobitz, W. C., Jr., J. B. Holyoke, and D. Brophy. 1956. Response of the human eccrine sweat duct to dermal injury. *J. Invest. Dermatol.* **26**: 247–262.

MacIntyre, B. A., R. W. Bullard, M. Banerjee, and R. Elizondo. 1968. Mechanism of enhancement of eccrine sweating by localized heating. *J. Appl. Physiol.* **25**: 255–260.

Montagna, W. 1962. "The Structure and Function of Skin." 2nd ed. Academic Press, New York.

Montagna, W., and R. A. Ellis. 1958. L'histologie et la cytologie de la peau humaine. XVI. Repartition et concentration des esterase carboxyliques. *Ann. Histochim.* **3**: 1–17.

Montagna, W., and V. Formisano. 1955. Histology and cytochemistry of human skin. VII. The distribution of succinic dehydrogenase activity. *Anat. Rec.* **122**: 65–77.

Montagna, W., and L. Giacometti. 1969. Histology and cytochemistry of human skin. XXXII. The external ear. *Arch Dermatol.* **99**: 757–767.

Montagna, W., H. B. Chase, and W. C. Lobitz, Jr. 1953. Histology and cytochemistry of human skin. IV. The eccrine sweat glands. *J. Invest. Dermatol.* **20**: 415–423.

Morel, F. 1969. Cyclic adenosine monophosphate, intracellular mediator of the action of numerous hormones. *Triangle* **9**: 119–126.

Ogawa, T. 1970. Local effect of skin temperature on threshold concentration of sudorific agents. *J. Appl. Physiol.* **28**: 18–22.

Pinkus, F. 1927. Die normale Anatomie der Haut. *In* "Handbuch der Haut-und Geschlechtskrankheiten" (J. Jadasohn, ed.), Bd. 1, S.1–378. Springer-Verlag, Berlin.

Pinkus, H. 1939. Notes on the anatomy and pathology of the skin appendages. I. The wall of the intra-epidermal part of the sweat duct. *J. Invest. Dermatol.* **2**: 175–186.

Randall, W. C., and K. K. Kimura. 1955. The pharmacology of sweating. *Pharmacol. Rev.* **7**: 365–397.

Randall, W. C., R. Deering, and I. Dougherty. 1948. Reflex sweating by prolonged arterial occlusion. *J. Appl. Physiol.* **1**: 53–59.

Roberts, D. F., F. M. Salzano, and J. O. C. Wilson. 1970. Active sweat gland distribution in Caingang Indians. *Amer. J. Physiol. Anthrop.* **32**: 395–400.

Robison, G. A., R. W. Butcher, and E. W. Sutherland. 1967. Adenyl cyclase as an adrenergic receptor. *Ann. N.Y. Acad. Sci.* **139**: 703–723.

Rothman, S. 1954. "Physiology and Biochemistry of the Skin." Univ. of Chicago Press, Chicago, Illinois.

Salt, B., K. Kostial, and H. Loskovic. 1965. The influence of temperature on acetylcholine synthesis in the superior cervical ganglion of the cat. *Arch. Intern. Physiol.* **73**: 627–632.

Sato, K., and R. L. Dobson. 1970. Enzymatic basis for the active transport of sodium in the duct and secretory portion of the eccrine sweat gland. *J. Invest. Dermatol.* **55**: 53–56.

Sato, K., and R. L. Dobson. 1971a. Unpublished data.

Sato, K., and R. L. Dobson. 1971b. Glucose metabolism of the isolated eccrine sweat gland. I. The effects of mecholyl, epinephrine and ouabain. *J. Invest. Dermatol.* **56**: 272–280.

Sato, K., R. L. Dobson, and J. W. H. Mali. 1971. Enzymatic basis for the active transport of sodium in the eccrine sweat gland. Localization and characterization of Na-K-adenosine triphosphatase. *J. Invest. Dermatol.* **57**: 10–16.

Schiefferdecker, P. 1922. Die Hautdrüsen des Menschen und des Säugetieres, ihre Bedeutung sowie die Muscularis sexualis. *Zoologica* **72**: 1–154.

Schneyer, L. H., and C. A. Schneyer. 1963. Influence of pilocarpine on transport by salivary gland slices. *Amer. J. Physiol.* **205**: 1058–1062.

Schulz, I. J. 1969. Micropuncture studies of the sweat formation in cystic fibrosis patients. *J. Clin. Invest.* **48**: 1470–1477.

Schulz, I., K. J. Ullrich, E. Fromter, H. E. Emrich, A. Frich, V. Hegel, and H. Holzgrove. 1965. Micropuncture experiments on human sweat gland. *In* "Research on Pathogenesis of Cystic Fibrosis" (P. A. di Sant 'Agnese, ed.), pp. 136–146. Wickersham, Washington, D.C.

Silver, A., W. Montagna, and I. Karacan. 1964. Age and sex differences in spontaneous, adrenergic and cholinergic human sweating. *J. Invest. Dermatol.* **43**: 255–265.

Slegers, J. F. G. 1968. Ionic secretion by epithelial membranes. *In* "Cystic Fibrosis" (R. Porter and M. O'Connor, eds.), pp. 68–85. Little, Brown, & Co., Boston, Massachusetts.

Slegers, J. F. G., and W. M. Moon. 1968. Effect of acetazolamide on the chloride shift and the sodium pump in secretory cells. *Nature (London)* **220**: 181–182.

Smith, A. A. 1970. Histochemical differentiation of two forms of glycogen synthetase. *J. Histochem. Cytochem.* **18**: 756–759.

Smith, A. A., and R. L. Dobson. 1966. Sweating and glycogenolysis in the palmar eccrine sweat glands of the rhesus monkey. *J. Invest. Dermatol.* **47**: 313–316.

Spearman, R. I. C. 1968. Branched eccrine sweat glands in normal human skin. *Nature (London)* **219**: 84–85.

Szabó, G. 1967. The regional anatomy of the human integument with special reference to the distribution of hair follicles, sweat glands and melanocytes. *Phil. Trans. Roy. Soc. Lond., Series B.* **252**: 447–485.

Terebinsky, W. J. 1908. Contribution à l'étude de la structure histologique de la peau ches les singes. *Ann. Dermatol. Syphiligr.* **9**: 692–704.

Thaysen, J. H., and I. L. Schwartz. 1955. Fatigue of the sweat glands. *J. Clin. Invest.* **34**: 1719–1725.

Thomson, M. L. 1954. A comparison between the number and distribution of functioning eccrine sweat glands in Europeans and Africans. *J. Physiol. (London)* **123**: 225–233.

Travill, A. A., and M. F. Hill. 1963. Histochemical demonstration of myoepithelial cell activity. *Quart. J. Exp. Physiol.* **48**: 423–427.

Tsuchiya, F. 1954. Über die ekkrine Schweissdrüse des menschlichen Embryo, mit besonderer Berücksichtigung ihre Histo- und Cytogenese. *Arch. Histol.* **6**: 403–432. (In Japanese.)

Van Heyningen, R., and J. S. Weiner. 1952. The effect of arterial occlusion on sweat composition. *J. Physiol. (London)* **116**: 404–413.

Warndorff, J. A. 1970. The response of the sweat glands to acetylcholine in atopic subjects. *Brit. J. Dermatol.* **83**: 306–311.

Wells, T. R., and B. H. Landing. 1968. The helical course of the human eccrine sweat duct. *J. Invest. Dermatol.* **51**: 177–185.

Wolfe, S., G. Cage, M. Epstein, L. Tice, H. Miller, and R. S. Gordon, Jr. 1970. Metabolic studies of isolated human eccrine sweat glands. *J. Clin. Invest.* **49**: 1880–1884.

Yasuda, K., and W. Montagna. 1960. Histology and cytochemistry of human skin. XX. The distribution of monoamine oxidase. *J. Histochem. Cytochem.* **8**: 356–366.

Zelickson, A. S. 1961. Electron microscopic study of epidermal sweat duct. *Arch. Dermatol.* **83**: 106–111.

Author Index

Numbers in italics refer to the pages on which the complete references are listed.

Subject Index